KU-646-826

Contents

The editorial team

Author

Ruth Hull is a freelance writer who specialises in natural health. Born and educated in Zimbabwe, she completed a degree in Philosophy and Literature before studying and practising complementary therapies in London.

She now lives in South Africa where she lectures in complementary therapies at the renowned Jill Farquharson College in Durban. Ruth is married with two children.

General editor

Greta Couldridge has worked within the industry for many years, starting her career in a salon environment, before moving to teaching posts in further education colleges where she progressed to a management role. Throughout this time she worked as an external verifier and examiner before accepting a full time position with an awarding body developing qualifications and teaching and learning materials.

Greta is continually developing her skills, technical knowledge and keeping abreast of changes within the industry by attending workshops, courses and conferences.

Consultant

Sally Todd is the owner of UK Holistics, an award-winning training school for holistic and complementary therapies based in Grantham, Lincs. She teaches a full range of courses, holds a Certificate in Education and endorses this book as one of the most comprehensive and essential learning books available.

Publisher

Andy Wilson has worked in the publishing industry for 23 years, setting up The Write Idea in 1991. Most recently he has worked with one of the leading international therapy awarding bodies, helping to set up their publishing division and co-producing a number of successful text books.

ANATOMY & PHYSIOLOGY

for therapists and healthcare professionals

Ruth Hull

thewriteidea

Published by The Write Idea Ltd
10 Holland Street
Cambridge
CB4 3DL
01223 847765

First published August 2009
Reprinted July 2013
ISBN 978-0-9559011-1-9

Set in 10/12.5 Sabon.

Printed by Warners Midlands PLC.

Acknowledgements

The publishers would like to acknowledge the professionalism and dedication of the editorial team without whose hard work and enthusiasm this book would not have been possible. We would also like to thank the following for their valuable contributions to the development of this book:

Vicky Slegg of Liquorice Design for many of the illustrations

Lisa Kirkham for the page and cover design

The Wellcome Trust Photo Library for the pathology images at the end of each chapter

Hazel Godwin of Blackpool and the Fylde College for her assistance with preparing the text for reprint

Alf Welgemoed of The Academy of Beauty Therapy in Durban, South Africa for his assistance with preparing the text for reprint

From the author
I would like to thank Elizabeth Tarpey for her inspiration and generosity and Jenny and Clive Hull for their countless hours of babysitting. Finally, I would like to thank my husband, Douglas, without whose strength and support I would never have been able to write this book or even do many of the other things I have done in my life. Most importantly, I would like to thank him for being my soulmate.

Introduction

This book is for anyone studying anatomy and physiology at level 2 or 3 with any of the major awarding bodies. It presents all the information necessary to gain a thorough understanding of the subject in a clear, accurate and easily absorbed format. We have tried to strike a balance between a friendly, informal tone and serious academic content.

We hope you will enjoy using this book and would welcome any feedback, good or bad, which will help us to improve it in subsequent editions.

All of the illustrations in this book are available to order in large format for use as teaching and revision aids. For further information on these and forthcoming publications please contact us on 01223 847765 or email enquiries@writeidea.co.uk.

1 Before You Begin

Introduction

If someone asked you to name an organ superior to the bladder would you understand exactly what they meant? Before you even begin to look at the structure and function of the body, let's learn a few of the basic anatomical terms and their meanings.

What is anatomy? Anatomy is the study of the *structure* of the body. It looks at what the body is made of, for example, bones and organs.

What is physiology? Physiology is the study of the *functions* of the body. It looks at how the body works, for example, how the blood is pumped around the body.

What is pathology? Pathology is the study of the *diseases* of the body. It looks at what can go wrong in the body.

Anatomical position

The anatomical position is a basic position that can always be used as a reference point. It allows you to describe or name areas of the body in terms of a specific reference that all anatomists will know. This avoids any confusion.

The man in this image is standing in the anatomical position:
* his head is facing forward
* his feet are parallel
* his arms are hanging at his sides
* most importantly, his palms are facing forward.

The front of his body (where his face is) is called the anterior or ventral and the back of his body is the posterior or dorsal. These terms can also be used to indicate when something is in front of or behind something else. For example, the heart is anterior to the spine and the spine is posterior to the heart.

Directional terms

Imagine a line running right through the centre of the body: between your eyes, through the middle of your nose and mouth, down through the centre of your neck, chest, stomach and pelvis and ending between your feet. This is the midline or median line. It is an imaginary line that acts as a reference for many anatomical terms, including those describing certain movements such as adduction (which you will learn about later).

In relation to the midline are the following directional terms:

Superior – towards the head, above
Inferior – away from the head, below
Medial – towards the midline, on the inner side

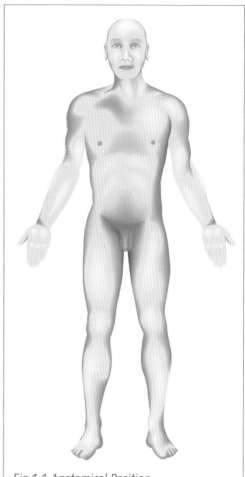

Fig 1.1 Anatomical Position

Lateral – away from the midline, on the outer side
Proximal – closer to its origin or point of attachment of a limb
Distal – farther from its origin or point of attachment of a limb
Superficial – toward the surface of the body
Deep – away from the surface of the body
Peripheral – at the surface or outer part of the body

Anterior – at the front of the body, in front of
Posterior – at the back of the body, behind
Cephalad – towards the head, above
Cranial – towards the head, above
Caudal – away from the head, below
Ventral – at the front of the body, in front of
Dorsal – at the back of the body, behind

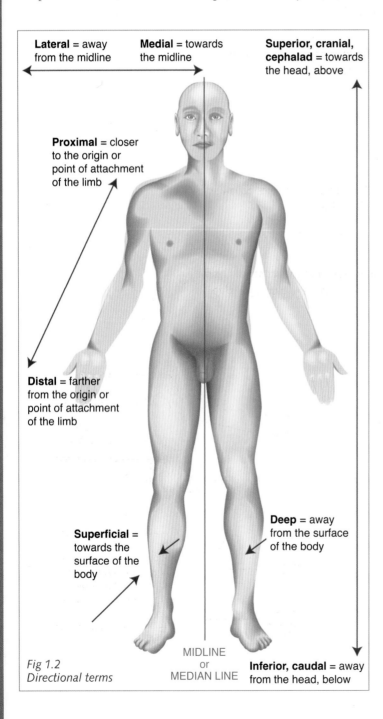

Lateral = away from the midline

Medial = towards the midline

Superior, cranial, cephalad = towards the head, above

Proximal = closer to the origin or point of attachment of the limb

Distal = farther from the origin or point of attachment of the limb

Superficial = towards the surface of the body

Deep = away from the surface of the body

MIDLINE or MEDIAN LINE

Inferior, caudal = away from the head, below

Fig 1.2 Directional terms

Here are a few examples to help you get used to the anatomical terms:

The nose is superior to the mouth.
The mouth is inferior to the nose.
The nose is medial to the ears.
The ears are lateral to the nose.
The elbow is proximal to the wrist (the arm's point of attachment is the shoulder).
The wrist is distal to the elbow (the arm's point of attachment is the shoulder).
The skin is superficial to the muscles.
The muscles are deep to the skin.
The hands and feet are at the periphery of the body.

Regions and cavities of the body

When studying the body it makes sense to divide it into smaller areas or regions. This enables us to discuss parts of the body without having to continually describe where they are. The body can be divided into:

- **Anatomical regions**

These relate to specific areas of the body. For example, the neck is the cervical region. When we discuss cervical nerves or cervical vertebrae we are talking about the nerves and vertebrae of the neck.

- **Body cavities**

These are spaces in the body that contain and protect the organs of the body. For example, the cranial cavity contains the brain.

Anatomical Regions Of The Body

Study tips

Certain things in anatomy simply have to be memorised and this is especially true for the anatomical regions of the body. The charts below will help you learn the correct terminology for each region.

Learn one chart at a time and touch the areas on your body as you memorise the words.

Make flash cards. On the front of the card will be the anatomical term. On the back of the card will be the area described. Use these cards in a study group.

GENERAL	
Term	**Region described**
Cutaneous	Skin

THE REGIONS OF THE HEAD AND NECK	
Term	**Region described**
Cephalic	Head
Cranial	Skull
Facial	Face
Frontal	Forehead
Ophthalmic/ Orbital	Eye
Otic	Ear
Buccal	Cheek
Nasal	Nose
Occipital	Back of the head
Cervical	Neck

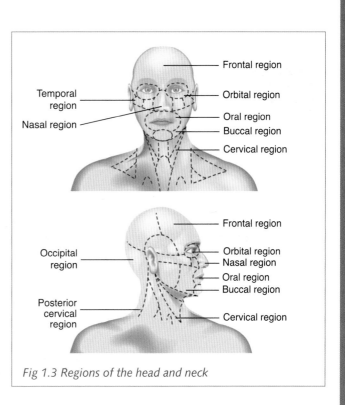

Fig 1.3 Regions of the head and neck

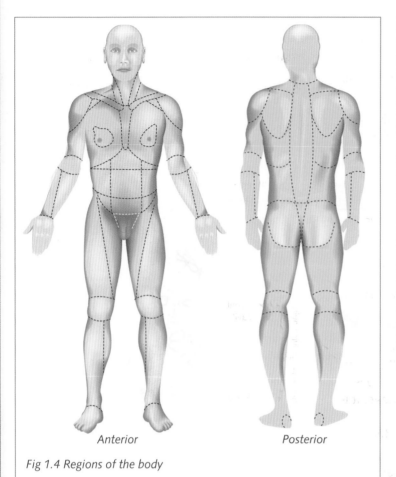

Anterior *Posterior*

Fig 1.4 Regions of the body

THE REGIONS OF THE TRUNK

Term	Region described
Thoracic	Chest
Costal	Ribs
Pericardial	Heart
Mammary	Breast
Abdominal	Abdomen
Umbilical	Navel
Dorsal	Back
Lumbar	Lower back
Coxal	Hip
Pelvic	Pelvis
Inguinal	Groin
Pubic	Pubis
Perineal	Area between anus and urethral opening

THE REGIONS OF THE ARM

Term	Region described
Acromial	Shoulder
Scapular	Shoulder blade
Axillary	Armpit
Brachial	Arm, upper limb
Cubital	Elbow or forearm
Antecubital	Front of the elbow
Olecranal	Back of the elbow
Antebrachial	Forearm
Carpal	Wrist
Manual	Hand
Palmar/ Metacarpal	Palm, inner surface of the hand
Digital/ Phalangeal	Fingers (also refers to toes)

THE REGIONS OF THE LEG

Term	Region described
Gluteal	Buttock
Crural	Leg, lower limb
Femoral	Thigh
Patellar	Front of the knee
Popliteal	Hollow behind the knee
Sural	Calf
Tarsal	Ankle
Pedal	Foot
Calcaneal	Heel
Plantar	Sole of the foot
Digital/ Phalangeal	Toes (also refers to fingers)

Body Cavities

Body cavities are spaces within the body that contain and protect the internal organs. There are two main cavities:

- The dorsal body cavity (at the back of the body)
- The ventral body cavity (at the front of the body).

These are subdivided as follows:

DORSAL BODY CAVITY	
Cranial cavity	Contains the brain and is protected by the bony skull (cranium).
Spinal cavity/ canal	Contains the spinal cord and is protected by the vertebrae.
VENTRAL BODY CAVITY	
Thoracic cavity	Contains the trachea, 2 bronchi, 2 lungs, the heart and the oesophagus and is protected by the ribcage. It is separated from the abdominal cavity by the diaphragm muscle.
Abdominopelvic cavity consisting of the:	
Abdominal cavity	Contains the stomach, spleen, liver, gall bladder, pancreas, small intestine and most of the large intestine. These organs are covered by a serous membrane called the peritoneum. The abdominal cavity is mainly protected by the muscles of the abdominal wall and partially by the ribcage and diaphragm.
Pelvic cavity	Contains a portion of the large intestine, the urinary bladder and the reproductive organs. It is protected by the pelvic bones.

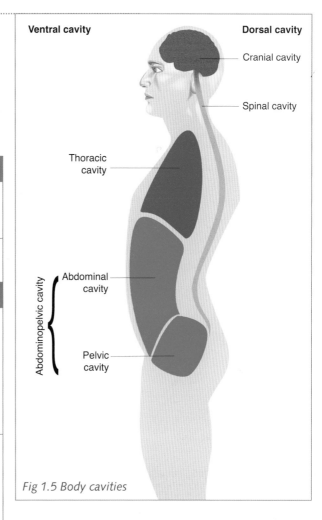

Fig 1.5 Body cavities

Note: For ease of learning, blood vessels, lymph vessels, lymph nodes and nerves have not been included in the chart on the left.

When learning about the body cavities it is important to know the following terms:

- Parietal – relating to the inner walls of a body cavity
- Visceral – relating to the internal organs of the body.

Dividing The Body

The body is often divided by:

Planes
These are imaginary lines used to cut the body or organs into sections. They can be used anywhere on the body or on any organ.

Quadrants
These are imaginary lines that specifically divide the abdominopelvic cavity into four parts so that it is easier to locate the organs found in this large cavity.

Planes of the body

Anatomists often divide the body or an organ into sections so that they can study its internal structures. These sections are made along imaginary flat surfaces called planes and are as follows:

- Sagittal plane – divides the body vertically into right and left portions.
- Frontal/coronal plane – divides the body vertically (longitudinally) into posterior and anterior portions.
- Transverse plane (cross-section) – divides the body horizontally into inferior and superior portions.
- Oblique plane – divides the body at an angle between the transverse plane and the frontal/sagittal plane.

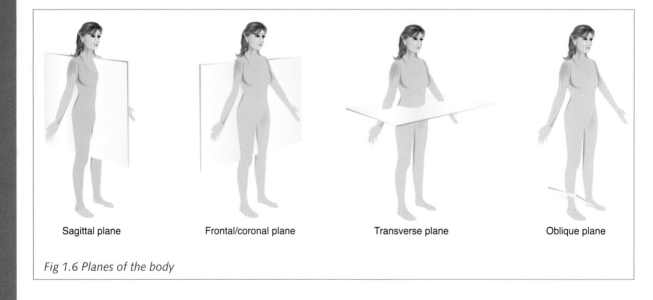

Sagittal plane   Frontal/coronal plane   Transverse plane   Oblique plane

Fig 1.6 Planes of the body

Quadrants of the body

The abdominopelvic cavity is large and contains many organs. It is, therefore, helpful to divide it into smaller regions that can be named according to their relative positions. These regions are quadrants and include the right upper quadrant (RUQ), the left upper quadrant (LUQ), the right lower quadrant (RLQ) and the left lower quadrant (LLQ).

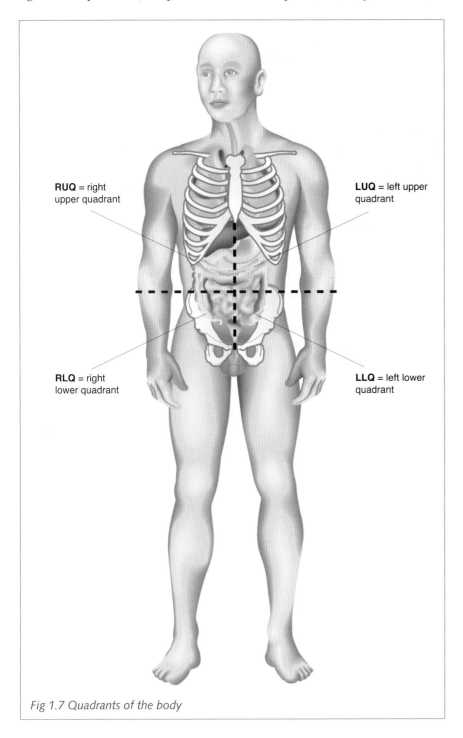

RUQ = right upper quadrant

LUQ = left upper quadrant

RLQ = right lower quadrant

LLQ = left lower quadrant

Fig 1.7 Quadrants of the body

NEW WORDS

Anatomical position	position in which the body is standing erect with the feet parallel, the arms hanging down by the side and the face and palms facing forward
Anatomy	the study of the structure of the body
Anterior	at the front of the body, in front of
Caudal	away from the head, below
Cephalad	towards the head, above
Coronal plane	divides vertically into anterior and posterior portions
Cranial	towards the head, above
Cross-section	divides horizontally into inferior and superior portions
Deep	away from the surface of the body
Distal	farther from its origin or point of attachment of a limb
Dorsal	at the back of the body, behind
Frontal plane	divides vertically (longitudinally) into anterior and posterior portions
Inferior	away from the head, below
Lateral	away from the midline, on the outer side
Medial	towards the midline, on the inner side
Median line	line through the middle of the body
Midline	line through the middle of the body
Oblique plane	divides at an angle between a transverse plane and a frontal or sagittal plane
Parietal	relating to the inner walls of a body cavity
Pathology	the study of the diseases of the body
Peripheral	at the surface or outer part of the body
Physiology	the study of the functions of the body
Plane	imaginary flat surface that divides the body or organs into parts
Posterior	at the back of the body, behind
Proximal	closer to its origin or point of attachment of a limb
Quadrant	a region of the abdominopelvic cavity
Sagittal plane	divides vertically into right and left portions
Superficial	towards the surface of the body
Superior	towards the head, above
Transverse plane	divides horizontally into inferior and superior portions
Ventral	at the front of the body, in front of
Visceral	relating to the internal organs of the body

Multiple choice questions

1. Superficial means:
   a. Towards the centre of the body
   b. Towards the surface of the body
   c. Towards the head
   d. Towards the feet.
2. The front of the body is referred to as:
   a. Dorsal
   b. Proximal
   c. Superficial
   d. Ventral.
3. In the anatomical position the palms are:
   a. Facing backwards
   b. Facing forwards
   c. In front of the body
   d. Behind the body.
4. The shoulders are _____ to the neck.
   a. Medial
   b. Superior
   c. Lateral
   d. Inferior.
5. The skeleton is _____ to the skin.
   a. Proximal
   b. Distal
   c. Superficial
   d. Deep.
6. The knee is _____to the thigh.
   a. Proximal
   b. Distal
   c. Superficial
   d. Deep.
7. An imaginary line that divides the body horizontally into superior and inferior portions is called:
   a. Sagittal plane
   b. Frontal plane
   c. Transverse plane
   d. Oblique plane.
8. The feet are _____ to the knees.
   a. Caudal
   b. Cephalad
   c. Medial
   d. Lateral.

9. Anatomy is the study of the:
   a. Diseases of the body
   b. Structure of the body
   c. Function of the body
   d. None of the above.
10. The study of how the heart works forms part of:
    a. Anatomy
    b. Pathology
    c. Physiology
    d. Herbology.
11. Imaginary straight lines used to divide the body into sections are called:
    a. Planes
    b. Cavities
    c. Quadrants
    d. Regions.
12. The left upper quadrant is commonly referred to as the:
    a. RUQ
    b. LUQ
    c. RLQ
    d. LLQ.
13. Oblique planes cut the body:
    a. At angles to the other planes
    b. Perpendicular to the other planes
    c. Parallel to the other planes
    d. None of the above.
14. The right lower quadrant is found in the:
    a. Thoracic cavity
    b. Spinal cavity
    c. Cranial cavity
    d. Abdominopelvic cavity.
15. Sagittal planes divide the body into:
    a. Superior and inferior portions
    b. Right and left portions
    c. Anterior and posterior portions
    d. Angles.

Now that you have a basic idea of anatomical language, it is time to start looking at the body in more depth. The following chapter describes the organisation of the body and introduces you to the different systems that you will need to learn. Each of these systems is then covered in depth in its own chapter.

As you work through this book you will find the following:

- *Italic text* has been used to emphasise key words or phrases.
- **Infoboxes** – these boxes will help you put your learning into perspective and bring it to life for you.
- New words – you may want to memorise these new words to help you improve your vocabulary. Any words you see in green are defined at the end of each chapter under the New Words section as well as in the glossary at the end of the book.
- **In the classroom** – these are simple ideas and exercises that can be used in a classroom environment or with a group of friends in a study group.
- **Study tips** – these will help you revise previous sections that are important to what you are currently learning and will also give you simple techniques to help you grasp a topic.
- **Revision and multiple choice questions** – by covering these at the end of every chapter you will be able to see how well you have understood the topic.
- **Superscript numbers**, e.g. [i] or [ii] – these are references showing where the information was obtained. You can find the details for each of these references on page 452.

Anatomy and physiology is a fascinating subject and easy to learn because you are everything you need to know – simply look at yourself when learning the subject. Think about what is going on inside your body and learn about yourself!

2 Organisation Of The Body

Introduction

The body is a complex and intricate machine. When you break it down into its smallest parts and put it together again you will be fascinated at how clever it really is. In this chapter we look firstly at the organisation of the body and how it all fits together and then we look at exactly what makes up the body: a cell. We will then discover how cells function, divide and unite to form tissues such as skin, muscles and nerves.

Student objectives

By the end of this chapter you will be able to:

- Identify the different levels of organisation of the body.
- Describe the basic chemical make-up of the body.
- Describe the structure and function of cells and discuss how they reproduce.
- Identify the major tissues that make up the body and be able to describe their structure, function and location.
- Identify the major systems of the body.

Anatomy and physiology in perspective

You may be asking yourself why you should study the chemistry of the body. There is one simple answer. You are made up of thousands of tiny chemicals continuously reacting with one another. What you put into your body (everything you eat or drink) and what you put on top of it (such as creams or lotions) is also made up of chemicals. These chemicals react with those in your body and change it. Now, isn't it worth knowing a little about chemistry?

Levels Of Structural Organisation Of The Body

Look at yourself: you are an organism, a living individual who breathes, moves, eats and functions. But do you know what you are really made of?

Cellular level: Cells are the basic structural and functional unit of the body. There are many different types of cells and they all have very specific functions. For example, there are nerve cells and blood cells. Cells combine to form tissues.

Tissue level: Tissues are groups of cells and the materials surrounding them. There are four tissue types which perform particular functions. These types are epithelial, muscular, connective and nervous tissue. Two or more different types of tissue combine to form organs.

Chemical level: The body is made up of tiny building blocks called atoms. Atoms such as carbon, hydrogen and oxygen, are essential for maintaining life and combine to form molecules such as fats, proteins and carbohydrates. These molecules then combine to form cells.

Organismic level: an organism is a living person – you.

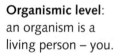
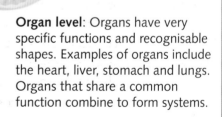

Organ level: Organs have very specific functions and recognisable shapes. Examples of organs include the heart, liver, stomach and lungs. Organs that share a common function combine to form systems.

System level: Systems are composed of related organs and they perform a particular function. For example, the digestive system is composed of organs such as the stomach, liver, pancreas, small and large intestines. Its function is to break down and digest food. Finally, systems combine to form the organism.

Fig 2.1 Levels of structural organisation of the body

Now that you can identify the different levels of structural organisation of the body it is time to take a closer look at each of them.

Chemical Organisation Of The Body

Atoms and elements

All living and non-living things consist of matter and the smallest unit of matter is the atom. Atoms combine to form either elements or compounds. Elements, such as oxygen, hydrogen and carbon, are composed of only one type of atom and they cannot be broken down into smaller substances by ordinary chemical methods. Compounds, on the other hand, are made up of different elements combined in specific proportions and an example is water, which is made up of the elements hydrogen and oxygen. The chart below shows the common elements found in the body.

ELEMENTS OF THE BODY	
Element	**Role in the body**
There are 4 major elements in the body which make up 96% of the body's mass. They are:	
Oxygen	Is a component of water and organic molecules and is essential to cellular respiration which is a process in which cellular energy, adenosine triphosphate (ATP), is produced.
Carbon	The main component of all organic molecules (e.g. carbohydrates, lipids, proteins and nucleic acids).
Hydrogen	A component of water, all foods and most organic molecules. It also influences the pH of body fluids.
Nitrogen	A component of all proteins and nucleic acids.
There are 9 lesser elements in the body which make up 3.9% of the body's mass. They are:	
Calcium	Found in bones and teeth. It is also necessary for muscle contraction, nerve transmission, release of hormones and blood clotting.
Phosphorous	Found in bones and teeth as well as in nucleic acids and many proteins. It also forms part of ATP.
Potassium	Necessary for many chemical reactions within the cell. It is also important for nerve impulses and muscle contraction.
Sulphur	A component of some vitamins and many proteins.
Sodium	Necessary for many chemical reactions in the extracellular fluid (fluid outside of the cell). It also plays a role in water balance, nerve impulses and muscle contraction.
Chlorine	Necessary for many chemical reactions in the extracellular fluid.
Magnesium	Found in bone and in some enzyme activity.
Iodine	Necessary for thyroid hormones.
Iron	A component of the haemoglobin molecule which transports oxygen within red blood cells.
	There are 13 other elements in the body that are present in such small quantities that they are known as trace elements. They make up 0.1% of the body's mass and are Aluminium, Boron, Chromium, Cobalt, Copper, Fluorine, Manganese, Molybdenum, Selenium, Silicon, Tin, Vanadium and Zinc.

Atoms and elements in a nutshell
- Atoms are the smallest unit of matter.
- Elements consist of uniquely structured atoms.
- The four major elements in the body are carbon, hydrogen, oxygen and nitrogen.

Chemical reactions

When atoms combine with or break apart from other atoms they react to form new products with new properties. This reaction is called a chemical reaction and chemical reactions are the foundation of all life processes. When chemicals react they either use or absorb energy or they give off or release energy. For example, reactions involving growth and tissue repair use energy while reactions involving the breakdown of food give off energy that the body can then use.

Molecules and compounds

When two or more atoms combine they form molecules and when atoms of different elements combine they form compounds. Most of the chemicals in the body exist in the form of compounds. The chart on pages 22 and 23 shows the major compounds present in the body.

Nutrients

Nutrients are found in the food we eat and are essential for the growth, maintenance and repair of the body. Essential nutrients include macronutrients (carbohydrates, proteins, fats, essential minerals and water) and micronutrients (vitamins and trace minerals). The chart on pages 22 and 23 discusses some macronutrients. Other nutrients include:

Vitamins: These are organic compounds required in minute amounts. They are essential for the normal functioning of the body and they help convert food into energy; help form tissues such as bones, muscles, nerves, blood and skin; and help the body resist infection. There are two groups of vitamins:

Fat Soluble Vitamins: A, D, E, K – these vitamins are stored by the body in the liver and fatty tissues.
Water Soluble Vitamins: B, C – these are not stored by the body and are easily lost through excretion. They are also more sensitive to the effects of storing and cooking.

Minerals: Minerals play a role in body growth and maintenance and form bones, teeth, hair, nails, red blood cells, body fluids, hormones and enzymes. They are categorised as follows:

Macro minerals: calcium, phosphorous, sodium, potassium, chloride, magnesium – essential minerals needed daily.
Trace minerals: iron, zinc, copper, fluoride, iodine, selenium, cobalt, molybdenum, manganese.

Essential fatty acids (EFAs): Essential fatty acids are fats that cannot be synthesised by the body and are vital for the proper functioning of the body.

They are necessary for brain and nerve function, the cardiovascular and immune systems, the hormones and the joints of the body.

Free radicals: Free radicals are highly unstable, reactive molecules that damage cells and are implicated in many diseases and ageing.

Antioxidants: Antioxidants are substances that combat or neutralise free radicals. They include vitamins E, C and beta-carotene; and the minerals selenium, zinc, manganese and copper.

The appendix on page 453 details some important sources and functions of nutrients.

Molecules and compounds in a nutshell
- Water is the solvent in body fluids.
- Carbohydrates are the fuel of the body.
- Fats have many roles such as insulation, protection and energy storage.
- Proteins are the building blocks of the body.
- Nucleic acids build our genes.
- ATP gives our cells energy.
- Nutrients are needed for growth, maintenance and repair.

Anatomy and physiology in perspective
One molecule you will learn about in the chart on the next two pages is DNA. DNA fingerprinting can be used to identify a child's parents or to convict criminals. Only a minute quantity of DNA is needed, for example from a single drop of blood or a strand of hair.

MAJOR COMPOUNDS OF THE BODY

Inorganic compounds – *These compounds generally do not contain the element carbon and are usually simpler and smaller than organic compounds.*

Compound	Elements present	Role in the body
Water	Hydrogen, Oxygen	This is the most abundant substance in the body because it is the **solvent** in body fluids. This means that different materials and substances can dissolve in it.

Water has many important functions in the body, including:

- *Maintaining body temperature* – water can absorb and give off large amounts of heat without its temperature changing too much. Therefore, it is able to maintain a normal internal temperature despite hot sun, cold winds and other external changes.

- *Acting as a lubricant* – water acts as a lubricant where internal organs touch and slide over one another or where bones, ligaments and tendons meet and rub together. It also lubricates the gastrointestinal tract so that food can easily move through it and faeces out of it.

- *Providing cushioning* – water creates a protective cushion around certain organs. For example, it forms cerebrospinal fluid which cushions the brain and protects it from external trauma.

- *Being a 'universal solvent'* – water can dissolve or suspend many different substances and is therefore an ideal medium in which chemical reactions can take place. Water can also transport substances such as nutrients, respiratory gases and waste around the body.

Organic compounds – *These are compounds that contain the element carbon. Carbon is a very useful element because it reacts easily to form large molecules that do not dissolve easily in water. These molecules build body structures and also break down to give off energy when the body needs it. The compounds below are all organic compounds and are important nutrient chemicals to the body.*

Compound	Elements present	Role in the body
Carbohydrates	Carbon, Hydrogen, Oxygen	Carbohydrates are the fuel of the body and include:

- *Monosaccharides* (simple sugars) – simple sugars are the building blocks of carbohydrates and a source of energy for chemical reactions. Glucose is the main form in which sugar is used by your cells. Fructose is found in fruits and galactose is present in milk. Some sugars also form parts of structural units. For example, deoxyribose is a sugar that forms part of the DNA molecule, which carries hereditary information.

- *Starch* – this is a large molecule and is the main carbohydrate found in food.

- *Glycogen* – glycogen is an energy reserve and is stored in the liver and skeletal muscles.

- *Cellulose* – is a carbohydrate built by plants. We eat it but cannot digest it and so it creates bulk which aids the movement of food and waste through our intestines. Cellulose is commonly referred to as fibre.

Compound	Elements present	Role in the body
Lipids (fats)	Carbon, Hydrogen, Oxygen	Lipids have over double the energy value of carbohydrates and proteins and are easily converted to body fat. They are composed of glycerol and fatty acids and most of them are **hydrophobic** (insoluble in water). There are many different types of lipids with very diverse roles, the main ones being:

MAJOR COMPOUNDS OF THE BODY

Compound	Elements present	Role in the body

- *Triglycerides* (neutral fats) – these are the most plentiful fats in the body and in your diet and are solids (fats) or liquids (oils) at room temperature. They are found in fat deposits beneath the skin and around organs. They protect and insulate the organs and are also a major source of stored energy.
- *Phospholipids* – these lipids form an integral part of the cell membrane and are also found in high concentrations in the nervous system.
- *Steroids* – steroids are a type of lipid and many different types of steroids are found in the body. For example, sex hormones and cholesterol (a steroid-alcohol or sterol).
- Other types of lipids include *eicosanoids* which have diverse effects on inflammation, immunity, blood clotting and other bodily responses; *fatty acids* which are important energy-supplying molecules; *carotenes* needed for the synthesis of vitamin A; *vitamin E* which contributes to the functioning of the nervous system, wound healing and is also an antioxidant; *vitamin K* which is necessary for blood clotting; and *lipoproteins* which help transport lipids in the body.

Proteins	Carbon, Hydrogen, Oxygen, Nitrogen. May contain Sulphur.	Proteins are the main family of molecules from which the body is built and are themselves built from amino acids. They are diverse in size, shape and function and play the following roles in the body:

- *Structural* – proteins are the building blocks of the body. For example, bone is built from collagen and skin, hair and nails are built from keratin.
- *Regulatory* – hormones are made from proteins and they regulate the bodily functions. For example, insulin helps regulate blood glucose levels.
- *Contractile* – the proteins myosin and actin enable muscles to shorten which allows for movement.
- *Immunity* – antibodies are proteins which protect against invading microbes.
- *Transport* – some carrier molecules are proteins. For example, haemoglobin transports oxygen in the blood.
- *Catalytic* – enzymes are proteins which help speed up biochemical reactions.
- *Energy source* – proteins can also act as a source of energy in times of dietary inadequacy.

Proteins are the main family of molecules from which the body is built and are themselves built from amino acids. The proteins in the body are built from 20 different amino acids, 10 of which the body synthesises and 10 of which are called essential amino acids because they must be obtained from proteins in the diet.

Nucleic acids	Carbon, Hydrogen, Oxygen, Nitrogen, Phosphorous	Nucleic acids are very important molecules which are found inside cells. There are two types:

- *Deoxyribonucleic Acid (DNA)* – DNA is found inside the cell's nucleus and makes up chromosomes, which contain our genes. DNA is the inherited genetic material inside every cell. It provides instructions for building every protein in the body and it replicates itself before a cell develops to ensure that the genetic information inside every body cell is identical.
- *Ribonucleic Acid (RNA)* – RNA is the 'molecular slave[i]' to DNA. It transports the orders of DNA from the nucleus to ribosomes where it is used to create specific proteins as per the genetic code.

Adenosine triphosphate (ATP)	Carbon, Hydrogen, Oxygen, Nitrogen, Phosphorous	ATP is the main energy-transferring molecule in the body and provides a form of chemical energy that can be used by all body cells.

Cellular Organisation Of The Body

Now that you understand the basic chemistry of the body, let us take a look at the next level up: the cellular level of organisation. Cytology is the study of cells, which are the basic structural and functional unit of the body. Cells vary greatly in size, shape and structure according to their function. For example, red blood cells are shaped like saucers to enable them to efficiently carry oxygen while nerve cells have long threadlike extensions for transmitting messages. In this chapter you will study a generalised animal cell.

Despite their differences, most cells are made up of a nucleus, cytoplasm and a plasma membrane and they are all bathed in interstitial fluid, a dilute saline solution derived from the blood.

Interstitial fluid is outside of the cell and is also known as extracellular fluid, intercellular fluid or tissue fluid. The fluid inside the cells is called intracellular fluid. Both the interstitial fluid and the intracellular fluid are made up of oxygen, nutrients, waste and other particles dissolved in water. The diagram below shows the structure of a generalised animal cell. Please refer to pages 30–31 for descriptions of the cell structures.

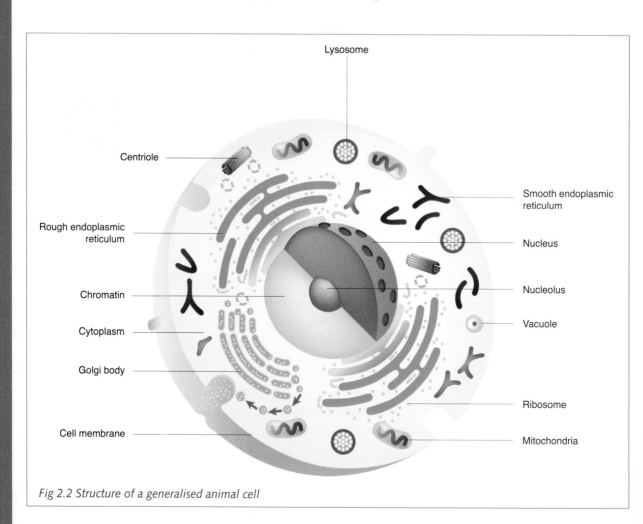

Fig 2.2 Structure of a generalised animal cell

Anatomy and physiology in perspective
Take a moment to put the chemistry you have just studied into perspective. A living cell is about 60% water and its structure is generally made up of the four elements carbon, oxygen, hydrogen and nitrogen. A cell also contains small amounts of other elements which are essential to its functioning, for example, nerve cells need sodium and potassium in order to transmit messages.

Plasma membrane

Structure and function of the plasma membrane
The plasma membrane is a thin and flexible barrier that surrounds the cell and regulates the movement of all substances into and out of it. It is made up of lipids and proteins which are organic compounds (please refer to the chart on major compounds of the body on pages 22 and 23 for further information on lipids and proteins).

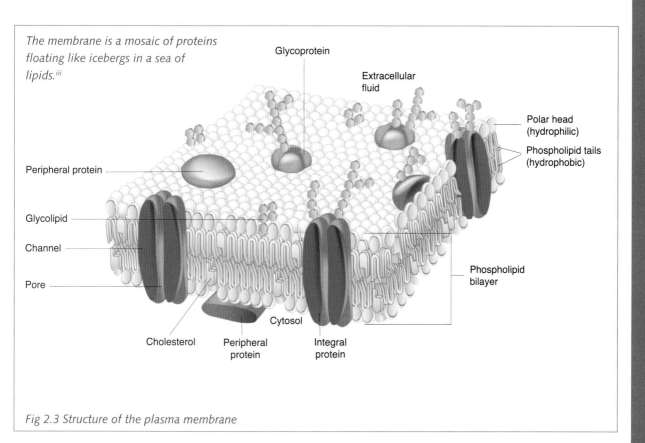

Fig 2.3 Structure of the plasma membrane

Lipids

Three types of lipids are present in the plasma membrane:

Phospholipids: approximately 75% of the lipids in the plasma membrane are phospholipids. Each phospholipid molecule has a head and a tail (like a tadpole). The head is hydrophilic (water-loving) while the tail is hydrophobic (water-hating) and the phospholipids lie tail to tail in two parallel layers. This is called a phospholipid bilayer and it forms the framework of the membrane. The hydrophilic heads are on the outside of the layers while the hydrophobic tails meet in the inside of the layers. Having the hydrophobic tails inside the membrane makes it impermeable to most water-soluble molecules.

Glycolipids: the functions of these lipids are still being discovered, but they do play a part in cell communication, growth and development. They are cell identity markers and so enable the cell to recognise whether other cells are of its own type or if they are potentially harmful foreign cells. They are found in the phospholipid layer that faces the extracellular fluid.

Cholesterol: these lipids are found only in animal cells and help to strengthen the membrane.

Proteins

Two types of proteins are scattered in the phospholipid bilayer:

Integral proteins: these extend all the way through the membrane to create channels which allow for the passage of materials in and out of the cell. Glycoproteins are proteins with attached sugar groups and (like glycolipids) are cell identity markers.

Peripheral proteins: these are loosely attached to the surfaces of the membrane and they can separate easily from it.

The proteins in the cells determine the functions of the cell and they play a variety of roles. Proteins can act as:

- **Channels** – some proteins have a pore (hole) in them through which other substances can move into and out of the cells.
- **Transporters** – some proteins carry substances within the membrane.
- **Receptors** – some proteins are able to identify and attach to specific molecules such as hormones or nutrients and thus guide them through the plasma membrane.

Other proteins act as *enzymes* or *cytoskeleton anchors*.

Transport across the plasma membrane

The life of a cell is dependent on the materials that are moved in and out of the cell: it needs energy and nutrition to function and it also needs to get rid of waste products and harmful substances.

Materials are transported into and out of the cell via the plasma membrane and there are two types of transport: passive processes and active processes.

Anatomy and physiology in perspective
Next time you pour yourself a cup of coffee put some sugar in it but don't stir it. After a while, taste the coffee (without having stirred the sugar). You will notice that it is sweet. Why? The particles of sugar will have diffused from an area of high concentration (the sugar) into an area of low concentration (the coffee).

Passive processes

In passive processes substances are moved without using cellular energy. These processes include:

Simple diffusion – the word *'diffus'* means spreading and in diffusion substances move from areas of high concentration to areas of low concentration. Molecules that diffuse easily through the plasma membrane include oxygen, carbon dioxide, nitrogen, steroids, fat-soluble vitamins (A, D, E and K), water, urea and small alcohols.

Diffusion is a very simple process:
- All substances have kinetic energy and are continually moving about.
- If there are many particles of a substance in one area then it is termed an area of 'high concentration'.
- An area with few particles of the substance will be called the area of 'low concentration'.
- The difference between the two areas forms the 'concentration gradient'.
- The particles from the area of high concentration will diffuse or spread into the area of low concentration until both areas have the same number of particles and equilibrium is reached.
- The rate of diffusion is determined by the temperature of the substance and the concentration gradient.

In the classroom
Demonstrate simple diffusion by putting a drop of food colouring into a glass of water and leaving it until all the water has changed colour. This is diffusion.

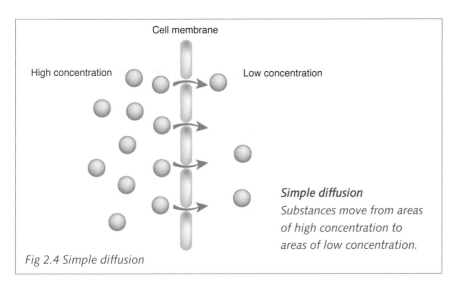

Cell membrane

High concentration

Low concentration

Simple diffusion
Substances move from areas of high concentration to areas of low concentration.

Fig 2.4 Simple diffusion

Osmosis – osmosis is the diffusion of water through a selectively permeable membrane. A selectively permeable membrane allows only water molecules to move across it.

In osmosis, water molecules move from an area of high water concentration to an area of lower water concentration. It must be remembered, that where there is a high concentration of water molecules there will be a low concentration of solute molecules. Therefore, osmosis is also the movement of water molecules from an area of low solute concentration to an area of higher solute concentration.

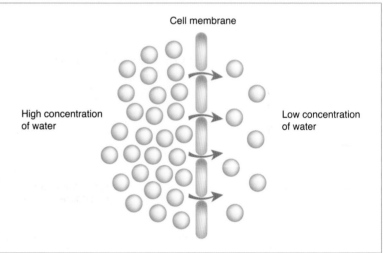

Osmosis

Water moves from areas of high water concentration to areas of lower water concentration, or… water moves from areas of low solute concentration to areas of high solute concentration. Both of these processes are always across a selectively permeable membrane.

Fig 2.5 Osmosis

Facilitated diffusion – molecules that are not lipid-soluble need help across the plasma membrane. These molecules include urea, glucose, fructose, galactose and certain vitamins.

Channel or transporter proteins within the membrane help these molecules across the plasma membrane. The rate of facilitated diffusion is determined by the number of channels/transporters and the concentration difference on the two sides of the membrane.

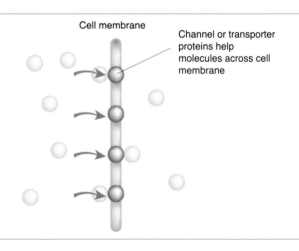

Facilitated diffusion

Substances are helped across the plasma membrane.

Fig 2.6 Facilitated diffusion

Active processes

In active processes cells need to use some of their own energy gained from the splitting of ATP to transport materials across the plasma membrane. Active processes usually involve moving substances against the concentration gradient. These processes include:

Active transport – integral membrane proteins act as pumps to push the molecules across the plasma membrane. These pumps are powered either directly or indirectly by ATP. Substances transported include ions, amino acids and monosaccharides.

Transport in a nutshell
- Substances move across the plasma membrane through *passive processes* – these do not use cellular energy and *active processes* – these use cellular energy.

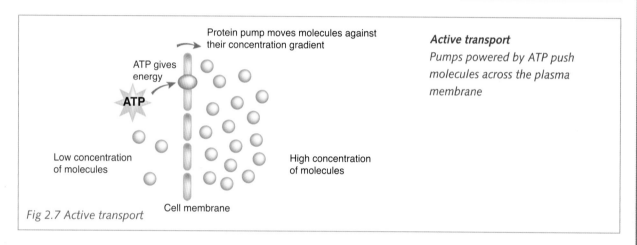

Protein pump moves molecules against their concentration gradient

ATP gives energy

ATP

Low concentration of molecules

High concentration of molecules

Cell membrane

Fig 2.7 Active transport

Active transport
Pumps powered by ATP push molecules across the plasma membrane

Vesicular transport – vesicles are small liquid containing sacs and in vesicular transport these sacs carry large particles such as bacteria, white blood cells, polysaccharides and proteins across the cell membrane.

Substances *enter* the cells through endocytosis, a process in which a segment of the plasma membrane surrounds the substance, encloses it and brings it into the cell.

Substances *exit* the cells through exocytosis, a process in which the vesicles fuse with the plasma membrane to release their contents into the extracellular fluid.

Cell membrane

Vesicle

Molecules to be secreted

Fig 2.8 Vesicular transport

Vesicular transport
Sacs called vesicles carry particles across the plasma membrane

Cytoplasm and cell structures

The cytoplasm is all the cellular material inside the plasma membrane, excluding the nucleus. It is the site of most cellular activities and includes cytosol, organelles and inclusions.

Cytosol

Cytosol is a thick, transparent, gel-like fluid made up of mainly water. It also contains solids and solutes, as well as spaces called vacuoles which house cellular wastes and secretions.

Organelles

Organelles are the 'little organs' of the cell and they play a specific role in maintaining the life of the cell. They are very specialised structures and differ according to the cell type and its functions. Some of the important organelles are:

- **Mitochondria** – these are the 'powerhouses' of the cell and are where ATP is generated through the process of cellular respiration.

Anatomy and physiology in perspective
Active cells that need a lot of energy, for example muscle cells, contain many mitochondria.

- **Ribosomes** – these tiny granules are the actual sites of protein synthesis in the cell – they are where proteins are made. Some ribosomes float freely in the cytoplasm and are called free ribosomes. Others are attached to the endoplasmic reticulum.
- **Endoplasmic reticulum (ER)** – this is a network of fluid-filled cisterns (channels or tubules) that coils through the cytoplasm. It provides a large surface area for chemical reactions and also transports molecules within the cell. There are two types of endoplasmic reticulum:
  Rough ER – this has ribosomes attached to it and provides a site for protein synthesis. It also temporarily stores new protein molecules and participates in the formation of glycoproteins. In addition, Rough ER works together with the Golgi complex to make and package molecules that are to be secreted from the cell.
  Smooth ER – this has no ribosomes attached to it and therefore no proteins are made here. Instead, it provides a site for the synthesis of certain lipids (fatty acids, phospholipids and steroids) and the detoxification of various chemicals such as alcohol, pesticides and carcinogens.
- **Golgi complex** – the Golgi complex is sometimes called the Golgi apparatus and is located near the nucleus. It is made of flattened cisterns with tiny vesicles attached to their edges and it processes, sorts and packages proteins and lipids for delivery to the plasma membrane. It also forms lysosomes and secretory vesicles.
- **Lysosomes** – these vesicles are formed inside the Golgi complex. They contain powerful digestive enzymes and are able to break down and recycle many different molecules. They help recycle the cell's own worn-out structures as well as foreign substances.

 **Anatomy and physiology in perspective**
Lysosomal enzymes help digest any cellular debris at sites of injury, thus preparing the area for repair.

- **Peroxisomes** – these are vesicles containing enzymes that detoxify any potentially harmful substances in the cell.
- **Centrosomes** – these are found near the nucleus and have two roles. In non-dividing cells they organise the microtubules which help support and shape the cells and move substances. In dividing cells they form the mitotic spindle (see section on cell division). Centrosomes contain centrioles.
- **Centrioles** – these are found within the centrosomes and play a role in cell division and also in the formation and regeneration of flagella and cilia. Flagella are projections on the outside of the cells which enable them to move, for example, a sperm cell has flagella. Cilia are hair-like projections which enable the cell to move substances across its surface. For example, ciliated cells of the respiratory tract move mucus.

Nucleus
The nucleus is the largest structure in the cell and controls all cellular structure and activities. It also contains most of the hereditary units known as genes (mitochondria also contain a few genes). Almost all body cells except for red blood cells contain a nucleus, while some cells such as skeletal muscle cells contain several nuclei.

The nucleus is constructed as follows:
- **Nuclear envelope** – the nucleus is surrounded by a double membrane of phospholipid bilayers (similar to the plasma membrane). This membrane contains nuclear pores (channels) which allow for the movement of molecules into and out of the nucleus.
- **Nucleolus** – inside the nucleus is a spherical body made up of protein, some DNA and RNA. This is where ribosomes are assembled.
- **Chromatin** – chromatins are only present in cells that are not dividing. They are a mass of 46 chromosomes all tangled together. During cell division the chromatin separates (see section on cell division).
- **Genes** – these are the hereditary unit of the cell which control the structure and activity of the cell. Genes are arranged along structures called chromosomes.

Inclusions
Inclusions form a diverse group of substances that are temporarily produced by some cells, for example, melanin and glycogen.

The cell in a nutshell
Imagine the cell as a bakery:
- The plasma membrane is the building walls and doors
- The cytoplasm is the bakery floor
- The nucleus is the manager
- Mitochondria are the generators
- Ribosomes are the ovens
- Endoplasmic reticulum is the production line
- Golgi complex is the packing and distribution department
- Lysosomes are the recycling site
- Peroxisomes are the cleaners.

Cell Division

So far you have been learning how your cells function on a day-to-day basis. But what happens if they become damaged, diseased or worn-out and need to be replaced? They reproduce themselves through a process called cell division. There are two types of cell division:

- **Somatic cell division** – this occurs in all cells, except bacteria and some cells of the reproductive system, and it takes place when the body needs to replace dead and injured cells or produce new cells for growth. Through a process of nuclear division called mitosis, a single diploid parent cell duplicates itself to produce two identical daughter cells.
- **Reproductive cell division** – this occurs when a new organism is to be produced. Through a process of nuclear division called meiosis, sperm and egg cells are produced and these form a new organism.

In this chapter we will only study somatic cell division in depth but, before we go any further, you need to know the following terms:

- Somatic cell = any cell except the reproductive cells.
- Diploid = having two complete sets of chromosomes per cell.

Human cells have 46 chromosomes. You inherit 23 chromosomes from your mother and 23 from your father.

Lifecycle Of A Cell

A cell has two distinct periods in its life cycle: interphase and cell division.

Interphase and DNA replication

Interphase is the time in which a cell grows and carries out its usual metabolic activities. It is also the time in which the cell prepares itself for division through DNA replication.

The DNA molecule is made up of two strands or chains of nucleotides twisted around each other into a double helix. This is a spiral curve that looks like a rubber ladder twisted around its axis. During DNA replication, the molecule uncoils and separates into its two nucleotide strands. Each strand now serves as a template for building a new nucleotide strand.

Note: each chromosome molecule is now made up of two strands, each called a sister chromatid and held together by a centromere.

Cell division

Two processes take place during cell division: mitosis and cytokinesis.

Mitosis
Mitosis is the period in which the mother cell divides into two daughter cells, each daughter cell containing the same genes as the mother cell. Although mitosis is a continuous process, it is easier to categorise it into the following four stages:

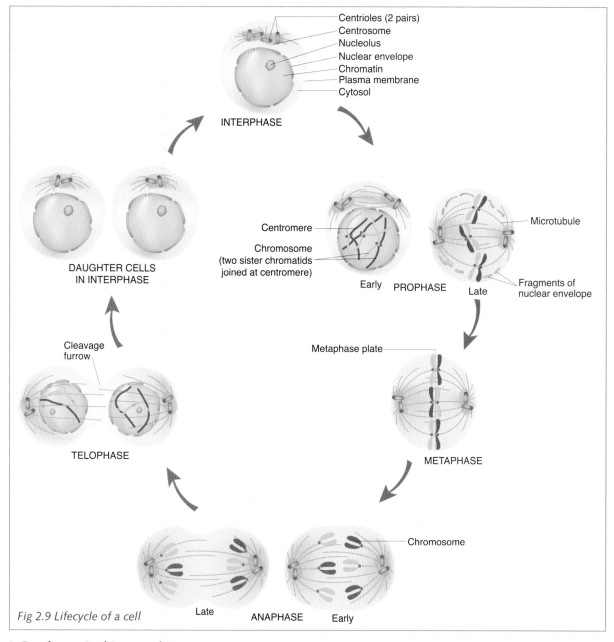

Fig 2.9 Lifecycle of a cell

1. **Prophase** – in this stage the:
   - *chromatin fibres condense and shorten into chromosomes*. This condensing of fibres is thought to prevent the entangling of DNA strands as they move during mitosis.
   - *the centrosomes and centrioles form the mitotic spindle* through separating from each other and moving to opposite sides of the cell. The mitotic spindle is made of microtubules and is shaped like a football. It forms a site of attachment for the chromosomes and distributes them to opposite ends of the cell.
   - *the nuclear envelope and nuclei break down and are absorbed into the cytosol.*

2. **Metaphase** – this is a short phase in which the *centromeres line up at the centre of the spindle.*

3. **Anaphase** – *centromeres split and the chromatids/chromosomes move apart towards opposite ends of the cell.* Note that once they are separated the sister chromatids are now referred to as daughter chromosomes.

4. **Telophase** – *this is prophase in reverse* and is the period in which:
   - the chromosomes go to the opposite ends of the cell, uncoil and become threadlike chromatin again
   - the spindle breaks down and disappears
   - a nuclear envelope forms around each chromatin mass
   - nucleoli appear in each of the daughter nuclei.

Remember: pro = before, meta=after, ana=upward, telo=far/end

Cytokinesis (Cytoplasmic Division)

Cytokinesis actually happens during mitosis and is the process by which the cell splits into two new cells. During the late anaphase, a cleavage furrow forms in the plasma membrane. This is a slight indentation that extends around the centre of the cell. The furrow deepens until the opposite surfaces of the cell make contact. The cell then splits into two daughter cells, each with their own separate portions of cytoplasm and organelles. The daughter cells are smaller than their mother cells, but genetically identical. They will soon grow and carry out normal cellular activities until it is their turn to divide – and so continues the cycle of life.

Infobox

Anatomy and physiology in perspective
Cancer is the uncontrolled division of body cells.

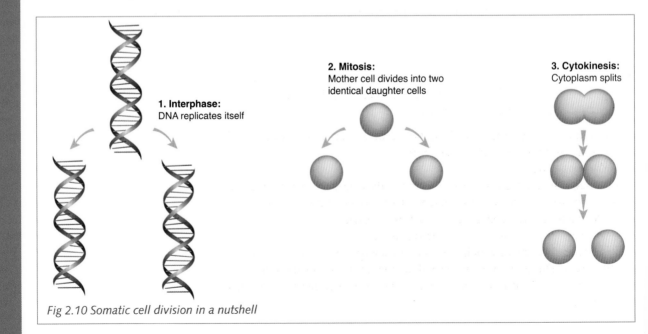

Fig 2.10 Somatic cell division in a nutshell

1. **Interphase:**
DNA replicates itself

2. **Mitosis:**
Mother cell divides into two identical daughter cells

3. **Cytokinesis:**
Cytoplasm splits

Tissue Level Of Organisation Of The Body

Now that you know what a cell is, it is time to look at what happens when similar cells work together to perform a common function: they form groups called tissues. For example, nerve cells work together to form nervous tissue.

In this section you will learn about the different types of tissues, the study of which is known as histology.

 **Anatomy and physiology in perspective**
A biopsy is the removal of a sample of tissue for examination. It enables medical professionals to help diagnose diseases such as cancer.

There are four types of tissue in the body: epithelial, connective, muscle and nervous. The table below gives a brief overview of them.

TYPES OF TISSUE IN THE BODY		
Name	**Function**	**Example**
Epithelial	• Covers body surfaces and therefore functions in protection, absorption and filtration • Lines hollow organs, cavities and ducts • Forms glands and therefore functions in secretion	Skin
Connective	• Protects and supports the body and its organs • Binds organs together • Stores energy reserves as fat • Provides immunity	Blood, adipose tissue, bone
Muscle	• Provides movement and force	Skeletal muscle
Nervous	• Initiates and transmits nerve impulses	Nerves

Epithelial tissue

Epithelial tissue, also called epithelium, forms body linings and glands. It is a large group of tissue that includes many different types, but they all generally share the same characteristics:

- Cells fit closely together with little extracellular material between them.
- Cells are arranged in continuous sheets.
- Cells always have:
  – a free surface which is exposed to the body's exterior or a body cavity
  – an interior surface which is attached to a basement membrane.
- Cells are constantly being regenerated by mitosis.
- Tissues are firmly joined to connective tissue.
- Tissues are avascular, which means that they have no blood supply of their own. Therefore, they receive all nutrients and remove waste through the process of diffusion with the blood vessels located in the connective tissue below.

There are four different cell shapes that make up epithelial tissue: squamous, cuboidal, columnar and transitional. Epithelial tissue is classified according to the numbers of layers it has and the shape of its cells. The following chart details this classification.

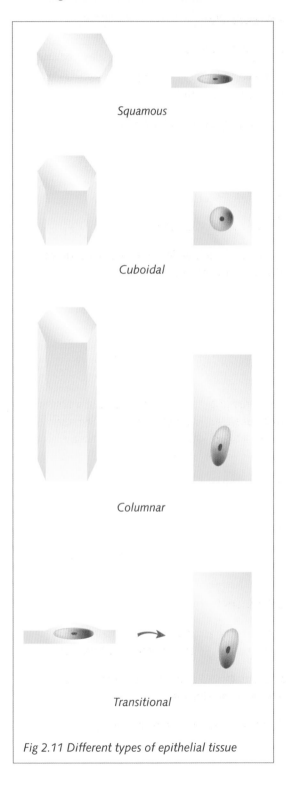

Squamous

Cuboidal

Columnar

Transitional

Fig 2.11 Different types of epithelial tissue

For each category, a whole cell is shown on the left and a longitudinal section is shown on the right.

CLASSIFICATION OF EPITHELIAL TISSUE

Simple epithelium – simple epithelium is a single layer of cells. It is usually very thin and functions in absorption, secretion and filtration. It includes:

Simple squamous (pavement) epithelium

Description: A single layer of flat squamous cells.

Location example: Lines blood vessels, lymphatic vessels, air-sacs of lungs and forms serous membranes.

Function: Filtration, diffusion, osmosis and secretion.

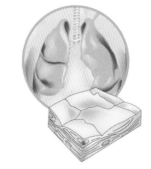

Simple cuboidal epithelium

Description: A single layer of cube-shaped cells.

Location example: Covers the surface of the ovaries, lines kidney tubules and forms the ducts of many glands.

Function: Secretion and absorption.

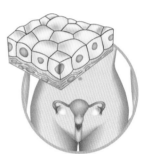

Simple columnar epithelium

Description: A single layer of rectangular cells. May contain goblet cells which produce a lubricating mucus and microvilli which are finger-like projections that increase the layer of the plasma membrane.

Location example: Lines the gastrointestinal tract.

Function: Secretion and absorption.

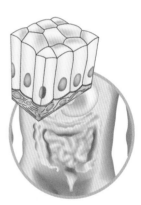

Ciliated simple columnar epithelium

Description: A single layer of rectangular cells that contain hair-like projections called cilia, which help move substances.

Location example: Lines part of the upper respiratory tract, the Fallopian tubes, and some of the sinuses.

Function: Moves fluids or particles along a passageway.

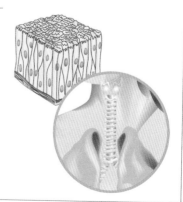

CLASSIFICATION OF EPITHELIAL TISSUE

Stratified epithelium – *Stratified epithelium consists of two or more layers of cells. It is durable and functions in protecting underlying tissues in areas of wear and tear. Some stratified epithelia also produce secretions. Stratified epithelium includes:*

Stratified squamous epithelium

Description:	Consists of several layers of cells which are squamous in the superficial layer and cuboidal to columnar in the deep layers. It exists in keratinised and non-keratinised forms. Keratin is a tough, waterproof protein that is resistant to friction and helps repel bacteria.
Location example:	The superficial layer of the skin is a keratinised form while wet surfaces such as the mouth and tongue are non-keratinised.
Function:	Protection.

Transitional epithelium

Description:	Consists of layers of cells that change shape when the tissue is stretched.
Location example:	Organs of the urinary system, for example the bladder.
Function:	Allows for distension.

Connective tissue

Connective tissue is the most abundant and widely distributed tissue in the body and takes many forms. For example, fat, blood and bone are all types of connective tissue.

The characteristics common to all types of connective tissue are:
- Cells are surrounded and separated by a matrix made of protein fibres and a fluid, gel or solid ground substance. This ground substance is usually produced by the connective tissue cells and deposited in the space between the cells.
- Tissue has a rich blood supply except for cartilage and tendons.
- Tissue has a nerve supply except for cartilage.

Classification of connective tissue is not always clear-cut but it can be classified as in the chart on the next page.

CLASSIFICATION OF CONNECTIVE TISSUE

Loose connective tissue – *loose connective tissue has loosely woven fibres and many cells. It is a softer connective tissue and includes:*

Areolar tissue

Description: This is a loose, soft, pliable, semifluid tissue and is the most widely distributed type of connective tissue in the body. It contains collagen, elastic and reticular fibres which give areolar tissue its elasticity and extensibility (see Chapter 3 for more information on these fibres).

Location example: Surrounds body organs and is found in the subcutaneous layer of the skin.

Function: Strength, elasticity and support (it is often the glue that holds the internal organs in their positions).

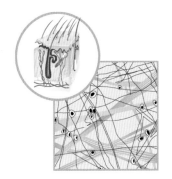

Adipose tissue

Description: This is an areolar tissue that contains mainly fat cells.

Location example: Subcutaneous layer of the skin, surrounding organs such as the kidneys and heart and found in the mammary glands.

Function: Insulation, energy reserve, support and protection.

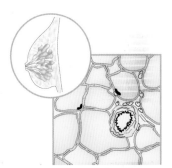

Lymphoid (reticular) tissue

Description: Contains thin fibres that form branching networks to form the stroma (a support framework for many soft organs).

Location example: Lymph nodes, the spleen, red bone marrow.

Function: Support.

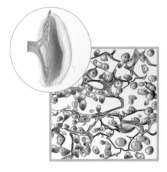

CLASSIFICATION OF CONNECTIVE TISSUE

Dense connective (fibrous) tissue – *dense connective tissue contains thick, densely packed fibres and fewer cells than loose connective tissue. It forms strong structures and there are different types of dense connective tissue, including elastic connective tissue and dense regular connective tissue.*

Elastic connective tissue (yellow elastic tissue)

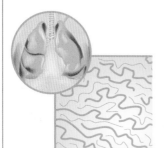

Description: Consists of many freely branching elastic fibres, few cells and little matrix. The elastic fibres give the tissue a yellowish colour.

Location example: Lung tissue, walls of arteries, trachea, bronchial tubes.

Function: Elasticity for stretching and strength.

Dense regular connective tissue (white fibrous tissue)

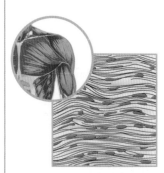

Description: Consists mainly of collagen fibres arranged in parallel bundles with cells in between the bundles. The matrix is a shiny white colour.

Location example: Tendons, ligaments, aponeuroses.

Function: Attachment.

Bone (osseous tissue) – *bone is an exceptionally hard connective tissue that protects and supports other organs of the body. It consists of bone cells sitting in lacunae (cavities) surrounded by layers of a very hard matrix that has been strengthened by inorganic salts such as calcium and phosphate. It includes compact and cancellous bone.*

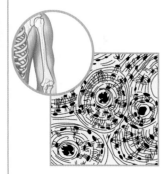

Description: Compact bone is the hard, outer more dense type of bone while cancellous bone lies internal to the compact bone and is spongy.

Location example: Bones.

Function: Support, protection, movement and storage.

CLASSIFICATION OF CONNECTIVE TISSUE

Cartilage – *Cartilage is a resilient, strong connective tissue that is less hard and more flexible than bone. It consists of a dense network of collagen and elastic fibres embedded in a rubbery ground substance. Cartilage has no blood vessels or nerves and includes:*

Hyaline cartilage

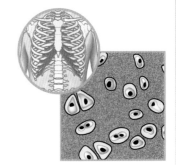

Description: This is the most abundant cartilage in the body but is also the weakest. It consists of a resilient gel-like ground substance, fine collagen fibres and cells.

Location example: Ends of long bones, ends of ribs, nose.

Function: Movement, flexibility and support.

Fibrocartilage

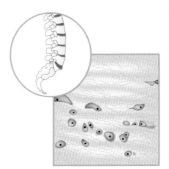

Description: This is the strongest of the three types of cartilage and consists of cells scattered between bundles of collagen fibres. It is a strong and rigid cartilage.

Location example: Intervertebral discs.

Function: Support and strength.

Elastic cartilage

Description: This is a strong and elastic cartilage consisting of cells in a threadlike network of elastic fibres within the matrix.

Location example: Eustachian tubes, supports external ear and epiglottis.

Function: Supports and maintains shape.

Blood (vascular tissue) – *blood is an unusual type of connective tissue whose matrix is a fluid called blood plasma. Plasma consists mainly of water with a variety of dissolved substances such as nutrients, wastes, enzymes, gases and hormones in it. Blood also contains:*

- *Red blood cells (erythrocytes) – these transport oxygen to body cells and remove carbon dioxide from them.*

- *White blood cells (leucocytes) – these are involved in phagocytosis (the destruction of microbes, cell debris and foreign matter), immunity and allergic reactions.*

- *Platelets – these function in blood clotting.*

Muscle tissue

The fibres of muscle tissue are actually elongated cells that provide a long axis for contraction. Thus, muscle tissue is able to shorten (contract) to produce movement. Muscle tissue also maintains posture and generates heat. Muscle tissue is characterised according to its location and function. There are three types: skeletal, cardiac and smooth (visceral).

CLASSIFICATION OF MUSCLE TISSUE

Skeletal muscle tissue

Description: Skeletal muscle tissue is attached to bones and composed of long, cylindrical shaped fibres with multiple nuclei. The fibres are also striated which means that when looked at under a microscope alternating light and dark bands are seen. Skeletal muscles are under conscious control and are therefore called voluntary muscles.

Location example: Attached to bones by tendons.

Function: Motion, posture and heat production.

Cardiac muscle tissue

Description: Cardiac muscle tissue forms most of the wall of the heart and consists of branched, striated fibres. Cardiac tissue is not under conscious control. It is regulated by its own pacemaker and the autonomic nervous system and is therefore an involuntary muscle.

Location example: Heart wall.

Function: Pumps blood.

Smooth (visceral) muscle tissue

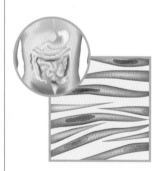

Description: Smooth muscle tissue is so called because it contains non-striated (smooth) fibres. It is not under conscious control. It is regulated by the autonomic nervous system and is therefore an involuntary muscle.

Location example: Walls of hollow structures such as blood vessels and intestines.

Function: Constricts and dilates structures to move substances within the body (eg. blood in blood vessels, foods through gastrointestinal tract).

Nervous tissue

Nervous tissue is made up of neurons and neuroglia and it is found in the brain, spinal cord and nerves. Its function is that of communication – of being sensitive to stimuli, generating and conducting impulses.

Neurons are composed of a cell body whose cytoplasm is drawn out into a long extension. These extensions are called dendrites or axons and they receive and conduct electrochemical impulses. Neuroglia are the supporting cells that insulate, support and protect the neurons. They do not generate or conduct nerve impulses.

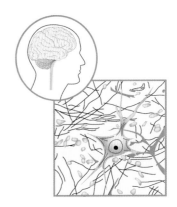

Membranes

Membranes are thin, flexible sheets made up of different tissue layers. They cover surfaces, line body cavities and form protective sheets around organs. They are categorised as epithelial and connective tissue (synovial) membranes.

Epithelial membranes

Epithelial membranes consist of an epithelial layer and an underlying connective tissue layer and include mucous, serous and cutaneous membranes.

EPITHELIAL MEMBRANES

Mucous membranes

Description:	Mucous membranes line body cavities that open directly to the exterior. They are 'wet' membranes whose cells secrete mucous which prevents cavities from drying out, lubricates food as it moves in the gastrointestinal tract and traps particles in the respiratory tract (note: urine, not mucous, moistens the urinary tract). Mucous membranes also act as a barrier that is difficult for pathogens to penetrate.
Location example:	The hollow organs of the digestive, respiratory, urinary and reproductive systems.
Function:	Lubrication, movement and protection.

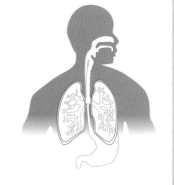

EPITHELIAL MEMBRANES

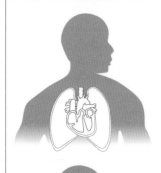

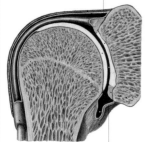

Serous membranes

Description: Serous membranes line body cavities that do not open directly to the exterior and it covers the organs that lie within those cavities. They are composed of two layers: the parietal layer is attached to the cavity wall and the visceral layer is attached to the organs inside the cavity. Between these two layers is a watery lubricating fluid called the serous fluid. This fluid enables the organs to move and glide against each other easily. The layers of the serous membrane consist of thin layers of areolar connective tissue covered by a layer of simple squamous epithelium.

Location example: There are three serous membranes in the body:
- *Pleura* – lines the thoracic cavity and covers the lungs.
- *Pericardium* – lines the cardiac cavity and covers the heart.
- *Peritoneum* – lines the abdominal cavity and covers the abdominal organs and some of the pelvic organs.

Function: Lubrication and protection.

Cutaneous membrane

Description: This is the skin and is composed of a superficial keratinising stratified squamous epithelium and an underlying dense connective tissue layer. It will be discussed in more detail in the following chapter.

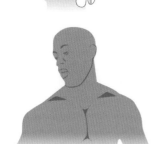

Synovial membrane (synovial sheath)

Synovial membrane (synovial sheath)

Fig 2.12 Synovial membrane

Connective tissue membranes (synovial membranes)

Connective tissue membranes are known as synovial membranes and are composed of areolar connective tissue with elastic fibres and fat. They line:

- The cavities of freely movable joints and secrete synovial fluid which lubricates the joints.
- Bursae and tendon sheaths which provide cushioning.

Tissues in a nutshell
- Epithelial tissue lines and protects.
- Connective tissue connects.
- Muscle tissue moves.
- Nervous tissue communicates.
- Membranes cover.

System Level Of Organisation Of The Body

You have studied the body at the chemical, cellular and tissue levels and now it is time to briefly look at the system level of organisation. You will learn more about the systems and their organs in the following chapters.

There are certain things you need to do in order to survive. For example, you need to be able to do such things as eat, breathe and move. The systems of our body carry out these crucial life processes and each system is composed of several related organs that share a common function. These systems are the:

- Integumentary system (the skin, hair and nails)
- Skeletal system
- Muscular system
- Nervous system
- Endocrine system
- Respiratory system
- Cardiovascular system
- Lymphatic and immune system
- Digestive system
- Urinary system
- Reproductive system

These systems work together to ensure that the body maintains a stable internal environment despite any changes in its external environment. This process is known as homeostasis.

Integumentary system – *protects the body; regulates body temperature; eliminates waste; helps produce vitamin D; contains nerve endings sensitive to pain, temperature and touch.*

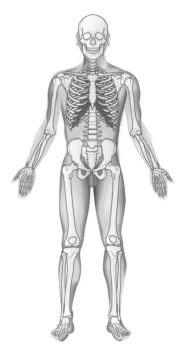

Skeletal system – *provides movement, support and protection; stores minerals; houses the cells that create blood cells.*

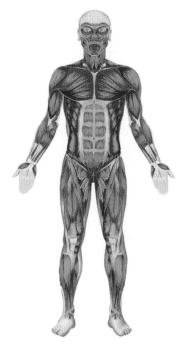

Muscular system – *powers movement; maintains posture; generates heat.*

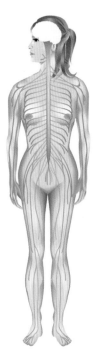

Nervous system – *regulates body activities through detecting, processing and responding to change in both the external and internal environments.*

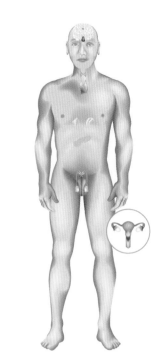

Endocrine system – *regulates body activities through hormones.*

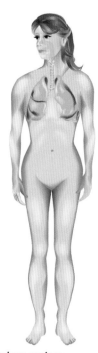

Respiratory system – *supplies oxygen and removes carbon dioxide; helps produce vocal sounds.*

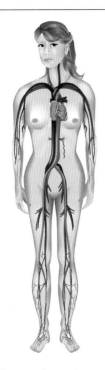

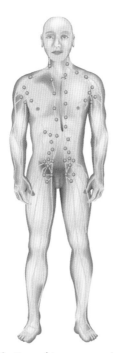

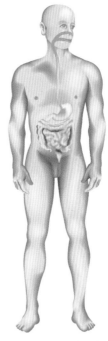

Cardiovascular system – transports blood which carries oxygen, carbon dioxide, nutrients and waste to and from cells; regulates body temperature.

Lymphatic and immune system – functions in immunity, protection and waste removal; returns proteins and plasma to the cardiovascular system; transports fats.

Digestive system – breaks down food and absorbs nutrients; eliminates waste.

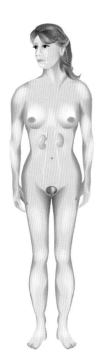

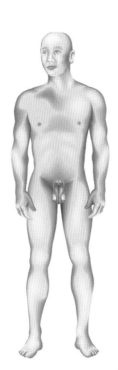

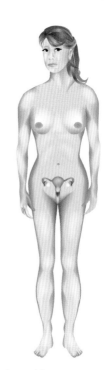

Urinary system – eliminates waste; regulates water, electrolyte and acid-base balance of blood.

Reproductive system – reproduces life.

NEW WORDS

Autonomic nervous system	part of the nervous system responsible for the control of functions that are not under conscious control (for example, the beating of the heart).
Cistern	a channel or tubule in the cell.
Cytology	the study of cells.
Diploid	having two complete sets of chromosomes per cell.
Enzyme	a protein that speeds up the rate of a reaction without itself being used in the reaction.
Equilibrium	balance.
Gene	the basic unit of genetic material.
Histology	the study of tissues.
Homeostasis	the process by which the body maintains a stable internal environment.
Hydrophilic	water-loving.
Hydrophobic	water-hating.
Kinetic energy	the energy of motion.
Metabolic	see *metabolism*.
Metabolism	the changes that take place within the body to enable its growth and function.
Phagocytosis	the engulfment and destruction of microbes, cell debris and foreign matter by phagocytes, which are a type of white blood cell.
Solvent	a liquid in which a solid is dissolved.
Somatic	any cell except the reproductive cells.
Vacuole	a space within the cytoplasm of a cell that contains material taken in by the cell.
Vesicle	a small, fluid-filled sac.

Study Outline

Levels of structural organisation of the body

The body can be divided into 6 levels of structural organisation. From the simplest to the most complex level they are: chemical, cellular, tissue, organ, system and organismic.

Chemical organisation of the body

1. Atoms are the smallest unit of matter.
2. Elements consist of uniquely structured atoms.
3. The four major elements in the body are carbon, hydrogen, oxygen and nitrogen.
4. The nine lesser elements in the body are calcium, phosphorous, potassium, sulphur, sodium, chlorine, magnesium, iodine and iron.
5. The thirteen trace elements in the body are aluminium, boron, chromium, cobalt, copper, fluorine, manganese, molybdenum, selenium, silicon, tin, vanadium and zinc.
6. Chemical reactions occur when atoms combine with or break apart from other atoms to form new products.
7. Molecules are the combination of two or more atoms.
8. Compounds are molecules of different types of atoms.
9. Water is the most abundant substance in the body. It functions in maintaining body temperature, acting as a lubricant, providing cushioning and being a solvent.
10. Carbohydrates are the fuel of the body and are made up of simple sugars (monosaccharides).
11. Lipids (fats) insulate and protect the body, are a source of stored energy and form an integral part of many structures and functions of the body. They are made up of glycerol and fatty acids.
12. Proteins are the building blocks of the body and are made up of amino acids.
13. Nucleic acids are found inside our cells. Deoxyribonucleic acid (DNA) makes up chromosomes which contain our genes. Ribonucleic acid (RNA) is used to create specific proteins as per the genetic code.
14. Adenosine triphosphate (ATP) is the main energy-transferring molecule in the body.
15. Nutrients are essential for the growth, maintenance and repair of the body. Macronutrients are carbohydrates, proteins, fats, essential minerals and water. Micronutrients are vitamins and trace minerals.
16. Vitamins are organic compounds required in minute amounts for the normal functioning of the body.
17. Minerals play a role in body growth and maintenance.
18. Essential fatty acids are fats that are vital for the proper functioning of the body.
19. Free radicals are highly unstable, reactive molecules that can damage cells.
20. Antioxidants are substances that combat or neutralise free radicals.

Cellular organisation of the body

1. Cytology is the study of cells.
2. Cells are the basic structural and functional unit of the body.
3. Cells are made up of a nucleus, cytoplasm, plasma membrane and various organelles.
4. The plasma membrane is a barrier that surrounds the cells and regulates the movements of substances into and out of it. It consists of lipids and proteins.
5. Lipids make up the structure of the membrane which is a phospholipid bilayer that also contains glycolipids and cholesterol.
6. Proteins are scattered in the phospholipid bilayer and act as channels, transporters and receptors.
7. Materials are moved in and out of the cell through two types of transport: passive processes and active processes. Passive processes are simple diffusion, osmosis and facilitated diffusion. Active processes are active transport and vesicular transport.
8. Simple diffusion is the movement of substances from areas of high concentration to areas of low concentration.
9. Osmosis is the movement of water from areas of high water concentration to areas of lower water concentration across a selectively permeable membrane.
10. In facilitated diffusion, substances are helped across the plasma membrane by channel or transporter proteins.
11. In active transport, pumps powered by ATP push molecules across the plasma membrane.
12. In vesicular transport, vesicles carry particles across the plasma membrane. Substances enter cells through endocytosis and leave cells through exocytosis.

> **Enter = endo**
> **Exit = exo**

13. The cytoplasm is all the cellular material inside the plasma membrane excluding the nucleus. Cytoplasm contains cytosol, organelles and inclusions.
14. Organelles are mitochondria, ribosomes, endoplasmic reticulum, Golgi complex, lysosomes, peroxisomes, centrosomes and centrioles.
15. Mitochondria are the powerhouses of the cells.
16. Ribosomes are where proteins are made.
17. Endoplasmic reticulum (ER) transports molecules and Rough ER contains ribosomes which are the site for protein synthesis. Smooth ER have no ribosomes and are the site for lipid synthesis and detoxification.
18. The Golgi complex processes, sorts and packages proteins and lipids.
19. Lysosomes break down and recycle molecules.
20. Peroxisomes detoxify harmful substances.
21. Centrosomes contain centrioles and function in cell division.
22. Centrioles function in cell division and play a role in the formation of flagella and cilia.
23. The nucleus is the control center of the cell. It is surrounded by a membrane called the nuclear envelope/membrane and it contains a nucleolus and chromatins.

24. The nucleolus is where ribosomes are made.
25. Chromatins contain 46 chromosomes and they separate during cell division. Chromosomes are made up of DNA which carries the genetic material of the cell.
26. Cells reproduce themselves through the process of cell division.
27. In somatic cell division, a single parent cell duplicates itself into two identical daughter cells. This process is called mitosis and it happens in four stages: prophase, metaphase, anaphase and telophase.

> **P-MAT**
> Prophase – Metaphase – Anaphase – Telophase

28. In reproductive cell division, sperm and egg cells are produced. This process is called meiosis.

Tissue level of organisation

1. Tissues are groups of cells that share a common function.
2. Histology is the study of tissues.
3. There are four types of tissue in the body: epithelial, connective, muscle and nervous.

> Emergencies Create Nervous Muscles
> Epithelial, Connective, Nervous, Muscle

4. Epithelial tissue forms body linings and glands and functions in protection, absorption and filtration.
5. There are four types of cells that make up epithelial tissue: squamous (thin and flat), cuboidal (cube shaped), columnar (column shaped) and transitional (changing shapes).
6. Simple epithelium is a single layer of cells. There are four types: squamous, cuboidal, columnar and ciliated. It is very thin and functions in absorption, secretion and filtration. It lines vessels such as blood and lymph vessels, and the respiratory and gastrointestinal tracts.
7. Stratified epithelium consists of two or more layers of cells. There are two types: stratified squamous and transitional. It is durable and functions in protecting underlying tissues in areas of wear and tear. Some also produce secretions. Examples include the skin and the organs of the urinary system.
8. Connective tissue is the most abundant and widely distributed tissue in the body. Cells are surrounded and separated by a matrix made of protein fibres and a fluid, gel or solid ground substance. There are many different types of connective tissue including: loose connective tissue (areolar, adipose, lymphoid), dense connective tissue (yellow elastic, white fibrous), bone, cartilage (hyaline, fibrocartilage, elastic) and blood.

> Lucy Always Ate Lizards Despite their Yellow and White Blood.
> Carl Hated the Fibres Enveloped in their Bones.
> Loose (Areolar, Adipose, Lymphoid), Dense (Yellow elastic, White fibrous), Blood. Cartilage (Hyaline, Fibrocartilage, Elastic), Bone.

9. Loose connective tissue has loosely woven fibres and many cells. It is a softer connective tissue that includes areolar tissue, adipose tissue and lymphoid tissue. Examples include the tissue that surrounds the organs of the body and lymph nodes.

10. Dense connective tissue contains thick, densely packed fibres and fewer cells than connective tissue. It forms strong structures and includes elastic connective tissue and dense regular connective tissue. Examples include lung tissue, tendons and ligaments.

11. Bone tissue consists of a very hard matrix that has been strengthened by inorganic salts. It includes compact and cancellous bone and it functions in protection and support.

12. Cartilage consists of a dense network of collagen and elastic fibres embedded in a rubbery ground substance. It has no blood vessels or nerves and it functions in movement, flexibility and support. It includes hyaline cartilage, fibrocartilage and elastic cartilage.

13. Blood consists of a fluid matrix called plasma and red blood cells, white blood cells and platelets.

14. Muscle tissue is made of elongated cells and is able to contract and produce movement. It also helps maintain posture and generates heat. There are three types of muscle tissue: skeletal, cardiac and smooth (visceral).

15. Skeletal muscle is attached to bones and composed of long, cylindrical shaped, striated fibres with multiple nuclei. They are voluntary muscles.

16. Cardiac muscle forms most of the wall of the heart and consists of branched, striated fibres. It is involuntary muscle tissue.

17. Smooth muscle moves substances within the body and consists of non-striated or smooth fibres. It is an involuntary tissue.

18. Nervous tissue is made up of neurons and neuroglia and functions in communication. It is found in the brain, spinal cord and nerves.

Membranes

1. Membranes are thin, flexible sheets made up of different tissue layers. They cover surfaces, line body cavities and form protective sheets. There are two groups of membranes: epithelial and connective tissue.

2. Epithelial membranes consist of epithelial tissue and an underlying layer of connective tissue. They include mucous membranes which line body cavities that open directly to the exterior, for example, the hollow organs of the respiratory system; and serous membranes which line body cavities that do not open directly to the exterior, for example, the pleura.

3. Connective tissue membranes are synovial membranes and are composed of areolar connective tissue with elastic fibres and fat. They line joints, bursae and tendon sheaths.

System level of organisation

1. Systems are groups of organs that share a common function.

2. The systems of the body are the Integumentary System, Skeletal System, Muscular System, Nervous System, Endocrine System, Respiratory System, Cardiovascular System, Lymphatic System, Digestive System, Urinary System and the Reproductive System.

3. Homeostasis is the process by which the body maintains internal equilibrium despite changes in its external environment.

Revision

1. Name the six levels of organisation of the body.
2. Define the following:
   - Atom
   - Element
   - Molecule
   - Compound
3. Name the four major elements that make up the body.
4. Describe the importance of water in the body.
5. Describe the term carbohydrate. List the functions of carbohydrates.
6. Describe the term lipid. List the functions of lipids.
7. Describe the term protein. List the functions of proteins.
8. Explain what a cell is.
9. List the functions of the following:
   - Nucleus
   - Mitochondria
   - Ribosomes
   - Golgi complex
   - Lysosomes
   - Peroxisomes
10. Describe how somatic cells reproduce.
11. Put the following processes into their chronological order: telophase, metaphase, anaphase, prophase.
12. Describe cytokinesis.
13. Describe the term 'tissue'.
14. Name the four major types of tissues in the body. For each one give an example.
15. Name the major systems of the body and briefly describe their functions.

Multiple choice questions

1. **Which of the following statements is correct?**
   a. Oxygen, carbon, cholesterol and nitrogen are major elements in the body.
   b. Selenium, lipids, chromium and nitrogen are major elements in the body.
   c. Oxygen, carbon, hydrogen and nitrogen are major elements in the body.
   d. Ergosterol, carbon, hydrogen and chromium are major elements in the body.

2. **Water is the:**
   a. Solvent in body fluids.
   b. Solute in body fluids.
   c. Solution in body fluids.
   d. None of the above.

3. **Carbohydrates are:**
   a. Atoms
   b. Elements
   c. Compounds
   d. Matter.

4. **Glycogen and cellulose are:**
   a. Proteins
   b. Lipids
   c. Carbohydrates
   d. None of the above.

5. **Triglycerides are:**
   a. Proteins
   b. Lipids
   c. Carbohydrates
   d. None of the above.

6. **The building blocks of the body are:**
   a. Proteins
   b. Lipids
   c. Carbohydrates
   d. None of the above.

7. **Which of the following statements is false?**
   a. Nucleic acids are found inside cells.
   b. Nucleic acids make up our genes.
   c. DNA and RNA are nucleic acids.
   d. Nucleic acids are vitamins.

8. **The fat soluble vitamins are:**
   a. B, C
   b. A, D, E, K
   c. B, C, D, K
   d. A, D, C.

9. **Cytology is the study of:**
   a. Tissue
   b. Plants
   c. Cells
   d. Chemicals.

10. **The plasma membrane:**
    a. Surrounds the cell.
    b. Surrounds the nucleus.
    c. Surrounds the organ.
    d. Surrounds the mitochondria.

11. **Which of the following are passive forms of transport across the cell membrane:**
    a. Simple diffusion and vesicular transport.
    b. Osmosis and facilitated diffusion.
    c. Active transport and simple diffusion.
    d. Facilitated transport and vesicular transport.

12. **Which of the following is an organelle:**
    a. Plasma membrane
    b. Cytosol
    c. Cytoplasm
    d. Ribosome.

13. **The process by which a parent cell duplicates itself into two identical daughter cells is called:**
    a. Meiosis
    b. Mitosis
    c. Somosis
    d. Somotosis.

14. **What happens during interphase?**
    a. A cell grows and DNA replicates itself.
    b. A cell grows and divides itself.
    c. A cell grows and dies.
    d. A cell grows and DNA does not replicate itself.

15. **Which of the following is a type of connective tissue?**
    a. Smooth muscle tissue
    b. Stratified squamous epithelium
    c. Nerves
    d. Bone.

16. **Which type of tissue initiates and transmits impulses?**
    a. Muscle
    b. Connective
    c. Epithelial
    d. Nervous.

17. **Simple squamous epithelium consists of:**
    a. Two or more layers of flat cells.
    b. A single layer of flat cells.
    c. Two or more layers of rectangular cells.
    d. A single layer of rectangular cells.

18. **Which of the following statements is correct?**
    a. Bone tissue has a rubbery ground substance.
    b. Cartilage has a very hard matrix that has been strengthened by inorganic salts.
    c. Blood has a matrix called plasma.
    d. None of the above.

19. **Cardiac muscle is:**
    a. Striated and involuntary.
    b. Striated and voluntary.
    c. Non-striated and involuntary.
    d. Non-striated and voluntary.

20. **Membranes are:**
    a. Sheets made up of neurons and neuroglia.
    b. Sheets made up of the same tissue layers.
    c. Sheets made up of different tissue layers.
    d. Sheets made up of fibres.

3 The Skin, Hair and Nails

Introduction

Which is the largest organ in your body?

Here is a clue – in an average adult it can weigh almost 5kg and cover an area about the size of a large dining room table. Although it is so large, it can be as thin as ½ mm and yet it is the only solid protection we have against the environment. In addition, it reflects the state of our health and emotions. You guessed it… the skin![i]

In this chapter you will discover the intricacies that make up the skin, hair and nails and you will see how they combine to form the integumentary system.

Student objectives

By the end of this chapter you will be able to:

- Explain the functions of the skin.
- Describe the anatomy of the skin.
- Describe the structure and function of the hair.
- Describe the structure and function of the nail.
- Describe the structure and function of the cutaneous glands.
- Identify common pathologies of the skin, hair and nails.

Skin

Did you know?
It is normal to lose about 500ml of sweat per day – that's about 1.5 Coke cans.

Functions of the skin

The skin is the ultimate overcoat – it keeps us warm when it is cold and cool when it is hot; it protects us from sunlight, rain and injury; it secretes substances to keep itself in good condition; and, in addition to all of this, it even repairs itself when cut or torn.

Heat regulation

The body works hard at keeping its normal internal temperature around 37°C (98.6°F). It does this by adapting to environmental changes in temperature by either cooling itself down in a warm environment or warming itself up in a cool environment. The body also needs to adapt to internal changes in temperature in times of fever or exercise. It regulates its temperature by:

- Cooling itself down through:
    - **Sweating** – Tiny glands release sweat onto the skins surface. From here the sweat evaporates, drawing heat from the body and cooling it down.
    - Vasodilation – Blood vessels dilate and this causes blood to rush to the capillaries in the surface of the skin. From here the heat in the blood is lost through radiation. The skin also appears pinker and warmer.

- Warming itself up through:
    - Decreased sweat production.
    - Vasoconstriction – Blood vessels constrict and this reduces the flow of blood through the capillaries. Thus the heat in the blood is conserved. The skin also appears whiter and cooler.
    - **Goosebumps** – The arrector pili muscles contract and cause the hairs to stand on end in what is known as goose bumps. This traps a layer of warm air close to the surface of the body.
    - **Shivering** – Our muscles contract more frequently to create heat within the body.

Sensation

The dermis of the skin contains nerve endings that are known as cutaneous sensory receptors. They are sensitive to:

- Touch
- Temperature
- Pressure
- Pain.

Some areas have more receptors than others and are therefore more sensitive. For example, the lips and fingertips are extremely sensitive areas of the body.

Protection

What would happen to you if your skin was only a membrane that held your insides in? You would go swimming, absorb all the water in the pool and become waterlogged. The millions of bacteria in your environment would enter your body and make you ill. The moment you went outdoors into the sun you would burn. As you can see, the skin is a unique barrier that protects you from a range of things including:

- **Physical trauma and abrasion** – the skin insulates and cushions the organs and structures of the body.
- **Bacterial invasion** – the skin secretes acidic substances which form an acid mantle. This inhibits the growth of bacteria. The skin also contains phagocytes which ingest foreign substances and pathogens.
- **Dehydration** – keratin in the skin forms a waterproof barrier which prevents water from passing into and out of the body. This prevents us from literally drying-out.
- **Ultraviolet radiation** – melanin helps to protect us from the sun's damaging ultraviolet radiation.
- **Chemical damage** – the skin acts as a physical barrier to many potentially harmful substances.
- **Thermal damage** – the skin acts as a barrier which can help protect us from burns and scalds.

It is also important to remember that one of the main functions of the skin is that of sensation. The skin is sensitive to touch, pressure, pain and temperature. This gives it further protective properties in that it enables the body to react to stimuli and so protect itself from further injury.

Absorption
Although the skin is a waterproof, protective barrier, certain substances can still be absorbed through the skin. These substances include:

- Fat-soluble substances such as vitamins A, D, E and K.
- Topical steroids used to treat skin conditions such as eczema.
- Drugs used in transdermal patches. For example, hormones used in Hormone Replacement Therapy and nicotine used in nicotine patches.
- Some toxic chemicals such as mercury.
- Essential oils used in aromatherapy.

Excretion
The skin excretes wastes from the body. These wastes include:

- Urea
- Salts
- Water
- Aromatic substances (e.g. garlic).

Secretion
Sebaceous glands found in the skin secrete sebum which is a fatty substance that keeps the skin supple and waterproof.

Synthesis of vitamin D
The skin contains cholesterol molecules which, in the presence of sunlight, are converted to calcitriol, an active form of vitamin D. Vitamin D is a pro-hormone that helps regulate the calcium levels of the body and is necessary for the growth and maintenance of bones.

Did you know?
Your body only needs one hour per week of sunlight on the hands, arms and face to synthesise vitamin D. That is less than ten minutes per day! Any additional exposure to the sun increases your risk of skin damage and skin cancer.

Minor functions of the skin

The functions discussed on the previous page are the major functions of the skin, however the skin also functions in:

- **Immunity** – cells such as Langerhans cells play an important role in the immune system.
- **Blood reservoir** – 8–10% of the body's total blood flow is in the dermis of a resting adult. In times of moderate exercise the dermal capillaries dilate and blood flow to the area increases to help cool the body. However, in times of strenuous exercise, dermal blood vessels actually constrict so that more blood can circulate to the muscles.
- **Communication** – our skin communicates information about both our health and our emotions.

Anatomy of the skin

The skin is composed of two layers: the outer epidermis which is continuously being worn away and the dermis which contains the nerves, blood vessels, sweat glands and hair roots.

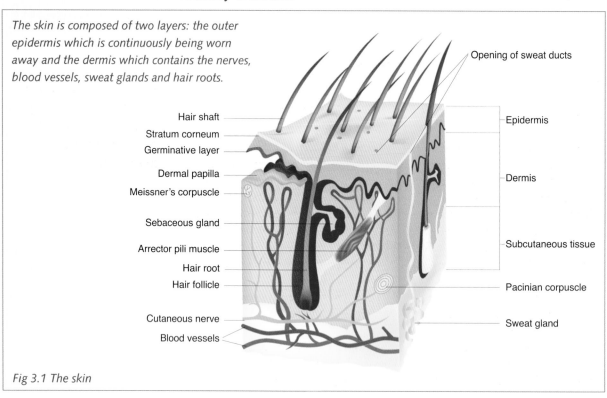

Hair shaft
Stratum corneum
Germinative layer
Dermal papilla
Meissner's corpuscle
Sebaceous gland
Arrector pili muscle
Hair root
Hair follicle
Cutaneous nerve
Blood vessels

Opening of sweat ducts
Epidermis
Dermis
Subcutaneous tissue
Pacinian corpuscle
Sweat gland

Fig 3.1 The skin

Dermatology is the study of the skin, which is a cutaneous membrane made of two distinct layers:

- the *epidermis* is a tough, waterproof outer layer that is continuously being worn away and
- the *dermis* lies beneath it and has a thicker layer that contains nerves, blood vessels, sweat glands and hair roots.

The dermis is attached to the subcutaneous layer which anchors the skin to the other organs in the body.

On the next page is a table briefly outlining the tissues that make up the skin.

NAME	TISSUE TYPE	GENERAL DESCRIPTION
Epidermis	Keratinised stratified squamous epithelium.	A thin layer of flat, dead cells that are continually being shed. This is a waterproof and protective layer.
Dermis	Connective tissue containing collagen and elastic fibres.	A thicker layer that supports the epidermis above it by enabling the passage of nutrients and oxygen. It also allows the skin to move, absorbs shocks, and cools and warms the body.
Subcutaneous (also called superficial fascia or hypodermis)	Areolar and adipose tissue.	This is not considered part of the skin but it anchors the skin to other organs and provides shock absorption and insulation.

Epidermis

The epidermis is composed of five layers of stratified squamous epithelial tissue that becomes tough and hard through a process called keratinisation.

Before looking at the different layers of the epidermis it helps to know a little about the four cells that are found in these layers.

- **Keratinocytes** – these make up about 90% of epidermal cells and they produce a protein called keratin. Keratin helps waterproof and protect the skin and the word itself, *kerato*, means 'horny'.
- **Melanocytes** – the word *melan* means 'black' and melanocytes produce a brown-black pigment called melanin. Melanin contributes to skin colour and absorbs ultraviolet light. Melanin granules actually form a protective layer over the nuclei of cells. This layer is only over the side of the nuclei that faces the surface of the skin and the melanin is like a sun-hat, protecting the nuclei from ultraviolet light.
- **Langerhans cells** – these arise from bone marrow and move to the epidermis. They respond to foreign bodies and thus play a role in skin immunity.
- **Merkel cells** – these are only found in the stratum basale of hairless skin and are attached to keratinocytes. They make contact with nerve cells to form Merkel discs that function in the sensation of touch.

We will now look at the five layers of the epidermis and the meanings of their names. This will give you an idea of the appearance of each layer.

Did you know?
A blister from a burn or excess friction is simply an area where the epidermis and dermis have separated.

Anatomy and physiology in perspective
The thickness of the skin varies depending on the area of the body. For example, the skin on the soles of the feet and the palms of the hands is thickest. This is because these areas are subject to a great deal of pressure and friction. On the other hand, the skin on the eyelids is very thin as it needs to be flexible and move quickly.

From the deepest to the superficial layer they are the:

- **Stratum basale** – means 'base' (also called the stratum germinativum which means 'to sprout'). This is the deepest layer of the epidermis and is the base from where new cells germinate or sprout.
- **Stratum spinosum** – 'thornlike or prickly'. This layer consists of prickly cells that are beginning to go through the process of keratinisation.
- **Stratum granulosum** – 'little grains'. This layer is made up of degenerating cells that are becoming increasingly filled with little grains or granules of keratin.
- **Stratum lucidum** – 'clear'. This is a waterproof layer of dead, clear cells.
- **Stratum corneum** – 'horny'. This is the outermost layer of the skin. Its cells are dead and now completely filled with keratin. Thus they are tough, durable and horny.

The epidermis is composed of five layers. Its cells are formed in the deepest layer and as they move upwards they become keratinised until only dead cells are found in the most superficial layer.

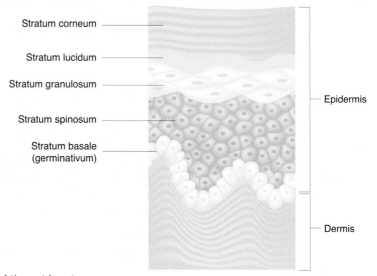

Stratum corneum

Stratum lucidum

Stratum granulosum

Stratum spinosum

Stratum basale (germinativum)

Epidermis

Dermis

Fig 3.2 The skin showing the main layers of the epidermis

Stratum basale (stratum germinativum) – basal cell layer

The stratum basale forms the deepest layer of the epidermis and is the layer closest to the dermis. Because it is close to the dermis it can receive nutrients and oxygen from it through the process of diffusion. This nourishment enables the stratum basale to produce millions of new cells each day.

The stratum basale has the following characteristics:

- It is composed of a single layer of cuboidal or columnar shaped cells that have nuclei.
- It is constantly producing new cells through cell division. This is the reason it is sometimes called the stratum germinativum.
- The cells are pushed upwards towards the superficial layers of the skin by newly produced cells below.
- The different cells present in the stratum basale are:
  - Stem cells for reproduction
  - Keratinocytes
  - Melanocytes
  - Merkel discs

Did you know?
When you shave the hairs off your skin, you are cutting off many layers of dead skin, yet you don't bleed. This is because the epidermis is avascular. That means it contains no blood vessels.

Stratum spinosum – prickle cell layer

The stratum spinosum consists of 8–10 layers of many-sided, irregular cells that appear to be covered with prickly thorns or spines. These spines join the cells tightly to one another. This is a transitional layer between the stratum basale and the stratum granulosum and it contains some dividing cells as well as cells that are beginning to go through the process of keratinisation.

Stratum granulosum – granular cell layer

The stratum granulosum consists of 3–5 layers of flattened cells whose nuclei are beginning to degenerate and die. This is because the cells are now far away from the nutrient-supplying dermis and they are becoming increasingly keratinised.

Stratum lucidum – clear cell layer

The stratum lucidum is composed of 3–5 layers of clear, flat, dead cells that have no nuclei and contain eleidin, the precursor to keratin. This layer is waterproof and most apparent in the thick skin on the palms of the hands and soles of the feet.

Stratum corneum – horny cell layer

The uppermost layer of the skin is the one that is exposed to all the harsh changes of the external environment as well as ultraviolet radiation, countless pathogens and chemicals. It is the layer that needs to protect the skin from the external environment and so it must be the strongest and toughest of all the layers. This is the stratum corneum which is made of 25–30 layers of flat, dead cells that are completely filled with keratin. This durable layer makes up about three quarters of the total thickness of the epidermis and it has the following characteristics:

- All cells are non-nucleated (i.e. they are dead).
- The cytoplasm of the cells has been replaced by the fibrous protein keratin.
- Cells are constantly shed through a process called desquamation.
- Cells are constantly replaced from beneath.

Think about it...
Is beauty only skin deep? When we look at a person we are really only seeing dead cells.

How does the sun affect your skin?

Ultraviolet (UV) light, emitted by the sun, is responsible for a great deal of damage to your skin. Excess exposure to it accelerates ageing of the skin, plays a significant role in the development of most types of skin cancer, suppresses the immune system and damages the retina of the eye. Parents of young children should be aware that most people receive 80% of their whole life's sun exposure before the age of twenty[ii] and it is therefore vital that they protect their children from the sun.

What is a sun tan?

Exposure to sunlight causes cells to produce more of the brown-black pigment melanin. This means the skin takes on a darker appearance or a sun tan. When those cells are finally shed through desquamation the melanin is lost and so the tan disappears.

Dermis

The dermis is the supportive layer beneath the epidermis and it is composed of connective tissue that contains both collagen and elastic fibres. It contains many different cells and structures, including:

- Fibroblasts – these are large, flat cells that synthesize the following fibres:
  - Collagen fibres – these are very tough, yet flexible, fibres that are resistant to a pulling force and give skin its extensibility (ability to stretch). Collagen fibres also attract and bind water and are thus responsible for keeping skin hydrated. They contain the protein collagen.
  - Elastic fibres – these are strong, thin fibres that give skin its elasticity (ability to return to its original shape after stretching). They can be stretched up to 150% of their relaxed length without breaking and contain the protein elastin.
  - Reticular fibres – these are thin fibres that form a branching network around other cells and give support and strength. They contain collagen protein coated with glycoprotein.
- Macrophages – these are cells that engulf and destroy bacteria and cell debris by a process called phagocytosis.
- Adipocytes – these are fat cells.
- Mast cells – these cells produce histamine which dilates small blood vessels during inflammation.
- Blood and lymphatic vessels.
- Nerves.
- Glands.
- Hair follicles.

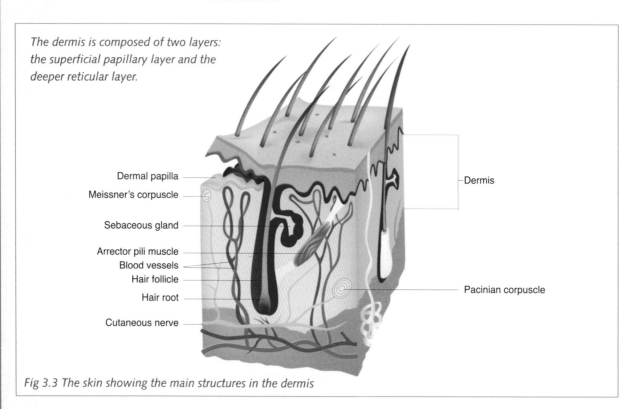

The dermis is composed of two layers:
the superficial papillary layer and the
deeper reticular layer.

Dermal papilla
Meissner's corpuscle
Sebaceous gland
Arrector pili muscle
Blood vessels
Hair follicle
Hair root
Cutaneous nerve

Dermis
Pacinian corpuscle

Fig 3.3 The skin showing the main structures in the dermis

The dermis can be divided into two layers:

- **Papillary layer** – so named because of its nipple-shaped projections that go into the epidermis (papilla means 'nipple').
- **Reticular layer** – whose varying thickness contributes to the varying thickness of the skin.

Papillary layer
The papillary layer is an undulating membrane that makes up approximately ⅕ of the thickness of the dermis. It is composed of areolar connective tissue and fine elastic fibres and has nipple-shaped fingerlike projections called papillae. These papillae go into the epidermis and:

- Greatly increase the surface area of the papillary layer.
- Contain loops of capillaries so that diffusion of nutrients and oxygen can take place between the dermis and epidermis.
- Some contain Meissner's corpuscles which are nerve endings that are sensitive to touch.

Anatomy and physiology in perspective
On the thick skin of the hand, the dermal papillae cause ridges in the overlying epidermis. These ridges are arranged in whorled and looped patterns that increase friction and act as suction cups so that you can pick up and grip objects. Sweat secretions into these ridges leave fingerprints on objects. Fingerprints are unique to each person and genetically determined. They develop in the embryo and do not change as a person grows – they only enlarge. Because of their uniqueness they are used to identify individuals.

Reticular layer
The reticular layer is deep to the papillary layer and is composed of dense, irregular connective tissue containing bundles of collagen and elastic fibres. These fibres give skin its extensibility, elasticity and strength. The reticular layer is the main support structure of the skin and it also contains:

- Hair follicles
- Nerves
- Oil glands
- Ducts of sweat glands
- Adipose tissue.

Subcutaneous layer
Although it is not part of the skin, it is important to learn about the subcutaneous layer as it is the tissue that attaches the reticular layer to the underlying organs. The subcutaneous layer contains:

- Areolar connective tissue
- Adipose tissue
- Lamellated corpuscles (Pacinian corpuscles) – these are nerve endings that are sensitive to pressure.

Skin types

Now that you know the structure and functions of the skin it is time to take a look at what you really see on a person and what distinguishes one person's skin from another. The common skin types are:

- **Normal (balanced) skin** – normal skin is a balanced skin in which there are no signs of oily or dry areas. It is actually a 'perfect' skin and is, of course, quite rare in adults. Normal skin:
  - has an even texture
  - has good elasticity
  - has small pores
  - feels soft and firm to the touch
  - is usually blemish free.
- **Oily skin** – in an oily skin there is an overproduction of sebum by the sebaceous glands. This can be caused by hormones, for example in puberty. Oily skin:
  - has an uneven texture
  - has normal elasticity
  - has large pores
  - feels thick and greasy to touch
  - often has blemishes such as comedones, papules, pustules and scars
  - appears sallow (slightly yellow) in colour and has a characteristic shine
  - ages slowly.
- **Dry skin** – in dry skin there is either an underproduction of sebum or a lack of moisture, or both together. Dry skin:
  - has flaky, dry patches and a thin, coarse texture
  - has poor elasticity
  - feels dry, coarse and papery to touch
  - looks like parchment and often has dilated capillaries around the cheek and nose areas
  - may also be sensitive
  - ages prematurely, especially around the eyes, mouth and neck.
- **Combination skin** – combination skin is a mixture of dry, normal and greasy skin and is the most common type of skin. Combination skin:
  - has an oily t-zone (forehead, nose and chin) which feels thick and greasy and is often blemished
  - has dry cheeks and neck which feel flaky and coarse and may have dilated capillaries
  - the tone and elasticity varies
  - may also have sensitive areas.
- **Sensitive skin** – sensitive skin often accompanies dry skin and is easily irritated. Sensitive skin:
  - is hypersensitive and reactions can include reddening, itching and chafing
  - often has dilated capillaries
  - is often dry and transparent
  - is usually warm to touch.

Anatomy and physiology in perspective

As we age our skin changes. We are born with soft, smooth skin that has a thick layer of fat and only a thin layer of protective keratin. It is not a very effective barrier against harmful substances.

As we grow our skin thickens and strengthens and by adulthood we have strong, supple skin that functions effectively in protecting us and regulating our body temperature.

Then, as we age, the skin begins to thin. In old age both the dermis and epidermis are thinner and a lot of the underlying fat layer that usually insulates us has disappeared. The number of sweat glands and blood vessels in our skin also decreases and this lowers our skin's ability to regulate our body temperature. Therefore, we become more susceptible to both the cold and the heat. The skin also loses its elasticity with old age and begins to wrinkle and sag and there is a decrease in melanocytes which causes an increase in sensitivity to the sun. Finally, a decrease in the number of nerve endings results in less sensitivity to external stimuli.

Hair

Hairs (or pili) grow over most of the body except the palms of the hands, the soles of the feet, the eyelids, lips and nipples. They come in different types:

- **Lanugo hair** – this is a soft hair that begins to cover a foetus from the third month of pregnancy. It is usually shed by the eighth month of pregnancy.
- **Vellus hair** – this is a soft and downy hair and it is found all over the body except the palms of the hands, soles of the feet, eyelids, lips and nipples.
- **Terminal hair** – this is a longer, coarser hair that is found on the head, eyebrows, eyelashes, under the arms and in the pubic area.

Structure of the hair

Hairs are columns of keratinised dead cells. When looked at longitudinally, they are made up of two parts:

- **Hair shaft** – the superficial end of the hair that projects from the surface of the skin. It is what we often call the hair 'strand'.
- **Hair root** – penetrates into the dermis.

Anatomy and physiology in perspective

It is the shape of the shaft that determines what type of hair you will have. For example, a round shaft produces straight hair, an oval shaft produces wavy hair and an elliptical or flat shaft produces curly hair.

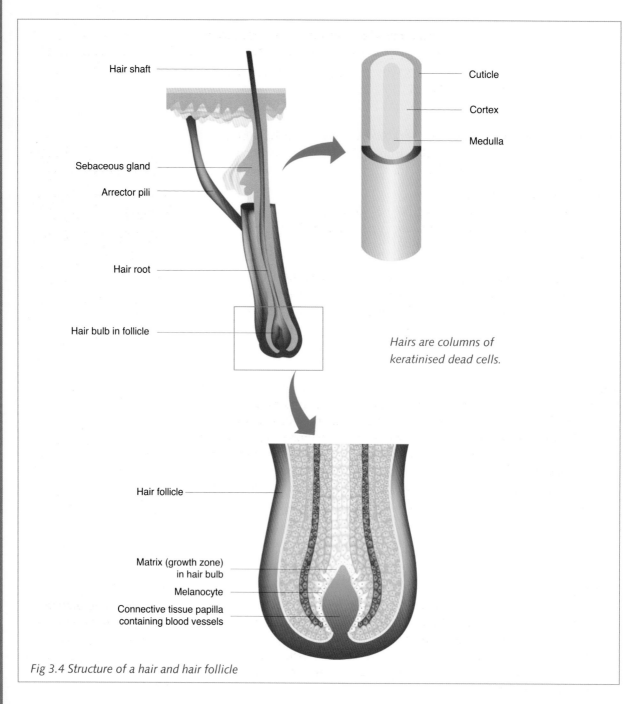

Hair shaft

Cuticle

Cortex

Medulla

Sebaceous gland

Arrector pili

Hair root

Hair bulb in follicle

Hairs are columns of keratinised dead cells.

Hair follicle

Matrix (growth zone) in hair bulb

Melanocyte

Connective tissue papilla containing blood vessels

Fig 3.4 Structure of a hair and hair follicle

Did you know?

It is normal to lose 70-100 hairs per day and hair loss can be accelerated by poor diet, illness, drugs, emotional stress and hormonal changes.

A cross-section of a hair shows that it consists of three concentric layers:

- **Inner medulla** – this consists of 2–3 rows of polyhedral cells and contains pigment granules and air spaces.
- **Middle cortex** – this is made up of elongated cells containing pigment granules in dark hair and air in white hair.
- **Outer cuticle** – this is a single layer of keratinised, flat, dead cells. These cells provide strength and help keep the inner layers of the hair tightly compacted.

Hairs grow out of hair follicles that surround the root of the hair. At the base of the follicle is the bulb which houses a nipple-shaped projection called the hair papilla. The papilla contains:

- Areolar and connective tissue.
- Blood vessels that provide nourishment to the cells.
- The matrix – a ring of cells which divide to create the hair.

Associated with hairs are:

- **Sebaceous oil glands** – secrete sebum which lubricates the hair.
- **Arrector pili muscles** – these are bands of smooth muscle cells that attach the side of the hair to the dermis. They contract to pull the hairs into a vertical position as a response to cold, fright or differing emotions.

Did you know?
Humans are born with all the hair follicles they will ever have and hair is one of the fastest growing tissues in the body.

Life cycle of a hair

The life cycle of a hair includes growing, transitional and resting stages. The growth stage is called *anagen* and in this stage a follicle reforms and the matrix divides to create a new hair. As with the skin, the older cells are pushed upwards as new cells develop below. The transitional stage is known as *catagen* and usually lasts 1–2 weeks. During this stage the hair separates from the base of the follicle as the dermal papilla breaks down. Finally, in the resting stage or *telogen*, the follicle is no longer attached to the dermal papilla and the hair moves up the follicle and is naturally shed.

Functions of the hair

Hair has two main functions:

- It helps to keep the body warm.
- It protects the body. For example:
  - Hair on the head guards the scalp against the harmful rays of the sun.
  - The eyebrows and eyelashes protect the eyes from foreign particles.
  - Nostril hairs prevent the inhalation of insects and other foreign bodies.
  - The hairs in the external ear canal protect the ears from insects and foreign particles.

Hair in a nutshell
Living cells in the matrix divide constantly and push the hair upwards. As the cells move upwards they fill with keratin and die, forming the hair shaft which warms and protects us.

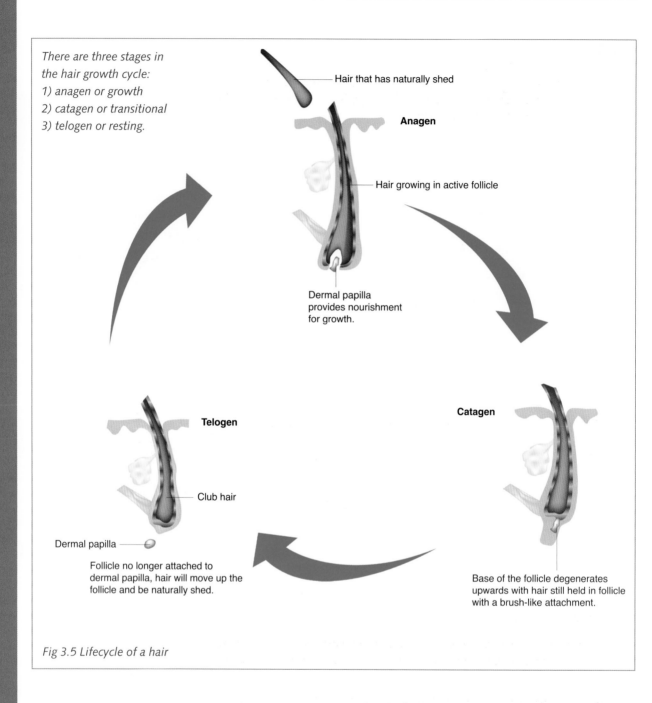

There are three stages in the hair growth cycle:
1) anagen or growth
2) catagen or transitional
3) telogen or resting.

Hair that has naturally shed

Anagen

Hair growing in active follicle

Dermal papilla provides nourishment for growth.

Telogen

Catagen

Club hair

Dermal papilla

Follicle no longer attached to dermal papilla, hair will move up the follicle and be naturally shed.

Base of the follicle degenerates upwards with hair still held in follicle with a brush-like attachment.

Fig 3.5 Lifecycle of a hair

Infobox

Anatomy and physiology in perspective

Men produce hormones called androgens that are responsible for the typical male-pattern of hair growth, such as beards and chest hair. These same hormones are, ironically, also responsible for male-pattern baldness. Women also produce small amounts of androgens and if, for some reason, they overproduce them, they can start to grow hair on their upper lip, chin, chest, inner thighs and abdomen.

Nails

Nails are hard plates that cover and protect the ends of the fingers and toes and enable us to pick up small objects easily. Onyx is a Latin word meaning nail and another term for a nail is unguis.

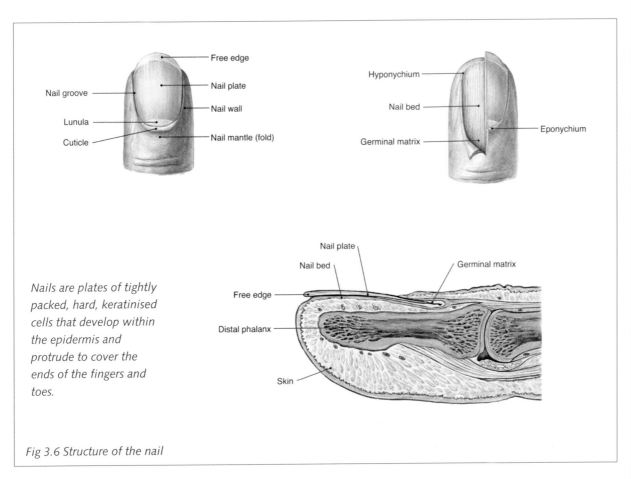

Nails are plates of tightly packed, hard, keratinised cells that develop within the epidermis and protrude to cover the ends of the fingers and toes.

Fig 3.6 Structure of the nail

Anatomy of the nail

For ease of learning, we will look at the nail in two parts. To begin with, we will look at the area beneath the nail plate. Starting at the proximal end is the:

- **Germinal matrix** – this area is rich in nerves and blood vessels and is the region where cell division takes place and growth occurs. Cells in this area are pushed upwards towards the lunula.
- **Nail groove** – grooves on the sides of the nail guide it up the fingers and toes.
- **Nail bed** – this area lies directly beneath the nail plate and secures the nail to the finger or toe. It contains blood vessels for nourishment and sensory nerve endings.

We will now look at the top of the nail, again starting from its proximal end:

- **Nail mantle** – this is the skin that lies directly above the germinal matrix, it is sometimes referred to as the nail fold.
- **Cuticle** – this is an extension of the horny layer of the epidermis from the nail mantle (fold) that overlaps the base of the nail plate. Its function is to protect the germinal matrix from infection.
- **Lunula** – the crescent-shaped white area at the proximal end of the nail.
- **Eponychium** – this is an extension of the cuticle from the base of the nail fold. Its function is to protect the matrix from infection.
- **Hyponychium** – located under the nail plate where the free edge forms. It protects the nail bed from infection.
- **Nail plate** – this is the visible body of the nail. It does not contain any blood vessels or nerves and it functions in protecting the nail bed beneath it.
- **Nail wall** – this is the skin that covers the sides of the nail plate and protects the nail grooves.
- **Free edge** – this is also called the distal edge and is the part of the nail that extends past the end of the finger or toe.

Functions of the nail

The key functions of the nail are:

- The nails protect the ends of the toes and fingers from physical trauma.
- The nails help us to grasp and pick up small objects.
- The nails enable us to scratch parts of the body.

Cutaneous glands

Glands are groups of specialised cells that secrete beneficial substances and also excrete waste products. There are several kinds of glands associated with the skin, the main ones being sebaceous and sweat glands. These glands are generally found in the dermis of the skin.

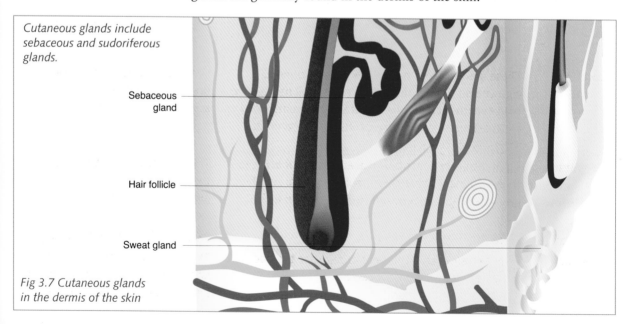

Cutaneous glands include sebaceous and sudoriferous glands.

Sebaceous gland

Hair follicle

Sweat gland

Fig 3.7 Cutaneous glands in the dermis of the skin

Sebaceous glands

The word *sebaceous* means 'oily' and sebaceous glands are oil glands that usually empty into hair follicles, although some do open directly onto the surface of the skin. Sebaceous glands secrete sebum which is an oily substance composed of fats, cholesterol, protein and inorganic salts. Sebum acts as a lubricant which:

- Prevents hair from drying out and becoming brittle.
- Prevents excessive evaporation of moisture from the skin, thus keeping it soft and moist.
- Inhibits the growth of certain bacteria.

Did you know?
The palms of your hands house more than 450 sweat glands per cm^2.

Anatomy and physiology in perspective
Whiteheads are enlarged sebaceous glands that have become blocked by accumulated sebum. Blackheads develop when the sebum oxidises, dries and darkens. Because sebum is nutritive to some bacteria, pimples can result from blackheads.

Sudoriferous (sweat) glands

Sudoriferous glands excrete sweat onto the surface of the skin. Sweat, or perspiration, is composed of water, ions, urea, uric acid, amino acids, ammonia, glucose, lactic acid and ascorbic acid. It aids heat-regulation of the body and also helps eliminate small amounts of waste. Sweat is acidic and therefore helps to inhibit the growth of bacteria on the skin's surface.

Sudoriferous glands are divided into two types depending on their location, structure, secretion and function: eccrine and apocrine. The table below highlights the differences between eccrine and apocrine glands.

Cutaneous glands in a nutshell
Cutaneous glands are groups of specialised cells that secrete substances such as sebum and excrete waste products such as sweat.

	ECCRINE	APOCRINE
Location	These are very common sweat glands found all over the body except for on the lips, nail beds, some of the reproductive organs and the eardrums. They are most numerous on the palms of the hands and the soles of the feet.	These are found in the axilla (armpits), pubic region and the areolae of the breasts.
Structure	Their secretory portion is located in the subcutaneous layer and their duct extends outwards through the skin, opening as a pore on the surface of the skin.	Their secretory portion is located in the dermis or subcutaneous layer and their duct opens into hair follicles.
Excretion	Sweat.	Sweat, fatty acids and proteins.
Function	They excrete waste and help to regulate the body's temperature by keeping it cool.	Their exact function is not yet known, but they are activated by pain, stress and sexual foreplay.

Common pathologies of the skin and hair

The condition of a person's skin and hair reflects their general health and emotional wellbeing. It is positively influenced by:

- Healthy nutrition.
- Exercise.
- Adequate water intake.
- Sufficient sleep.
- Rest and relaxation.
- Hygiene and care.

Factors that have a negative effect on one's skin and hair include:

- Poor nutrition and minimal water intake.
- Insufficient exercise.
- Lack of sleep, rest or relaxation.
- Emotional stress and tension.
- Excess alcohol and caffeine consumption.
- Smoking.
- Medication (some drugs have negative side-effects on the skin).
- Excess exposure to ultraviolet radiation.
- Use of harsh chemicals in detergents and soaps.
- Environmental pollution.
- Hormonal changes.
- Ageing.

Infobox

Anatomy and physiology in perspective
Remember that although some skin disorders have no medical consequences, they can have a traumatic effect on a person's confidence and self-image.

Terms for marks and growths on the skin

Although you may not need to learn the following terms, they will help you to visualise and thus learn some of the diseases described later.

Abrasion	A damaged area of the skin caused by the skin being scraped or worn away.
Bruise	A discolouration of the skin caused by the escape of blood from underlying vessels.
Bulla	A large, fluid filled spot.
Callus	A thick, protective layer of skin caused by repeated pressure or friction. Common on the fingers and feet.
Chilblains (perniosis)	Red, itchy swellings occurring on the extremities or legs due to exposure to the cold. They can occasionally blister.
Corn (clavus)	A small cone of compacted cells found either on the toes or between the toes. Caused by continual pressure and often accompanied by pain and inflammation.

Crust (scab)	An accumulation of dried blood, pus or skin fluids on the surface of the skin. Forms wherever the skin has been damaged.
Cyst	A semi-solid or fluid-filled lump above and below the skin.
Excoriation	The removal of the skin caused by scratching or scraping.
Fissure	A crack in the skin that penetrates into the dermis.
Macule	A small, flat, discoloured spot of any shape, e.g. freckles.
Mole	A small and dark skin growth. It is a concentrated area of melanin.
Nodule	A solid bump that may be raised.
Papule	A small, solid bump that does not contain fluid. For example, warts, insect bites and skin tags.
Plaque	A large, flat, raised bump or group of bumps.
Pustule	A lump containing pus.
Scales	Areas of dried, flaky cells. For example, in psoriasis or dandruff.
Scar	An area where normal skin has been replaced by fibrous tissue. Forms after an injury.
Stretchmarks	Small tears in the dermis caused by the skin stretching beyond its ability.
Telangiectasis	A localised collection of blood vessels in the skin. It is characterised by a red spot that can sometimes be spidery in appearance and that blanches under pressure.
Tubercle	A solid lump that is larger than a papule.
Tumour	An abnormal growth of tissue.
Ulcer	A deep, open lesion on the skin. Ulcers penetrate the dermis.
Vesicle	A small, fluid filled spot. For example, chickenpox and burns.
Wheal (hive)	A common allergic reaction in which there is swelling with an elevated, soft area.

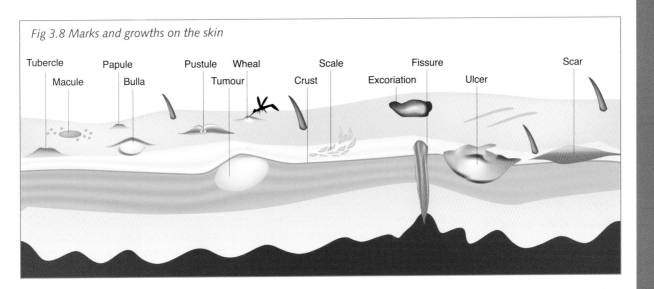

Fig 3.8 Marks and growths on the skin

Tubercle   Papule   Pustule   Wheal   Scale   Fissure   Scar

Macule   Bulla   Tumour   Crust   Excoriation   Ulcer

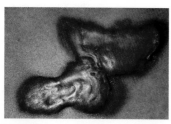

Keloids

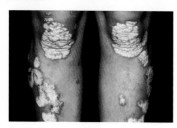

Psoriasis

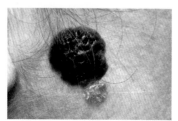

Seborrheic keratoses (senile warts)

Contact dermatitis

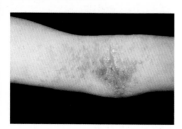

Eczema

Abnormal growth disorders

Keloids

Keloids form over healed wounds and are an overgrowth of scar tissue. They appear as smooth, shiny, flesh-coloured growths of fibrous tissue. Although they are not painful, they can itch or be sensitive to touch and can form in any scar, even acne scars.

Psoriasis

Psoriasis is a chronic disorder characterised by raised, red patches that have silvery scales. These patches are caused by unusually rapid cell growth, but the reason for this rapid cell growth is unknown. Some people with psoriasis will also have deformed, thickened nails. Psoriasis often runs in families and can be aggravated by sunburn, injuries or stress. It is sometimes improved through gentle exposure to the sun.

Seborrheic keratoses (senile warts)

Senile warts are common in middle-aged to elderly people and are harmless, flesh-coloured, brown or black growths. They are common on the torso and temples, but can appear anywhere on the skin.

Verrucae filliformis (skin tags)

Skin tags are harmless soft, small, flesh-coloured or dark growths. They generally appear on the neck, in the armpits or in the genital area.

Allergies

When a person has an allergic reaction to a substance, it is usually an inappropriate immune response to what is normally a harmless substance. When the skin is involved in an allergic reaction, the most obvious symptom is inflammation. Inflammation is a localised, protective response involving redness, swelling and pain.

Dermatitis and eczema

Dermatitis is a broad term for inflammation of the upper layers of the skin and it includes a variety of symptoms that usually involve itching, blistering, redness and swelling. Severe symptoms are oozing, scabbing and scaling. Dermatitis can be caused by allergens, irritants, dryness, scratching or fungi.

- **Contact Dermatitis** – an itchy rash that is confined to a specific area and the result of direct contact with a substance. It can be caused by cosmetics, metals such as nickel, certain plants such as poison ivy, drugs in some creams and certain chemicals used in clothing manufacturing.
- **Eczema** – a form of dermatitis that can be caused by either internal or external factors. It is characterised by itchiness, redness and blistering. It can be either dry or weeping and can cause scaly and thickened skin.

Urticaria (nettle rash or hives)

Urticaria is commonly called hives and is an allergic reaction triggered by certain foods, drugs or insect stings. It is identified by itchy wheals which are pale, slightly elevated swellings surrounded by an area of redness.

Bacterial infections of the skin

We come into contact with bacteria everyday, but most of the time our skin forms a barrier which prevents infection. However, some bacterial infections do occur. These are usually in people with a weakened immune system or with damaged skin (from sunburn, scratching or trauma). Many types of bacteria can affect the skin and the most common are *staphylococcus* and *streptococcus*.

Carbuncles
Carbuncles are a collection of small, shallow abscesses that are connected to one another under the skin. If a person has carbuncles, he or she will usually also have a fever and feel ill and tired.

Cellulitis
Cellulitis is a spreading bacterial infection of the skin. It begins as redness, pain and tenderness over an area and then develops into hot, swollen areas that appear slightly pitted. Cellulitis is caused by bacteria and can occur after being bitten by humans or dogs or injuring oneself in water or dirt.

Conjunctivitis
This is described in Chapter 6 under the heading of **Eye Problems**.

Folliculitis
Folliculitis is the bacterial infection of a hair follicle and appears as a small white pimple at the base of a hair. One or many follicles can be infected.

Furuncles (boils)
Furuncles are also known as boils or skin abscesses and are pus-filled pockets of infection beneath the skin. They are bacterial infections that are more common in people with poor hygiene or weakened immune systems.

Hordeolum (stye)
This is described in Chapter 6 under the heading of **Eye Problems**.

Impetigo
Impetigo is a very contagious bacterial infection that is common in children. Symptoms include itchiness, pain and scabby, yellow-crusted sores or fluid-filled blisters. It most commonly occurs on the face, arms and legs and is easily spread through scratching.

Paronychia
See section on Common Pathologies of the Nails on page 86.

Urticaria (nettle rash or hives)

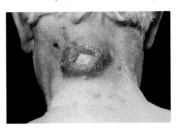

Carbuncle

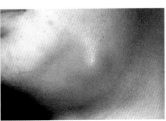

Cellulitis

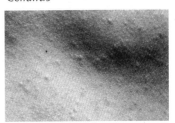

Folliculitis

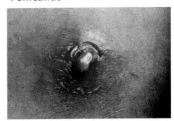

Furuncle (boil)

Impetigo

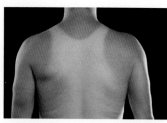

Sunburn

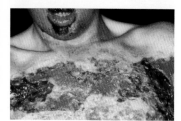

2nd and 3rd degree burns

Burns

Burns are injuries to the skin. They can be caused by heat, sunlight, chemicals, electricity or radiation. Burns differ according to their cause and the degree of the burn. For example, a first-degree burn is a shallow, superficial burn in which only the epidermis has been affected; a second-degree burn extends into the dermis; and a third-degree burn involves injury to all layers of the skin, including the subcutaneous layer. Symptoms of burns also vary according to the degree of the burn, from simple redness and swelling to blistering and even blackening and scarring of the skin.

Cancers

There are three main types of skin cancer: squamous cell carcinoma, basal cell carcinoma and melanoma. Long-term sun exposure generally plays a role in the development of skin cancer and fair-skinned people, who produce less melanin, are more susceptible to developing it.

Most skin cancers are curable if treated early enough and so it is vitally important that any unusual, persistent skin growth be examined by a doctor as soon as possible.

As a therapist it is important to have a basic understanding of the different treatments available for cancer. This will help you decide whether or not it is safe for you to work on a client.

Although there are many different types of cancer, most of them are treated with either one or a combination of the following:

Surgery – surgery is usually used to treat tumours that have not spread beyond their original site of growth. However, surgery cannot always be used if the tumour is in an inaccessible area or if the surgery would involve the removal of an entire vital organ.

Radiation therapy – radiation therapy involves the focusing of radiation (an emission of energy in the form of x-rays or similar rays) on a specific area or organ of the body. It can be given via an external beam focused on the body or via radioactive substances that are injected into the body. Radiation kills cells, such as cancer cells, that divide rapidly. However, it can also damage non-cancerous cells that divide rapidly, for example, those of the skin, hair follicles, the lining of the mouth, oesophagus and intestines. Possible side effects of radiation therapy can include: tiredness; changes in appetite; weight loss; skin reactions similar to sunburn; and skin sensitivity to the sun or to cold winds. Unfortunately, some types of cancer cells are resistant to radiation.

Chemotherapy – chemotherapy involves the use of drugs to destroy cancer cells by interfering with their ability to divide. The drugs often affect non-cancerous cells as well, especially those of the skin, the lining of the mouth, the bone marrow, the hair follicles and the digestive system (these are all cells that divide rapidly). However, damage to these cells is temporary. Chemotherapy drugs can be given via injections, as tablets or as creams. There are many possible side effects to chemotherapy including: a reduction in the number of blood cells produced by the bone marrow; fatigue, nausea and vomiting; loss of appetite, diarrhoea and/or constipation; mouth ulcers and changes in taste; hair loss; skin changes such as rashes, dryness, discoloration and an increased skin sensitivity to sunlight; brittle or flaky nails, slow nail growth and the appearance of white lines across the nails; tingling or numbness in the hands or feet; anxiety, restlessness, dizziness and sleeplessness; headaches and changes in hearing. Please note, this is an extensive list of the side effects of chemotherapy and someone receiving it will not necessarily have all of the above.

Immunotherapy – immunotherapy involves stimulating the body's immune system against cancer. Substances such as vaccines, antibodies and biologic response modifiers are used to encourage the immune system to destroy the cancer cells.

Basal cell carcinoma (rodent ulcers)

Basal cell carcinoma is the most common form of skin cancer. Although it is called 'basal cell', it does not necessarily originate in the basal cells of the epidermis. Basal cell carcinoma can appear in many forms. For example, it may appear as raised bumps that break open and form scabs; or as a flat pale or red patch; or it may be an enlarged papule with a thickened, pearly border. Basal cell carcinomas may also be mistaken for sores that constantly bleed, scab and heal. Basal cell carcinomas rarely metastasise (spread to distant parts of the body). However, they can invade surrounding tissues and this can be serious if the carcinomas are located close to the brain, eyes or mouth.

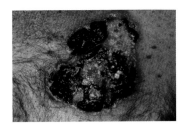

Basal cell carcinoma

Melanoma

Melanoma is a cancer that originates in the melanocytes of the skin. It can develop in sun-exposed areas or on moles and it can metastasise. Melanomas vary in appearance and signs to watch for include moles or freckles that are growing, changing in colour or changing in shape; moles that bleed or break open; or irregular black or gray lumps that appear on the skin.

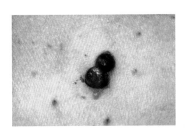

Melanoma

Squamous cell carcinoma (prickle-cell cancer)

Squamous cell carcinoma originates in the stratum spinosum (prickle-cell layer) of the epidermis and usually occurs on sun-exposed areas in fair-skinned people. Occasionally, it may develop in areas that are not exposed to the sun, for example, the mouth. It is characterised by a thick, scaly, warty appearance and if left untreated it can develop into an open sore that grows into the underlying tissue. In addition to prolonged sun exposure, causes of squamous cell carcinoma include certain chemicals, chronic sores, burns and scars. Squamous cell carcinomas do occasionally metastasise.

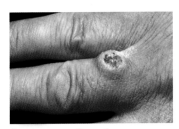

Squamous cell carcinoma

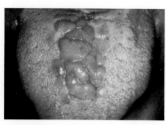
Candidiasis (thrush or yeast infection)

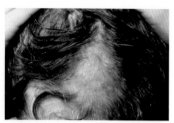
Tinea capitis (scalp ringworm)

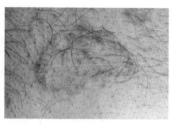
Tinea corporis (body ringworm)

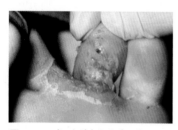
Tinea pedis (athlete's foot)

Pediculosis capitis (head lice)

Fungal infections of the skin

Fungi are organisms that often make their homes in moist areas of the skin, for example, between the toes or in the folds of skin.

Candidiasis (thrush or yeast infection)

Candidiasis, also called candidosis, is an infection by the yeast candida. This yeast is normally found in the mouth, digestive tract and vagina but it can infect other areas of the body if it is allowed to grow unchecked. Pregnant women, obese people and people with diabetes are more prone to candidiasis. So are those taking antibiotics. Different forms of candidiasis include:

- **Infections in skin folds** – characterised by a bright red rash with a softening and sometimes breaking down of the skin. The rash may itch or burn and small pustules may appear. Areas that are sometimes infected by candidiasis include the navel, the anus and the nappy area of a baby.
- **Vaginal candidiasis** – characterised by burning, itching and redness around the vagina that is accompanied with a white or yellow discharge.
- **Penile candidiasis** – characterised by a red, raw, painful rash on the penis.
- **Thrush** – this is candidiasis of the mouth and it is characterised by white, painful patches on the tongue and sides of the mouth.
- **Candidal paronychia** – this is candidiasis of the nail bed and it is characterised by redness and swelling that may lead to the nail turning white or yellow and separating from the nail bed.

Tinea (ringworm)

Tinea is a contagious fungal infection that is identified by ring-shaped patches on the skin. It is generally classified by its location on the body and includes:

- **Tinea capitis** (scalp ringworm) – scalp ringworm is an itchy, scaly, pink rash on the scalp that can cause patches of hair loss.
- **Tinea corporis** (body ringworm) – body ringworm appears as round patches that have pink scaly borders and clear centres. It can be very itchy.
- **Tinea pedis** (athlete's foot) – athlete's foot is a common infection found between the toes. Its symptoms include scaling with, or without, redness and itching. It is easily spread in moist areas where people walk barefoot, for example, communal showers.

Infestations of the skin

Most skin infestations are caused by skin parasites which are tiny insects or worms that burrow into the skin and live there.

Pediculosis (lice)

Lice are tiny, wingless insects which spread easily from person to person by physical contact. Although lice are difficult to find, their eggs (nits) are easily seen as shiny, grayish eggs that are firmly attached to hair shafts. Lice can occur on different parts of the body and generally cause severe itching. A lice infestation needs to be treated as soon as possible as intense scratching can lead to bacterial infections.

- **Pediculosis capitis** (head lice) – head lice infect scalp hair and are common in young children. They are spread through sharing personal items such as hair brushes and hats.
- **Pediculosis corporis** (body lice) – body lice are usually only found in people with poor hygiene or who live in overcrowded housing. The lice generally inhabit the seams of clothing.
- **Pediculosis pubis** (pubic lice) – pubic lice is sometimes called crabs. It is a lice infestation of the genital hair and is spread through sexual contact.

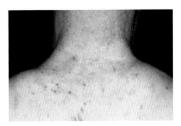

Pediculosis corporis (body lice)

Scabies (itch mites)

Scabies is a highly contagious infestation caused by the female itch mite who burrows her way into the horny layer of the skin. Here she lays her eggs and within a few days young mites have hatched. Tiny, itchy bumps which are usually worse at night develop.

Pigmentation disorders

Most pigmentation disorders are related to either an overproduction or underproduction of the pigment melanin. Melanin is produced by cells called melanocytes and it is the pigment responsible for the colour of a person's skin. A few pigment disorders can be related to other pigments, for example, jaundice is a buildup of the pigment bilirubin.

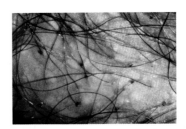

Pediculosis pubis (pubic lice)

Albinism

Albinism is a hereditary disorder caused by a lack of melanin in the body and can occur in people of all races. It is characterised by white hair, pale skin and pink or pale blue eyes. Albinism also results in abnormal vision, involuntary eye movements and extreme sensitivity to sunlight.

Chloasma

Chloasma, also known as melasma, is a blotchy, brownish patch caused by the overproduction of melanin. It is more common in women and usually affects the forehead, cheeks and around the lips. Chloasma can be triggered by hormonal changes in pregnancy or when taking the contraceptive pill. It can occasionally be caused by excessive sun exposure or as a reaction to some skin cosmetics and it usually fades with time.

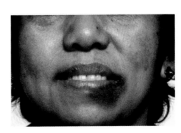

Chloasma

Dilated capillaries

Capillaries are the tiny blood vessels found in the dermis of the skin. When they dilate, more blood rushes to them and this appears as erythema.

Ephelides (freckles)

Freckles are flat, irregular patches of melanin produced in response to sunlight. They are harmless and common in fair-skinned people.

Lentigines

Erythema

Erythema is the term used to describe inflammation or redness of the skin. It is caused by the engorgement of skin capillaries.

Lentigines

Lentigines (singular *lentigo*) are small, pigmented patches that can be flat or slightly raised. They can occur anywhere on the body and vary in colour from tan-brown to black. They can appear at birth or in early childhood or can be due to sun exposure. In older people they are sometimes called 'liver spots' or 'sun spots'.

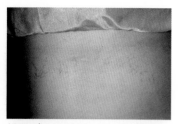

Vascular naevi

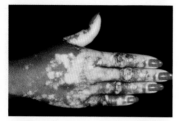

Vitiligo

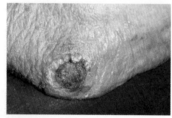

Pressure sores

Acne rosacea

Acne vulgaris

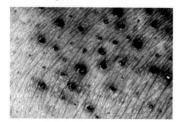

Comedones (blackheads)

Vascular naevi

Naevi (singular naevus) are malformed dilated blood vessels in the skin. They appear as red or purple blotches and can occur anywhere on the body. Birthmarks are a common type of naevi. There are different types of naevi, including:

- **Port-wine stains** – flat, red to purplish discolorations which are present at birth.
- **Spider naevi** – small, red spots surrounded by slender dilated capillaries that look like spider's legs. They are common in fair skinned people and are sometimes associated with sun exposure.
- **Strawberry naevi** – birthmarks that usually disappear during childhood. They are red and raised above the surface of the skin.

Vitiligo

Vitiligo is characterised by smooth, white patches of skin where there is a lack of melanocytes. Due to this lack of melanocytes, it is important to protect the skin from the sun with sunscreen and clothing. Vitiligo can sometimes be hereditary or caused by another disease or physical trauma and although it has no medical consequences, it can be very distressing for an individual.

Pressure sores (Decubitus ulcers)

Pressure sores, also called bed sores or decubitus ulcers, are caused by constant pressure to an area. This pressure prevents the flow of blood to the area and results in the death of the skin. The dead skin then breaks down and forms sores or ulcers. Pressure sores are common in people who are bed-ridden and are characterised by redness and inflammation.

Sebaceous gland disorders

Acne rosacea

Rosacea is sometimes called 'adult acne' and usually affects the central area of the face. It is characterised by redness, noticeable blood vessels and tiny pimples. Its cause is unknown, but it is aggravated by alcohol, spicy foods, menopause and stress.

Acne vulgaris

Acne vulgaris occurs mainly on the face, chest, shoulders and back. Its symptoms include comedones, papules, pustules, cysts and sometimes abscesses. Acne vulgaris is an inflammatory disorder of the sebaceous glands which is thought to be caused by an interaction between hormones, skin oils and bacteria. Symptoms can range from mild to severe.

Comedones (blackheads)

Enlarged sebaceous glands can become blocked by accumulated sebum. This is known as a whitehead. When the sebum oxidises and dries, it darkens and forms blackheads.

Milia

Milia are commonly found around the eyes and over the cheek area and appear as white, hard nodules under the skin. They are small, harmless keratin-filled cysts found just beneath the epidermis and are common in newborn babies.

Seborrhoea

Seborrhoea is characterised by an oily skin with enlarged, blocked pores, comedones and pustules. It is due to an excessive secretion of sebum by the sebaceous glands.

Steatomas (sebaceous cysts or wens)

Sebaceous cysts are usually found on the face, neck, scalp and back and are round lesions that have a smooth and shiny surface.

Sudoriferous (sweat) gland disorders

Anhidrosis

Anhidrosis, also called hypohidrosis, is a lack of sweating in the presence of an appropriate stimulus such as heat. It can be congenital or due to an illness.

Bromhidrosis (body odour)

Bromhidrosis, also called bromidrosis or osmidrosis, is body odour. It is caused by the breakdown of sweat by bacteria and yeasts that normally live on the skin. It can also be influenced by one's diet, genetics and general health.

Hyperhidrosis

Hyperhidrosis, also called hyperidrosis, is excessive or almost constant sweating. It can affect the entire surface of the skin but is usually limited to the armpits, genitals, palms and soles. It can be caused by an illness, a medical condition such as hyperthyroidism, or after the use of certain drugs.

Miliaria rubra (prickly heat)

Prickly heat is common in warm, humid climates and is caused by trapped sweat. The sweat is trapped in narrow ducts that carry it to the surface. This trapped sweat causes inflammation and a prickling, itching sensation. It can appear as an itchy rash of very tiny blisters or as large, reddened areas of the skin.

Viral infections of the skin

Viruses are tiny organisms that invade living cells and then multiply within the cells. They cause many common infections such as cold sores and warts. Diseases caused by viruses are often accompanied by rashes, spots or sores on the skin.

Herpes simplex (cold sore)

Cold sores are small, fluid-filled blisters that can keep recurring. They are normally found on the face and around the mouth. Cold sores usually begin with a tingling sensation at the site, followed by redness, swelling and the development of blisters. These blisters then break open leaving sores and scabs. This condition is passed through direct contact with the sores.

Milia

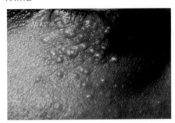

Seborrhoea

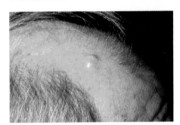

Steatoma (sebaceous cyst)

Herpes simplex (cold sore)

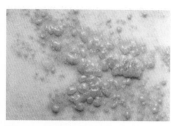

Herpes zoster (shingles)

Rubella (German measles)

Rubeola (measles)

Varicella (chickenpox)

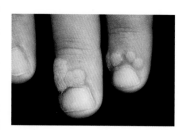

Warts

Herpes zoster (shingles)

Herpes zoster is a re-emergence of the chicken pox virus and it usually occurs when a person's immune system is weakened. It develops as a painful eruption of blisters limited to an area served by infected nerves. The infected person may feel unwell and feverish and/or may experience pain, tingling and itching.

Rubella (German measles)

Rubella is a contagious infection that begins with mild flu-like symptoms such as a runny nose and cough. Painless rose-coloured spots appear on the roof of the mouth and then merge and extend over the back of the throat. The lymph glands swell in the neck and a characteristic rash that begins on the face and neck and spreads down the body then appears. This rash is often accompanied by a reddening or flushing of the skin. Joint pain can also accompany rubella. Rubella is spread by airborne droplets of moisture.

Rubeola (measles)

Measles is a highly contagious infection which begins with flu-like symptoms such as a fever, runny nose, sore throat, cough and red eyes. Two to four days later, small white spots appear inside the mouth and then a few days later an itchy rash appears. This rash usually begins on the neck and then spreads to the rest of the body. Measles is spread by airborne droplets of moisture.

Varicella (chickenpox)

Chickenpox is a very contagious infection characterised by a mild fever that is followed by an itchy rash of small, raised spots that can blister, crust and scab. These scabs drop off after about 12 days. Chickenpox is spread by airborne droplets of moisture and after the infection the virus can remain dormant in the body and reactivate as shingles in later life.

Warts or verrucae

Warts are small, firm growths that have a rough surface. They are caused by a virus and can grow in clusters or as an isolated growth. Warts on the body are generally painless and common on areas that are frequently injured. For example, the knees, face, fingers and around the nails. Warts on the feet are called plantar warts or verrucae and can be painful due to the pressure of the weight of the body on them.

Common pathologies of nails

The nails can be affected by many skin conditions and they can also be indicators of internal imbalances, neglect or stress and anxiety. It is important to be able to recognise diseases and disorders of the nail as some can lead to cross-infection.

Factors that affect nail growth positively include:

- Good nutrition.
- A good supply of blood to the nails.

Factors that have negative affects on nail growth include:

- Poor health.
- Poor nutrition.
- Over exposure to chemicals such as detergents.
- Poor manicure and pedicure techniques.
- Injuries to the nail bed.
- Ageing.

Agnail (hang nail)
Hang nail is characterised by dry, split cuticles. It is often caused by poor treatment techniques, the use of detergents or soaking the hands in water for long periods. Although it is a harmless disorder, it can become infected.

Anonychia
Anonychia is the congenital absence of a nail.

Beau's lines
Beau's lines are transverse (horizontal) ridges or grooves on the nail plate. They can reflect a temporary retardation of growth due to ill health, a very high fever or a zinc deficiency in the body.

Egg shell nails (soft, thin nails)
Eggshell nails are unusually soft, thin, white nails that are curved over the free edge. They can be the result of poor diet, ill health, medication or some nervous disorders.

Koilonychias (spoon nail)
Koilonychias is the term given to spoon-shaped, concave nails resulting from abnormal growth. Spoon nails can be congenital, due to a lack of minerals or as a result of illness. The nails are also thin, soft and hollowed.

Leuconychia (white nails or white spots)
Leuconychia is the term given to white spots or streaks on the nail or to nails that are white or colourless. Leuconychia is often the result of trauma to the nail or air bubbles. It can also be an indication of poor health. The white spots will grow out with the nail.

Longitudinal furrows
Longitudinal (vertical) ridges can occur on the nail as a result of uneven nail tissue growth. To an extent, they are normal in adults and increase with age. However, severe ridges can result from injury to the nail matrix through poor treatment techniques or from the excess use of detergents and other harsh chemicals. They can also be caused by conditions such as psoriasis and poor circulation.

Onychauxis (thick nails) and onychogryphosis (ram's horn nail)
Onychauxis is an unusual thickness of the nail caused by trauma to the nail matrix, fungal infection or neglect. It can also be hereditary. Chronic thickening of the nail can lead to onychogryphosis, a condition in which the nail thickens and curves into a hooked nail. Onychogryphosis usually results from damage to the nail bed and can cause pain and injury to adjoining toes.

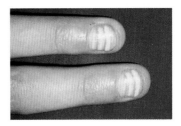
Beau's lines

Koilonychias (spoon nail)

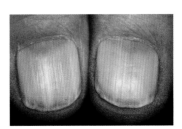
Leuconychia (white nails)

Longitudinal furrows

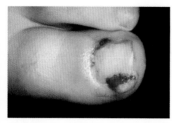

Onychocryptosis (ingrowing nail)

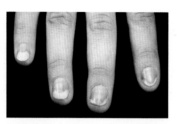

Onycholysis (separation of the nail from the nail bed)

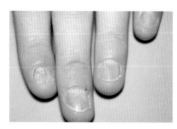

Onychophagy (nail biting)

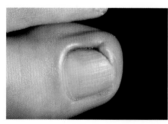

Paronychia (bacterial infection of the cuticle)

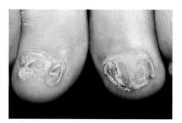

Tinea ungium (ringworm)

Onychocryptosis (ingrowing nail)

Onychocryptosis is an ingrowing nail: a condition in which the sides of the nail penetrate the skin. It is characterised by red, shiny skin around the nail and can be very sensitive and painful if touched. Ingrowing nails are most common on the large toe and can be caused by ill fitting shoes or improper nail cutting.

Onycholysis (separation of the nail from the nail bed)

Onycholysis is the separation of the nail from the nail bed. It usually begins to loosen at the free edge and continues up to the lunula, but does not fall off. It can be caused by illness, trauma, infection, certain drugs or abuse of the nails.

Onychophagy (nail biting)

Nail biting is common in people who suffer from stress or anxiety. It can result in an exposed nail bed which is inflamed and sore. Onychophagy is the term given to nails that have become deformed through excess nail biting.

Onychoptosis (nail shedding)

Onychoptosis is a condition in which parts of the nail shed. It usually occurs during ill health or as a reaction to some drugs. It can also be caused by trauma to the nail.

Onychorrhexis (brittle nails)

Onychorrhexis is the term given to dry, brittle, splitting nails that also have longitudinal ridges. It is common in old age or people with arthritis or anaemia. It can also be caused by poor treatment techniques and excess soaking of the hands in water or detergents.

Paronychia (bacterial infection of the cuticle)

Paronychia is a bacterial infection of the skin surrounding the nails. The tissues become red, swollen and painful. It is a common infection that can be the result of injury, nail biting or poor manicure techniques.

Pterygium (overgrowth of cuticle)

This is the condition in which the cuticle becomes overgrown and grows forward, sticking to the nail plate. The cuticle can become dry and split and it is usually a result of neglect or poor nail care.

Severely bruised nail

Bruised nails are a result of physical damage or trauma to the nail bed. A clot of blood forms under the nail plate and, in some cases, a severely bruised nail can fall off.

Tinea ungium (ringworm of the nail)

Tinea ungium is ringworm of the nail. It is a fungal infection characterised by thickened and deformed nails. Infected nails may separate from the nail bed, crumble or flake off.

NEW WORDS

Acid mantle	a film of sebum and sweat on the surface of the skin that protects against bacteria.
Avascular	lacking in blood vessels.
Cross-infection	the transfer of infection from one person to another.
Cutaneous	relating to the skin.
Dermatology	the study of the skin.
Desquamation	the process through which the skin is shed.
Elasticity	the ability to return to one's original shape after stretching.
Extensibility	the ability to stretch.
Integumentary system	the system of the skin and its derivatives (hair, nails and cutaneous glands).
Keratinisation	the process through which cells die and become full of the protein keratin.
Phagocytosis	the ingestion of bacteria.
Vasoconstriction	the constriction of blood vessels.
Vasodilation	the dilation of blood vessels.

Study Outline

Skin

Functions of the skin

1. **Sensation.** The skin contains cutaneous sensory receptors that are sensitive to touch, temperature, pressure and pain.
2. **Absorption.** The skin absorbs certain substances.
3. **Protection.** The skin protects against trauma, bacteria, dehydration, ultraviolet radiation, chemical damage and thermal damage.
4. **Heat regulation.** It cools the body through sweating and vasodilation. It warms the body through decreased sweat production, vasoconstriction, contraction of the arrector pili muscles, and shivering.
5. **Excretion.** The skin excretes wastes.
6. **Secretion.** The skin secretes sebum which keeps the skin supple and waterproof.
7. **Vitamin D synthesis.** The skin functions in the synthesis of vitamin D which helps regulate the calcium levels of the body and is necessary for the growth and maintenance of bones.

> Skin Always Protects Humans Every Single Day.
> Sensation, Absorption, Protection, Heat regulation, Excretion, Secretion, vitD.

Anatomy of the skin – main layers

1. The skin is a cutaneous membrane made of two distinct layers: the epidermis and the dermis.
2. The epidermis is the tough, waterproof outer layer that is continuously being worn away. It is made of keratinised stratified squamous epithelium.
3. The dermis is the thicker layer that contains nerves, blood vessels, sweat glands and hair roots. It is made of areolar connective tissue containing collagen and elastic fibres. It lies beneath the epidermis.
4. The dermis is attached to the subcutaneous layer which is made of areolar and adipose connective tissue.

Epidermis

1. The epidermis is made up of five layers: the stratum basale (basal-cell layer), stratum spinosum (prickle-cell layer), stratum granulosum (granular-cell layer), stratum lucidum (clear-cell layer) and stratum corneum (horny-cell layer).

> Bad Skin Grafts Look Comical.
>
> Basale, Spinosum, Granulosum, Lucidum, Corneum.

2. The basal-cell layer constantly produces new cells that are pushed upwards.
3. The prickle-cell layer is a transitional layer where some cells are dividing but others are starting to become keratinised.
4. In the granular-cell layer cells begin to die as they become increasingly keratinised.

5. The clear-cell layer is a waterproof layer of dead cells.
6. The horny-cell layer is the outermost layer and consists of strong, tough, dead cells that are completely full of keratin. The cells are constantly being shed through the process of desquamation.
7. Keratinocytes produce keratin which is a waterproof protein.
8. Melanocytes produce melanin which is a pigment that contributes to skin colour and absorbs ultraviolet light.

> Your body dresses itself for all seasons.
> It puts on a **raincoat made of keratin** and a **sunhat made of melanin**.

9. Skin is thickest on the palms of the hands and soles of the feet and thinnest on the eyelids.

Dermis

1. The dermis is the supportive layer beneath the epidermis and it is composed of two layers: the superficial papillary layer and the deeper reticular layer.
2. The papillary layer has finger-like projections that go into the epidermis. These projections contain loops of capillaries and nerve endings sensitive to touch.
3. The reticular layer houses hair follicles, nerves, oil glands, ducts of sweat glands and adipose tissue.

Subcutaneous layer

The subcutaneous layer attaches the reticular layer to the underlying organs and it contains areolar and adipose connective tissue and nerve endings that are sensitive to pressure.

Skin types

1. There are five basic skin types: normal, oily, dry, combination and sensitive.
2. Normal skin is a balanced skin in which there are no signs of oily or dry areas.
3. Oily skin has an overproduction of sebum by the sebaceous glands.
4. Dry skin has an underproduction of sebum or a lack of moisture, or both together.
5. Combination skin is a mixture of dry, normal and greasy skin and is the most common type of skin.
6. Sensitive skin often accompanies dry skin and is easily irritated.

Hair

There are three types of hair: lanugo, vellus and terminal.
1. Lanugo hair is a soft hair only found in a foetus.
2. Vellus hair is a soft hair found all over the body except the palms, soles, eyelids, lips and nipples.
3. Terminal hair is a long, coarse hair found on the head, eyebrows, eyelashes, under the arms and in the pubic area.

Structure of the hair

1. Hairs are columns of keratinised dead cells.
2. Hairs grow out of hair follicles.
3. Hairs are made up of a superficial end called the 'shaft' and a 'root' which penetrates into the dermis.
4. Arrector pili muscles pull the hair up to a vertical position as a response to cold, fright or differing emotions.

Functions of the hair

The hair provides warmth and protection.

Life cycle of a hair

1. There are three distinct phases in the life cyle of a hair: anagen, catagen and telogen.
2. Anagen is the growing or active phase in which a new hair develops at the matrix beneath the club hair.
3. Catagen is the transitional or changing stage in which a now fully grown hair detaches from the matrix.
4. Telogen is the resting or 'tired' phase in which a fully grown hair sits high up in the follicle and the hair bulb beneath it is inactive.

> ACT
> Anagen = Active
> Catagen = Changing
> Telogen = Tired

Nails

Structure of the nail

1. The nail covers the ends of the fingers and toes.
2. The nail plate is the visible body of the nail. Its sides are covered by the nail wall and at its proximal end is the lunula, which is a crescent-shaped white area.
3. The nail mantle, or proximal nail fold, is skin that lies above the matrix.
4. The cuticle is an extension of the horny layer of the epidermis from the nail mantle (fold). It protects the germinal matrix from infection.
5. The distal edge of the nail is called the free edge.
6. Beneath the nail mantle is the germinal matrix where cell division and nail growth occur.
7. The nail bed is beneath the nail plate and attaches it to the finger or toe.

Functions of the nail

Functions of the nail include protection, helping us pick up small objects and enabling us to scratch.

Cutaneous glands

1. Glands are specialised cells that secrete substances.
2. Cutaneous glands are usually found in the dermis of the skin.
3. Sebaceous glands secrete sebum.
4. Sebum is a lubricant which helps condition and protect the skin and hair.
5. Sudoriferous glands are sweat glands.
6. Sweat functions in heat-regulation and in the elimination of waste.
7. There are two types of sudoriferous glands: eccrine and apocrine.
8. Eccrine glands are found almost all over the body and open as pores on the surface of the skin. They secrete sweat and function in regulating the body's temperature and excreting waste.
9. Apocrine glands are only found in the armpits, pubic region and the areolae of the breasts. They open into hair follicles and secrete sweat, fatty acids and proteins. They are activated by pain, stress and sexual foreplay.

Revision

1. What type of membrane is the skin?
2. Name the two layers that make up the skin.
3. Name the tissue type that makes up the epidermis.
4. Name the tissue type that makes up the dermis.
5. Name the skin layer that contains the nerves, blood vessels and sweat glands.
6. Put the following layers of the epidermis into their correct order, starting from the deepest and ending with the most superficial layer: stratum spinosum, stratum granulosum, stratum basale, stratum corneum, stratum lucidum.
7. Identify the layer of the epidermis in which cells divide.
8. Identify the layer of the epidermis that consists of dead cells which are completely filled with keratin.
9. Name the process through which cells are constantly shed from the skin.
10. Name two types of fibres that are found in the dermis.
11. Name the two layers that make up the dermis.
12. Identify which layer of the dermis has finger-like projections that protrude into the epidermis.
13. Name the tissue type that makes up the subcutaneous layer.
14. Describe the main functions of the skin.
15. Explain how the body cools itself when it is too hot.
16. Explain how the body warms itself when it is too cold.
17. Name the type of cells found in hair.
18. Name the superficial end of the hair.
19. Name the end of the hair that penetrates the dermis.
20. Describe the two main functions of the hair.
21. Name the type of cells found in the nail.
22. Identify where in the nail the cells divide.
23. Give the name of the visible body of the nail.
24. Give the name of the white, crescent-shaped area on the nail.
25. Describe three functions of the nail.
26. Explain what a gland is.
27. Identify where cutaneous glands are found.
28. What do sebaceous glands secrete?
29. What do sudoriferous glands excrete?
30. Name the two types of sudoriferous glands.
31. Identify the type of sudoriferous gland found in the axilla.

Multiple choice questions

1. **Which of the following statements is correct?**
   a. The skin is one of the smallest organs of the body.
   b. The skin is one of the largest organs of the body.
   c. The skin is not an organ of the body.
   d. The skin is the smallest organ of the body.

2. **Keratinised stratified squamous epithelium tissue is found in the:**
   a. Reticular layer
   b. Dermis
   c. Epidermis
   d. Subcutaneous layer.

3. **Functions of the hair include:**
   a. Excretion
   b. Protection
   c. Sensation
   d. Absorption.

4. **In the nail, cell division takes place in:**
   a. The nail mantle
   b. The lunula
   c. The matrix
   d. The nail plate.

5. **Sebaceous glands secrete:**
   a. Sweat
   b. Sebum
   c. Uric acid
   d. Glucose.

6. **Which of the following statements is true?**
   a. Keratin helps waterproof and protect the skin.
   b. Keratin contributes to skin colour.
   c. Keratin absorbs ultraviolet light.
   d. Keratin functions in the sensation of touch.

7. **Which is the correct order of the layers of the epidermis, from the most superficial to the deepest layer?**
   a. Stratum spinosum, stratum basale, stratum lucidum, stratum granulosum, stratum corneum.
   b. Stratum basale, stratum spinosum, stratum granulosum, stratum lucidum, stratum corneum.
   c. Stratum corneum, stratum lucidum, stratum spinosum, stratum granulosum, stratum basale.
   d. Stratum corneum, stratum lucidum, stratum granulosum, stratum spinosum, stratum basale.

8. **Herpes simplex is a:**
   a. Fungal infection of the nail.
   b. Viral infection of the nail.
   c. Fungal infection of the skin.
   d. Viral infection of the skin.

9. **The skin functions in warming the body through which of the following processes:**
   a. Sweating
   b. Shivering
   c. Vasodilation
   d. None of the above.

10. **Which of the following is responsible for the colour of your skin?**
    a. Melanin
    b. Keratin
    c. Elastin
    d. Collagen.

11. **Tinea corporis is ringworm of the:**
    a. Feet
    b. Body
    c. Scalp
    d. Nails.

12. **New skin cells are constantly produced in the:**
    a. Stratum spinosum
    b. Stratum basale
    c. Stratum corneum
    d. Stratum granulosum.

13. **Where on the body are eccrine glands located?**
    a. In the axilla only.
    b. In the axilla, pubic region and areolae.
    c. Everywhere except the lips, nail beds, eardrums and on some of the reproductive organs.
    d. Everywhere.

14. **Keloids are:**
    a. Harmless soft, flesh-coloured growths.
    b. Thickened nails.
    c. Senile warts.
    d. An overgrowth of scar tissue.

15. **The clinical term for nail biting is:**
    a. Koilonychias
    b. Onychophagy
    c. Onycholysis
    d. Leuconychia.

4 The Skeletal System

Introduction

Imagine what we might look like with no bones. How would we move? How would we stand up straight? We would be a mass of soft tissues lying on the floor unable to move or sit up.

In this chapter you will learn about the skeletal system and discover what a unique and vital system it is.

Student objectives

By the end of this chapter you will be able to:

- Describe the functions of the skeletal system.
- Describe bone tissue and identify the different types of bones found in the body.
- Explain the structure of a long bone.
- Identify the organisation of the skeleton and name the bones of the body.
- Describe the joints of the body.
- Identify the common pathologies of the skeletal system.

Did you know?
- Weight for weight, bone is approximately five times stronger than steel.
- Bones are hard, yet flexible – your ribs are strong enough to protect your heart but flexible enough to allow movement for breathing.
- Bones are hard, yet light and account for only approximately 14% of the body's total weight.

Functions Of The Skeletal System

Bones are not dead materials that simply support your body. They are intelligent, living structures that are constantly changing by reshaping, rebuilding and repairing themselves and they play vital roles in both the structure and functioning of the body. Bones support and protect the body, allow for movement and mineral homeostasis and are a site of blood cell production as well as energy storage.

Support
Bones are the scaffolding of the body and provide a framework that supports and anchors soft tissues and organs.

Shape
Acting as a framework, bones form the basic shape of the body.

Protection
Bones are extremely hard and able to protect the body's vital organs. For example, the cranial bones protect the brain and the vertebrae protect the spinal cord.

Movement
Without bones we would not be able to stand, walk, run or even chew. Our bones are sites of attachment for skeletal muscles. When the muscles contract or shorten they pull on the bones and this generates movement.

Mineral homeostasis
In addition to forming bones, calcium is vital for nerve transmission, muscular contraction, blood clotting and the functioning of many enzymes in the body. Bones store most of the calcium present in our bodies. Depending on the blood calcium levels, they either release calcium into, or absorb it from, the blood to ensure there is always the correct amount of calcium present in the blood. Bones also store other minerals such as phosphorous. Mineral homeostasis in bones is controlled by hormones.

Site of blood cell production
Some bones in the body contain red bone marrow which, through the process of haemopoiesis, produces red blood cells, white blood cells and platelets.

Storage of energy
The yellow bone marrow found in some bones stores lipids which are an important energy reserve in the body.

Anatomy Of Bones

Bone tissue

Osteology is the study of bone. Osseous tissue (bone tissue) is a connective tissue whose matrix is composed of water, protein, fibres and mineral salts. The fibres are made of a protein called collagen which enables bone to resist being stretched or torn apart. This is known as 'tensile strength' and without collagen bones would be hard and brittle. The mineral salts are mainly calcium carbonate and a crystallised compound called *hydroxyapatite*. These salts give bone its hardness.

Before learning about the different types of bones, it helps to know the cells that make up bone tissue:

- **Osteoprogenitor cells** – these are stem cells derived from mesenchyme (the connective tissue found in an embryo). They have the ability to become osteoblasts.
- **Osteoblasts** – these cells secrete collagen and other organic components to form bones.
- **Osteocytes** – these are mature bone cells that maintain the daily activities of bone tissue. They are derived from osteoblasts and are the main cells found in bone tissue.
- **Osteoclasts** – these cells are found on the surface of bones and they destroy or resorb bone tissue.

Two types of bone tissue exist: compact and spongy.

- **Compact (dense) bone tissue** – This is a very hard, compact tissue that has few spaces within it. It is composed of a basic structural unit called an *osteon* or *Haversian system* which is made of concentric rings of a hard, calcified matrix called *lamellae*. Between the lamellae are small

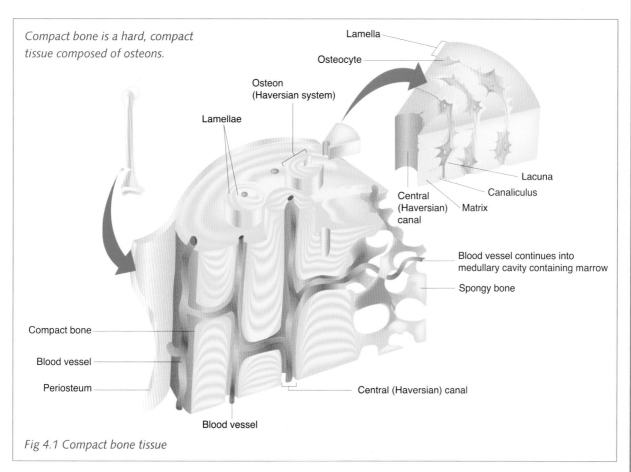

Compact bone is a hard, compact tissue composed of osteons.

Lamella
Osteocyte
Osteon (Haversian system)
Lamellae
Lacuna
Canaliculus
Central (Haversian) canal
Matrix
Blood vessel continues into medullary cavity containing marrow
Spongy bone
Compact bone
Blood vessel
Periosteum
Central (Haversian) canal
Blood vessel

Fig 4.1 Compact bone tissue

spaces called *lacunae* where osteocytes are housed. Through the centre of the lamellae run *Haversian canals* in which nerves, blood and lymph vessels are found. *Canaliculi* are tiny canals that radiate outward from the central canals to other lacunae. The main functions of compact bone tissue are protection and support. It forms the external layer of all bones.

Spongy bone is a light tissue made up of trabeculae.

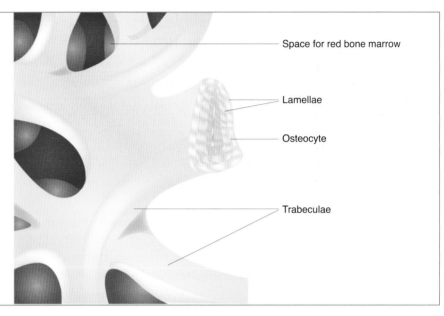

Space for red bone marrow

Lamellae

Osteocyte

Trabeculae

Fig 4.2 Spongy bone tissue

- **Spongy (cancellous) bone tissue** – this is a light tissue with many spaces within it and it has a sponge-like appearance. It does not contain osteons. Instead, it is made up of lamellae arranged in an irregular latticework of thin plates of bone called *trabeculae*. Within the trabeculae are lacunae containing osteocytes. Spongy bone tissue contains red bone marrow which is the site of blood cell production. It is found in the hip bones, ribs, sternum, vertebrae, skull and the ends of some long bones.

Bone formation and remodelling

Bone is a dynamic, living tissue which is constantly changing, repairing and reshaping itself. Most bones are formed through ossification, a process which begins somewhere between the sixth and seventh week of embryonic life and continues throughout adulthood. There are two types of ossification:

- **Intramembranous ossification** – bone forms on, or within, loose fibrous connective tissue membranes without first going through a cartilage stage.
- **Endochondral ossification** – bone forms on hyaline cartilage which has been produced by cells called chondroblasts.

New bone tissue constantly replaces old, worn-out or injured bone tissue through the process of remodelling. Mechanical stress, in the form of the pull of gravity and the pull of skeletal muscles, is integral to the process of remodelling. If these stresses are absent the bones weaken or if they are excessive the bones thicken abnormally. For example, the bones of people who are bedridden diminish while those of some athletes are found to be thicker than usual.

Did you know?
Every bone in your body is completely reformed approximately every ten years and the distal portion of your thigh bone is replaced approximately every four months.[i]

Anatomy and physiology in perspective
X-ray studies of the astronauts of the Gemini and Apollo space missions revealed that the heel bones of two of the three crewmen had decreased in density. This is thought to be due to their weightless experience in space.[ii]

What does ageing do to our bones?

Why do people often get shorter and smaller as they age? Why do their bones seem so brittle and easily broken? Why do they take so long to repair? Ageing has two specific effects on our bones. Firstly, ageing causes demineralisation of the bones. This is a loss of calcium and other minerals from the bone matrix and this process is especially evident in women after the age of 30. By the age of 70, a woman can easily have lost almost 30% of the calcium in her bones. Secondly, ageing decreases the body's ability to produce collagen which gives bones their tensile strength. Thus, bones become more brittle and susceptible to fracture.

Types Of Bones

The body contains many bones which are generally classified into the following types: long, short, flat, irregular or sesamoid.

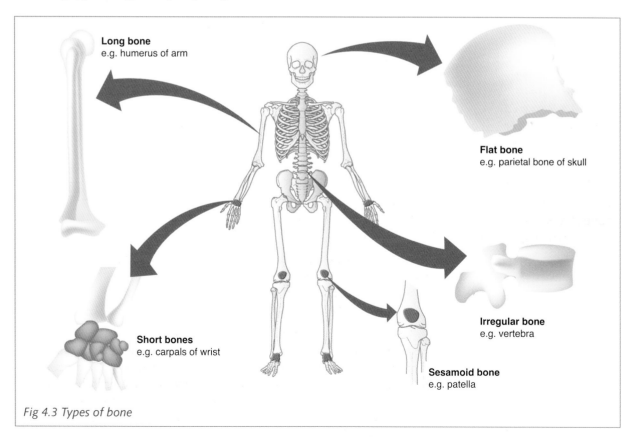

Fig 4.3 Types of bone

- **Long bones** – have a greater length than width and usually contain a longer shaft with two ends (see The Structure of a Long Bone for further details). Long bones are slightly curved to provide strength and are composed of mainly compact bone tissue with some spongy bone tissue. Examples of long bones include the femur, tibia, fibula, phalanges, humerus, ulna and radius.
- **Short bones** – cube-shaped and nearly equal in length and width. They are made up of mainly spongy bone with a thin surface of compact bone. Examples of short bones include the carpals and tarsals.
- **Flat bones** – very thin bones consisting of a layer of spongy bone enclosed by layers of compact bone. Flat bones act as areas of attachment for skeletal muscles and also provide protection. Examples of flat bones include the cranial bones, the sternum, ribs and scapulae.
- **Irregular bones** – most bones that cannot be classified as long, short or flat bones fall into the category of irregular bones. They have complex shapes and varying amounts of compact and spongy tissues. An example of an irregular bone is a vertebra.
- **Sesamoid bones** – oval bones that develop in tendons where there is considerable pressure, e.g. the patella (knee cap).

Structure of a long bone

To help you understand the structure of a bone, we will look at the structure of the humerus. This is the long bone in your arm.

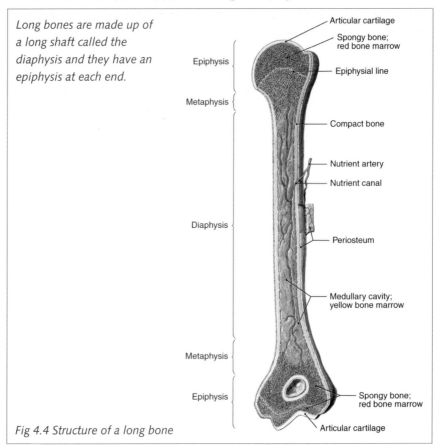

Long bones are made up of a long shaft called the diaphysis and they have an epiphysis at each end.

Epiphysis

Metaphysis

Diaphysis

Metaphysis

Epiphysis

Articular cartilage

Spongy bone; red bone marrow

Epiphysial line

Compact bone

Nutrient artery

Nutrient canal

Periosteum

Medullary cavity; yellow bone marrow

Spongy bone; red bone marrow

Articular cartilage

Fig 4.4 Structure of a long bone

A long bone has a main, central shaft called a *diaphysis*. The diaphysis is covered by a membrane known as the *periosteum*. The periosteum provides attachment for muscles, tendons and ligaments and is also essential for nutrition, repair and bone growth in diameter. The periosteum consists of two layers:

- **Outer fibrous layer,** made of dense irregular connective tissue that contains blood vessels, lymph vessels and nerves that pass into the bone.
- **Inner osteogenic layer,** made of elastic fibres and containing blood vessels and bone cells.

Each end of the diaphysis is called an *epiphysis*. Each epiphysis is covered by a thin layer of hyaline cartilage called *articular cartilage*. This cartilage reduces friction and absorbs shock at the area where the bone forms an articulation (joint) with the surface of another bone. The epiphysis is made of mainly spongy bone tissue and contains red bone marrow. This is where blood cells are produced.

The region where the diaphysis joins the epiphysis is the *metaphysis*. In a growing bone the metaphysis has a layer of hyaline cartilage that allows the diaphysis to grow in length. This is called the *epiphyseal plate*. In a mature bone that is no longer growing in length, the epiphyseal plate is replaced by the *epiphyseal line*.

Inside the diaphysis is a space known as the *medullary* or *marrow cavity*. The medullary cavity is lined by a membrane called the *endosteum*. This contains cells necessary for bone formation. In adults this cavity contains fatty yellow bone marrow which stores lipids.

Anatomy and physiology in perspective
If a child fractures a bone and damages the epiphyseal plate then the bone will always be shorter than it should be. However, if the epiphyseal plate has not been damaged, then the bone will be able to grow to its normal length.

Organisation Of The Skeleton

The skeleton is made up of a central axial skeleton which supports and protects the major organs of the head, neck and trunk; and an appendicular skeleton which forms the upper and lower extremities and their girdles.

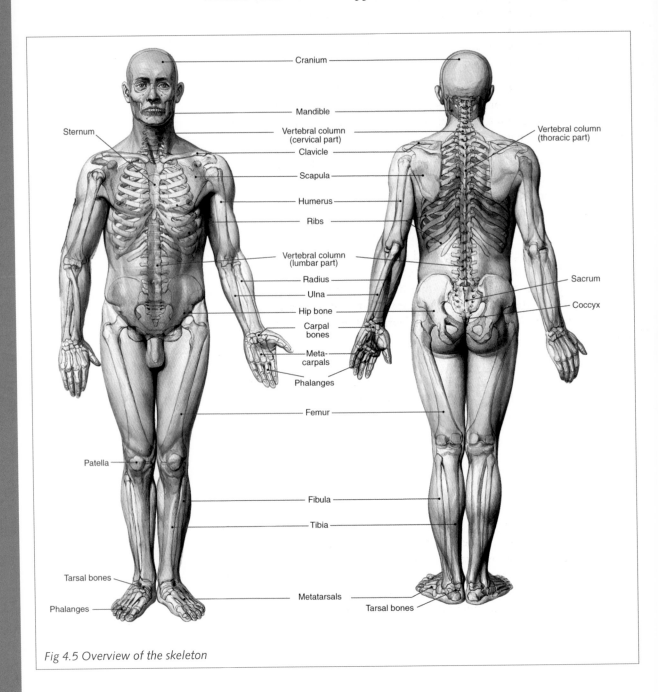

Fig 4.5 Overview of the skeleton

There are approximately 206 bones in the human body. This number can differ slightly depending on age and gender, for example, a baby has between 270 to 300 bones at birth. The skeleton is made up of:

- **The axial skeleton** – consists of the 80 bones that are found in the centre of the body, namely the skull, hyoid, ribs, sternum and vertebrae. The auditory ossicles are also usually included in the axial skeleton (these are discussed in Chapter 6).
- **The appendicular skeleton** – consists of the 126 bones of the upper and lower limbs and their girdles which connect them to the axial skeleton.

The bones of the skeleton are attached to one another by ligaments. Ligaments are tough, fibrous cords of connective tissue that contain both collagen and elastic fibres. They surround joints and bind them together, joining bones to bones.

The charts below identify the bones of the body and briefly mention some of the major joints of the body (joints and their movements will be discussed in more detail shortly).

Study tip
Learn the names of the bones well and you will be relieved you did so later. Many of the muscles, blood vessels and nerves that you also need to learn are named after the bones along which they run. For example the tibialis anterior is a muscle located on the anterior surface of the tibia bone, the radial artery is located near the radius bone and the ulnar nerve is found near the ulna bone. The time you spend learning the bones will not be wasted!

Please note
The tables from page 105 onwards mention joint types and movements. You may want to read the section on joints on page 122 before reading the following information.

Axial Skeleton

Bones of the skull

The skull contains twenty-two bones which can be divided into eight cranial and fourteen facial bones. Together these bones protect and support the brain and the special sense organs (of vision, taste, smell, hearing and equilibrium) and they form the framework of the face. They also protect the entrances to the digestive and respiratory systems and provide areas of attachment for muscles, including those of facial expression.

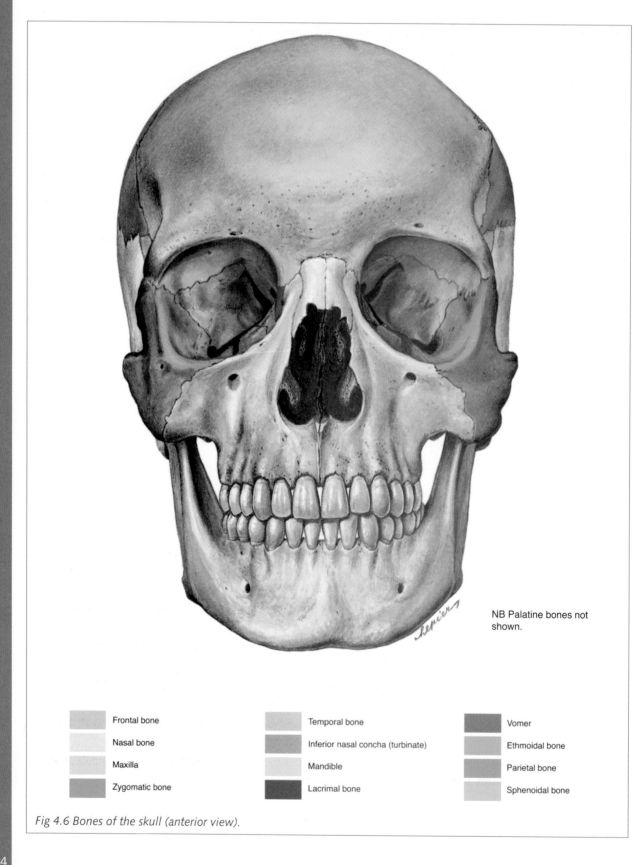

NB Palatine bones not shown.

Frontal bone	Temporal bone	Vomer
Nasal bone	Inferior nasal concha (turbinate)	Ethmoidal bone
Maxilla	Mandible	Parietal bone
Zygomatic bone	Lacrimal bone	Sphenoidal bone

Fig 4.6 Bones of the skull (anterior view).

Bones of the cranium

The cranium is a large cavity that encloses and protects the brain. It is made up of eight bones: the frontal bone, two parietal bones, two temporal bones, the occipital bone, the sphenoid bone and the ethmoid bone. Most of the bones of the cranium are held together by immovable joints called sutures.

BONES OF THE CRANIUM	
Frontal	Forms the forehead and the roofs of the orbits (eye sockets).
Parietals	Two parietal bones form the sides and roof of the cranium.
Temporals	Beneath the parietal bones are two temporal bones. They form the inferior lateral sides of the cranium and part of the cranial floor.
Occipital	Forms the back of the cranium and most of the base of the cranium.
Sphenoid	This butterfly-shaped bone articulates with all the other cranial bones and holds them together. It lies at the middle part of the base of the skull and forms part of the floor of the cranium, the sides of the cranium and parts of the eye orbits.
Ethmoid	This is the major supporting structure of the nasal cavity. It forms the roof of the nasal cavity and part of the medial walls of the eye orbits.

Bones of the face

There are fourteen bones that make up the face. They are: two nasal bones, two maxillae, two zygomatic bones, the mandible, two lacrimal bones, two palatine bones, two inferior nasal conchae (turbinates) and the vomer.

Most of the bones of the face are held together by immovable joints called sutures. Only the mandible (jawbone) is attached to the cranium by a freely movable joint called the temporo-mandibular joint (TMJ).

BONES OF THE FACE	
Nasal bones	These two bones form the bridge of the nose.
Maxillae	These two bones unite to form the upper jawbone, part of the floors of the orbits, part of the lateral walls and floor of the nasal cavity and most of the roof of the mouth. The maxillae articulate with every bone in the face except the mandible.
Zygomatics	These two bones are the cheekbones and form part of the lateral wall and floor of the orbits.
Mandible	This is the lower jaw. It is the only movable bone in the skull and is also the largest and strongest facial bone.
Lacrimals	These tiny bones form part of the medial wall of the eye orbit. They are about the same size and shape as a fingernail and are the smallest bones in the face.
Palatines	These two bones form part of the palate (roof of the mouth), part of the floor and lateral wall of the nasal cavity and part of the floors of the orbits.
Inferior nasal conchae (Turbinates)	These two bones form part of the lateral wall of the nasal cavity and help to circulate, filter and warm air before it passes into the lungs.
Vomer	This bone forms part of the nasal septum which divides the nose into left and right sides.

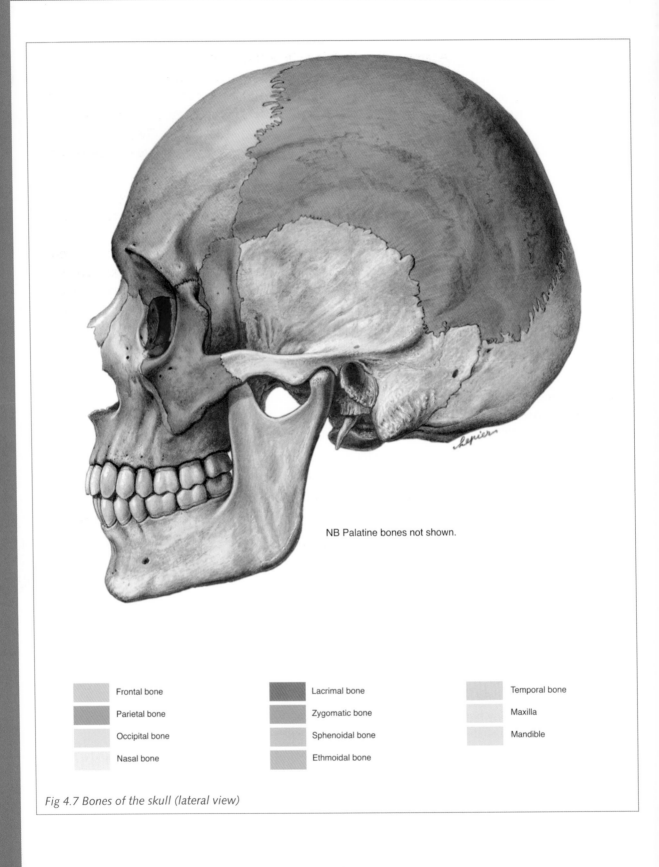

NB Palatine bones not shown.

	Frontal bone		Lacrimal bone		Temporal bone
	Parietal bone		Zygomatic bone		Maxilla
	Occipital bone		Sphenoidal bone		Mandible
	Nasal bone		Ethmoidal bone		

Fig 4.7 Bones of the skull (lateral view)

Anatomy and physiology in perspective

What exactly are sinuses and why can they cause us so much discomfort? Paranasal sinuses are air-filled spaces within the cranial and facial bones. They are located near the nasal cavity and serve as resonating chambers for sound when we speak. They are also lined with a mucous membrane and have tiny openings into the nasal cavity called ostia.

When you have a cold or an allergic reaction to something, the mucous membranes in the sinuses can swell and the mucous will be unable to pass through the tiny ostia and drain as it should normally do. This can lead to inflammation, infection and a great deal of pain and discomfort. There are three pairs of paranasal sinuses named after the bone in which they are located: the frontal, maxillary and sphenoid sinuses. There are also the ethmoid sinuses which consist of many spaces inside the ethmoid bone.

Sinuses are air-filled cavities that serve as resonating chambers for sound and that also produce mucous.

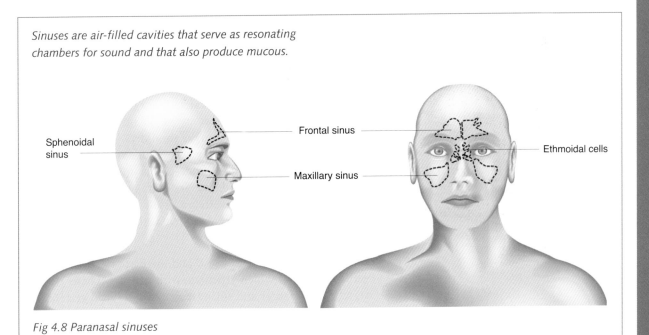

Fig 4.8 Paranasal sinuses

Bones of the neck and spine

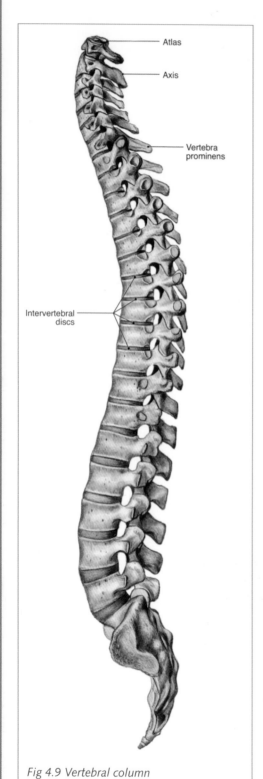

Atlas

Axis

Vertebra
prominens

Intervertebral
discs

*Fig 4.9 Vertebral column
viewed from the left*

Neck

The neck comprises the cervical vertebrae (to be discussed shortly), as well as a unique bone that does not articulate with any other bone in the body. This is the hyoid bone and it is suspended from the temporal bones by ligaments and muscles.

Hyoid

Supports the tongue and provides attachment for some of the muscles of the neck and pharynx.

Spine (vertebral column)

The spine is a strong, flexible structure that is able to bend and rotate in most directions. It supports the head, encloses and protects the spinal cord and is also a site of attachment for the ribs and the muscles of the back.

The spine is composed of thirty-three vertebrae as follows: seven cervical, twelve thoracic, five lumbar, five sacral and four coccygeal. The sacral vertebrae fuse to form the sacrum and the coccygeal vertebrae fuse to form the coccyx. Thus, there are 26 separate bones that make up the spine.

Cervical vertebrae (7)

The word cervix means 'neck' and seven cervical vertebrae form the neck. The first cervical vertebra (C1) is called the atlas and it supports the head, just as the mythological figure, Atlas, supported the world on his shoulders. The second (C2) is called the axis and it literally acts as an axis on which the atlas and head can rotate in a side-to-side movement. The third to sixth cervical vertebrae are quite normal, but the seventh (C7) is called the vertebra prominens and is the large prominence that can be seen and felt at the back of the neck.

Thoracic vertebrae (12)

The word thorax means 'chest' and there are twelve thoracic vertebrae, ten of which articulate with the ribs. These vertebrae are larger and stronger than the cervical vertebrae.

Lumbar vertebrae (5)

Five lumbar vertebrae support the lower back. They are the largest, strongest vertebrae and provide attachment for the large muscles of the back that support the weight of the upper body.

Sacral vertebrae (5)

Five sacral vertebrae fuse to form a triangular bone called the sacrum which is the strong foundation of the pelvic girdle.

Coccygeal vertebrae (4)

Four coccygeal vertebrae fuse to form a triangular shape called the coccyx or tailbone.

A closer look at the spine

The spine is a simple yet invaluable structure that needs to not only protect the spinal cord but also be able to move in a number of directions, withstand pressure, hold the weight of the upper body and maintain our bodies in an upright position. It is able to perform all these functions because it is made up of so many small, strong bones and joints.

Structure of a vertebra

A vertebra consists of a weight-bearing body and two vertebral arches which surround the vertebral foramen, a space through which the spinal cord runs.

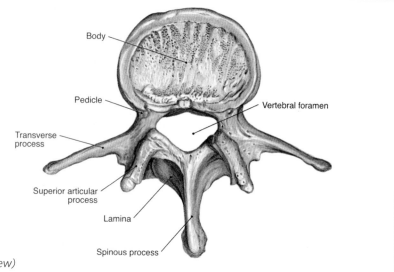

Body

Pedicle

Vertebral foramen

Transverse process

Superior articular process

Lamina

Spinous process

Fig 4.10 A typical vertebra (superior view)

Although vertebrae do differ in size and shape according to where in the spine they are found, they all share a similar structural pattern. They all have:

- **A body** – this is a thick, disc-shaped anterior portion that is weight-bearing.
- **A vertebral arch** – this is formed by two thick posterior processes (projections) called the pedicles and the laminae. The body and arches form a space called the vertebral foramen.
- **The vertebral foramen** – this is the space formed by the body and arches. It contains the spinal cord, adipose and areolar connective tissues and blood vessels. The vertebral foramina of all vertebrae join together to form the spinal canal.
- **Transverse processes** – these are two prominences that project laterally from either side of the vertebral arch, where the pedicles and lamina join. These processes serve as points of attachment for muscles.
- **Spinous process** – this is a single prominence that projects posteriorly from the vertebral arch where the laminae meet. It is also called the spine of a vertebra and serves as a point of attachment for muscles.
- **Superior and inferior articular processes** – these four processes form joints with the vertebrae above and below.

Intervertebral discs

Intervertebral discs are found between the vertebrae. They are composed of an outer ring of fibrocartilage called the annulus fibrosus and a soft, elastic inner nucleus pulposus. Intervertebral discs form strong joints that allow the spinal column to move in many directions and they also provide a cushioning that can flatten and absorb vertical shock when under compression.

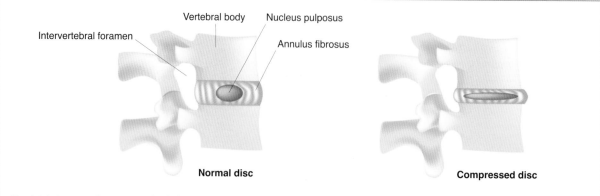

Fig 4.11 A normal intervertebral disc and a compressed intervertebral disc (lateral view)

 Infobox

Anatomy and physiology in perspective
A slipped disc is not a disc that has slipped out of the spinal column. It is actually a herniated disc that is often called a prolapsed intervertebral disc (PID). This means that the disc's outer ring of fibrocartilage has ruptured and its inner nucleus pulposus is protruding. This usually occurs if the ligaments surrounding the discs are injured or weakened and the discs are put under excess pressure.

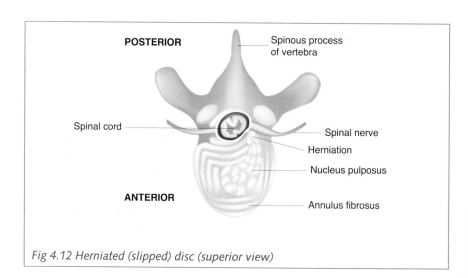

Fig 4.12 Herniated (slipped) disc (superior view)

Curves of the spine

When viewed from the side, the spine has four natural curves that increase its strength and flexibility, help maintain balance and absorb shock. They are named after the vertebrae that form them:

- Cervical curve
- Thoracic curve
- Lumbar curve
- Sacral curve

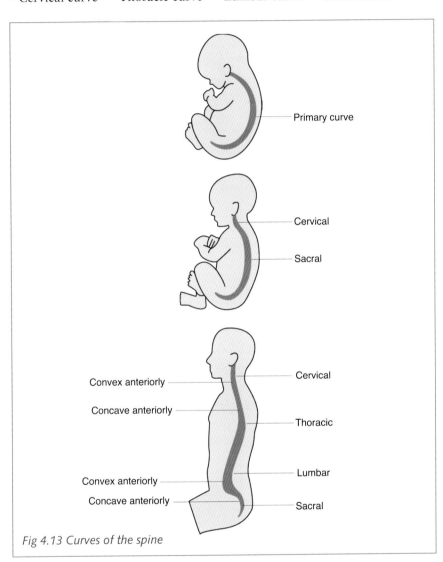

Fig 4.13 Curves of the spine

Anatomy and physiology in perspective

A foetus curled up inside the womb has only a single concave curve. When it is a baby and it starts to hold its head erect, a second curve, the cervical curve, then develops. When the baby starts to sit up straight, stand and walk, the lumbar curve develops. This curve separates the thoracic and sacral curves which are known as the primary curves. The cervical and lumbar curves are the secondary curves.

Thorax

The thorax, or chest, is a bony cage composed of the sternum, ribs and thoracic vertebrae. It encloses and protects the heart and lungs and supports the bones of the shoulder girdle and upper limbs.

BONES OF THE THORAX	
Sternum	This is the breastbone. It is a flat, narrow, long bone that is found in the middle of the anterior thoracic wall. The sternum is composed of three parts: The superior *manubrium* which articulates with the clavicle and the costal cartilages of the first two ribs. The long, middle *body* which articulates with the costal cartilages of the second through to seventh ribs. The inferior *xiphoid process* which provides attachment for some abdominal muscles.
Ribs	There are twelve pairs of ribs. The first seven pairs of ribs are attached to the sternum by a type of hyaline cartilage called costal cartilage. These seven pairs are known as *true ribs* because they are directly attached to the sternum. The remaining five pairs of ribs are *false ribs* because they are not directly attached to the sternum. The eighth, ninth and tenth pairs are attached to each other by their cartilages and then to the cartilages of the seventh pair of ribs. The eleventh and twelfth pairs of ribs are *floating ribs* that are only attached to abdominal muscles.
Thoracic vertebrae	Please refer to page 108 for more information on the thoracic vertebrae.

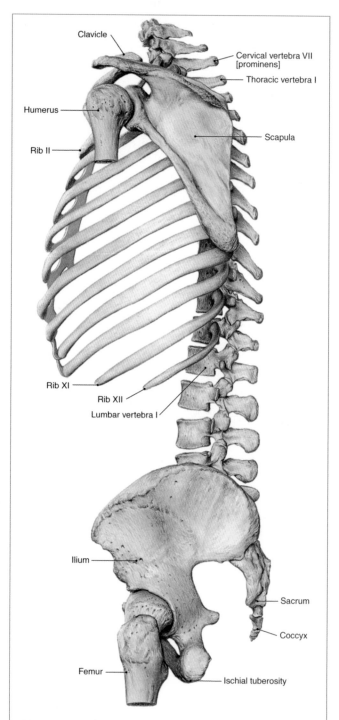

Fig 4.14 Vertebral column, pectoral and pelvic girdles (lateral view).

Labels on figure: Clavicle; Cervical vertebra VII [prominens]; Thoracic vertebra I; Humerus; Scapula; Rib II; Rib XI; Rib XII; Lumbar vertebra I; Ilium; Sacrum; Coccyx; Femur; Ischial tuberosity

Bones of the trunk, pectoral and pelvic girdles

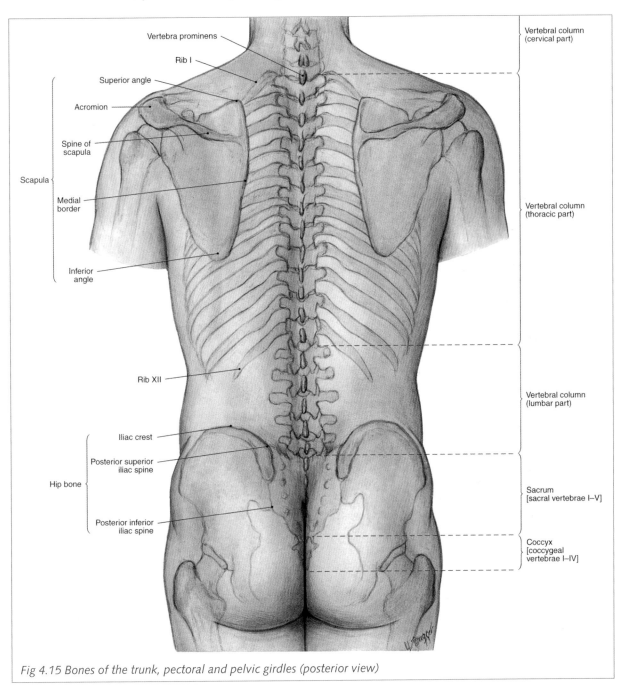

Fig 4.15 Bones of the trunk, pectoral and pelvic girdles (posterior view)

Appendicular Skeleton

The appendicular skeleton comprises the upper and lower extremities and their girdles – basically, the bones of the arms and legs, shoulders and pelvis.

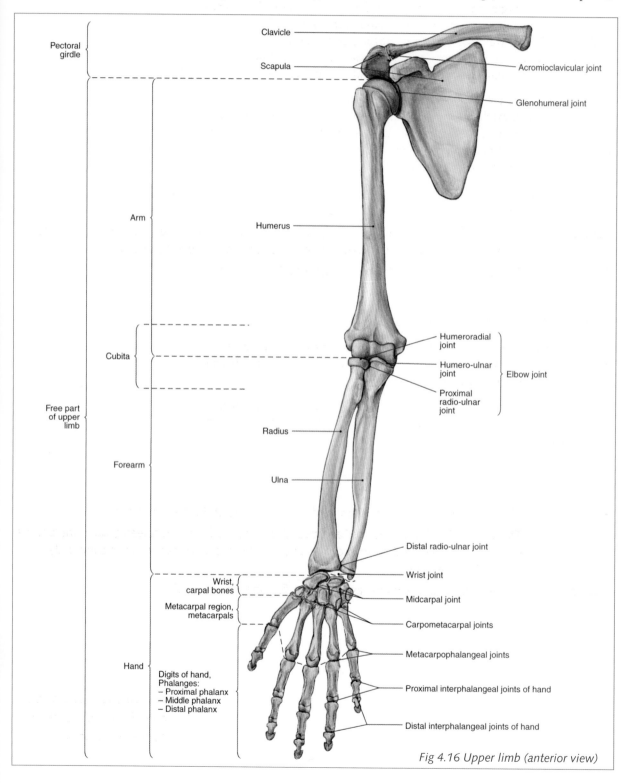

Fig 4.16 Upper limb (anterior view)

Pectoral girdle (the shoulder)

The shoulder is called the pectoral girdle and it attaches the arm to the trunk of the body. It consists of two bones: the clavicle and scapula.

The shoulder joint (glenohumeral joint) is a diarthrotic, synovial ball-and-socket joint formed by the humerus and scapula. It has exceptional flexibility and more freedom of movement than any other joint in the body. It allows the following movements: flexion, extension, abduction, adduction, medial rotation, lateral rotation and circumduction.

However, the shoulder's freedom of movement does make it prone to dislocation. It is strengthened by a group of muscles called the rotator cuff muscles. These surround the joint, joining the scapula to the humerus, and consist of the supraspinatus, infraspinatus, teres minor and subscapularis muscles (see Chapter 5 for more information on these muscles).

PECTORAL GIRDLE (THE SHOULDER)	
Clavicle	This is the collarbone and it is a long, slender, double curved bone that helps hold the arm away from the top of the thorax. It also helps prevent shoulder dislocation.
Scapula	This is the shoulder blade. It is a large, flat, triangular bone.

Upper limb (arm, forearm and hand)

Sixty bones make up the arm and hand. The arm consists of the humerus and the forearm is made up of the ulna and radius. The hand is made up of the carpals, metacarpals and phalanges.

The bones of the upper arm and forearm meet at the elbow joint which is classified as a hinge joint. This joint only allows for flexion and extension of the forearm.

BONES OF THE UPPER LIMB (ARM, FOREARM AND HAND)	
Humerus	This is the longest and largest bone of the upper limb and is the arm bone.
Ulna	This bone is located on the medial aspect (little finger side) of the forearm when the body is in the anatomical position. Its proximal end is called the olecranon which is commonly known as the elbow.
Radius	This is located on the lateral aspect of the forearm (thumb side).

Did you know?
The clavicle is one of the most frequently broken bones in the body.

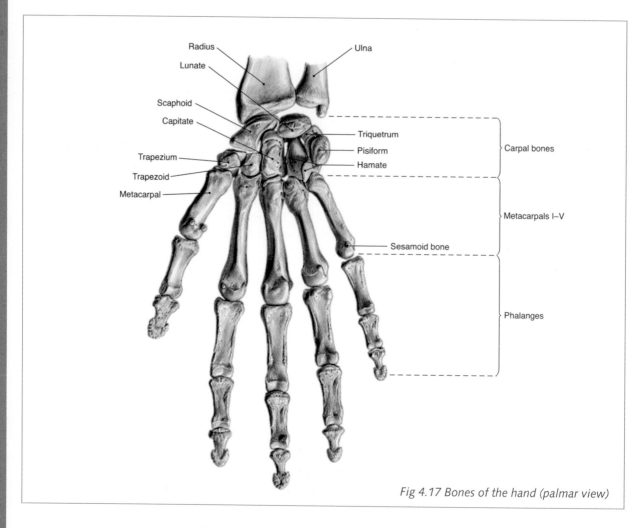

Fig 4.17 Bones of the hand (palmar view)

Hand

The bones of the hand fall into three regions: the carpals, metacarpals and phalanges. The carpals attach to the radius at the wrist joint. This is a diarthrotic, synovial condyloid joint that allows for flexion, extension, abduction, adduction and circumduction.

BONES OF THE HAND	
Carpals (8)	The wrist consists of eight small bones arranged in two irregular rows of four bones each. They are bound together by ligaments. These bones are the: Trapezium, Trapezoid, Capitate, Hamate, Scaphoid, Lunate, Triquetrum, Pisiform
Metacarpals (5)	Five metacarpals form the palm of the hand. They are numbered 1 to 5, starting with the thumb side of the hand.
Phalanges (14)	Fourteen phalanges make up the fingers. In each finger are a proximal, middle and distal phalange. In the thumb there are only a proximal and distal phalange. The thumb is sometimes called the *pollex*.

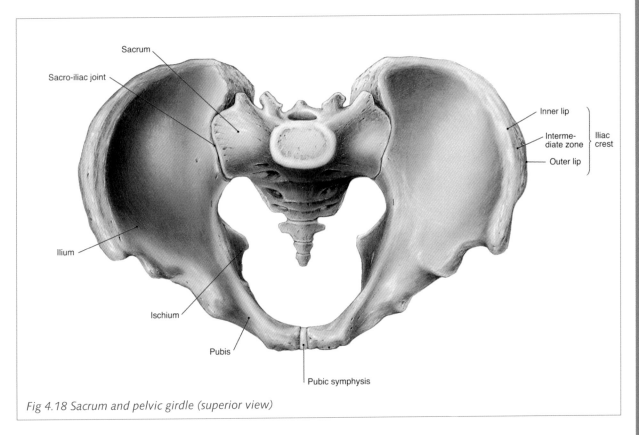

Fig 4.18 Sacrum and pelvic girdle (superior view)

The skeletal system

Pelvic girdle (hip)

The hip, or pelvic girdle, is a strong and stable structure upon which rests the total weight of the upper body. The pelvic girdle also supports both the spine and the visceral organs and is a site of attachment for many muscles.

The pelvic girdle attaches the leg to the trunk of the body and is composed of two hipbones called the *innominate bones*, *coxal bones* or *ossa coxae*. In a newborn baby, each of these bones is made up of three smaller bones – the ilium, ischium and pubis. In adults these bones are fused together.

Together with the sacrum and coccyx, the pelvic girdle forms the pelvis which is a bony, basin-like structure. The ilium, ischium and pubis fuse to form the *acetabulum*, a deep socket for the head of the femur. The joint formed by the femur and the acetabulum is called the hip joint (coxal joint) and is a diarthrotic, synovial ball-and-socket joint that allows for flexion, extension, abduction, adduction, circumduction and rotation.

The pubic bones are united anteriorly by a cartilaginous joint called the *pubic symphysis*. Posteriorly, the ilium of each hipbone connects with the sacrum at the *sacroiliac* joint.

BONES OF THE PELVIC GIRDLE (HIP)	
Ilium	This large, wing-like bone forms the superior portion of the hipbone. Its upper border serves as a site of attachment for many muscles and is called the iliac crest.
Ischium	This forms the inferior and posterior portion of the hipbone.
Pubis	This is the most anterior part of the hipbone.

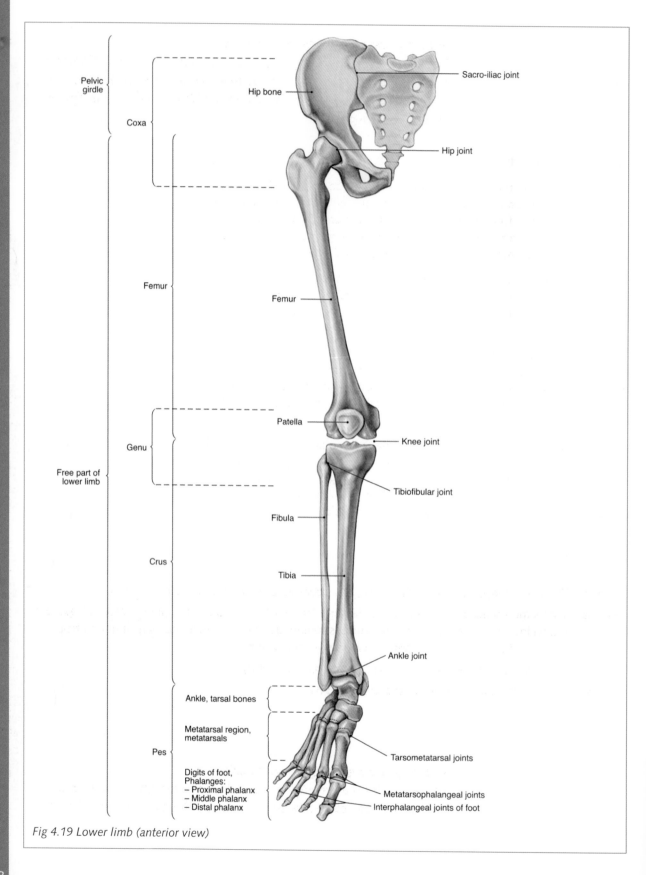

Pelvic girdle

Coxa

Hip bone

Sacro-iliac joint

Hip joint

Femur

Femur

Free part of lower limb

Genu

Patella

Knee joint

Tibiofibular joint

Fibula

Crus

Tibia

Ankle joint

Ankle, tarsal bones

Metatarsal region, metatarsals

Pes

Tarsometatarsal joints

Digits of foot,
Phalanges:
– Proximal phalanx
– Middle phalanx
– Distal phalanx

Metatarsophalangeal joints
Interphalangeal joints of foot

Fig 4.19 Lower limb (anterior view)

Anatomy and physiology in perspective
Because of her role in childbearing, a woman's pelvis differs from a man's pelvis. Her pelvis is larger and shallower and the bones are lighter and thinner. Her sacrum is also shorter and less curved and her pubic arch is more rounded.

Lower limb (thigh, knee, leg and foot)

Like the upper limb, the lower limb is composed of sixty bones. The thigh is made up of the femur, the knee is the patella, and the leg is composed of the fibula and tibia. The foot includes the tarsals, metatarsals and phalanges. The bones of the lower limb are all thicker and stronger than those of the upper limb because they need to bear the weight of the entire body.

The femur of the thigh and the fibula and tibia of the leg meet at the knee joint (tibiofemoral joint). This is classified as a diarthrotic synovial joint which allows for flexion, extension, slight medial rotation and lateral rotation when in a flexed position.

The knee joint is the largest joint in the body and it is actually made up of three smaller joints. These include an intermediate patellofemoral joint between the patella and the femur (this is a gliding joint) and a lateral and medial tibiofemoral joint between the tibia and femur (these are modified hinge joints).

There is no actual interlocking of bones in the knee joint and it is reinforced by tendons and ligaments only. These fibres connect the bones and also surround the joint to strengthen it. Between the tibia and femur are two menisci. These are fibrocartilage discs that help to compensate for the irregular shapes of the bones and that also help circulate synovial fluid.

BONES OF THE LEG	
Femur	This is the thigh bone and it is the longest, strongest and heaviest bone in the body.
Patella	This is the kneecap and it is a sesamoid bone that is attached to the tibia by the patellar ligament. It develops in the tendon of the quadriceps femoris muscle to protect the knee joint and help maintain the position of the tendon when the knee is bent.
Tibia	This is the shinbone and is the large, medial bone of the leg.
Fibula	This is a thin bone that runs parallel to the tibia.

Anatomy and physiology in perspective
The knee joint is a synovial joint and the common condition of 'water on the knee' is swelling due to an excessive production of synovial fluid.

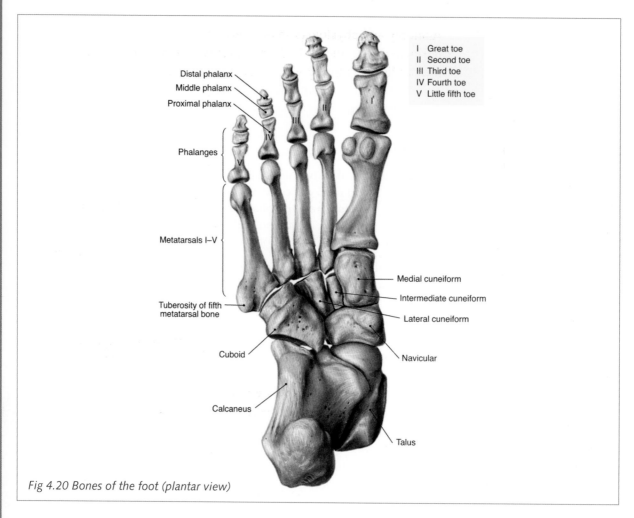

Fig 4.20 Bones of the foot (plantar view)

Foot

Like the bones of the hand, the bones of the foot fall into three regions: the tarsals, metatarsals and phalanges. These are extremely strong bones that support our body weight and enable us to walk and run. The tibia and fibula of the leg meet the talus of the foot at the ankle joint (talocrural joint). This is a diarthrotic, synovial hinge joint that allows for dorsiflexion and plantar flexion of the foot.

BONES OF THE FOOT	
Tarsals (7)	Seven tarsal bones make up the *tarsus*, or back portion, of the foot. They are the: Talus (ankle bone), Calcaneus (heel bone), Cuboid, Navicular, Three cuneiforms (medial, intermediate and lateral)
Metatarsals (5)	Five metatarsals form the metatarsus of the foot. They are numbered 1 to 5, starting with the large toe (medial) side of the foot.
Phalanges (14)	Fourteen phalanges make up the toes. In each toe are a proximal, middle and distal phalange. In the large toe there are only a proximal and distal phalange. The large toe is sometimes called the hallux.

Anatomy and physiology in perspective
A prosthesis is an artificial limb that is attached to the body as a substitute for a missing or non-functioning limb.

The arches of the foot distribute the weight of the body over the entire foot and give feet their springiness.

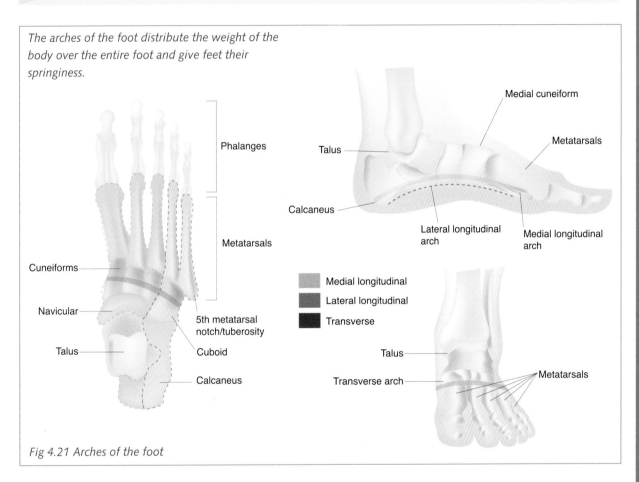

Fig 4.21 Arches of the foot

Arches of the foot

The bones of the foot are arranged into arches that distribute the weight of the body over the entire foot. They are strong arches that give the foot its 'springiness' and so enable the foot to support the weight of the body and provide leverage while walking. They include the:

- **Medial longitudinal arch** – this arch runs down the medial length of the foot (longitudinally) and is composed of the calcaneus, navicular, all three cuneiforms and the medial first three metatarsals.
- **Lateral longitudinal arch** – this arch runs down the lateral length of the foot (longitudinally) and is composed of the calcaneus, cuboid and the lateral two metatarsals.
- **Transverse arch** – this arch cuts transversely across the foot and is formed by the cuboid, the three cuneiforms and the bases of the five metatarsals.

Did you know?

If you are under 40 years of age, some of the bones in your body are still growing. An embryo has no bones and its 'skeleton' is composed entirely of fibrous connective tissue membranes, called mesenchyme, and hyaline cartilage. Around the sixth or seventh week of embryonic life, ossification begins and although bone growth in length is usually completed around 25 years of age, bones continue to thicken and be remodelled throughout life.

Take a quick look at some developmental aspects of the skeleton:

- **The head** – Babies are born with soft spots on the tops of their heads. These are called fontanelles and are membrane filled spaces between the cranial bones. They enable the baby's skull to change size and shape so that it can pass through the birth canal and also allow rapid growth of the brain during infancy. They ossify and become sutures.
- **The face** – During the first two years of life, the shape of the face changes dramatically while the brain and cranial bones expand, the teeth form and the sinuses increase in size. The face only stops growing when you reach about 16 years of age.
- **The back** – Not only are you born with one curve in your spine that develops into four, fusion of your sacral vertebrae only begins around the age of 16–18 years and is completed by about 30 years of age. In addition, fusion of your coccyx takes place somewhere between 20–30 years of age.
- **The feet** – The arches of the feet are developed by 12 or 13 years of age.

Finally, the xiphoid process of the sternum does not ossify until about age 40.

Joints

The bones of the skeleton are held together by joints. Joints are also called articulations and are the points of contact between bones, cartilage and bones or teeth and bones. Without joints, our bodies would be solid structures and we would be unable to move. Thus, they function not only in holding the bones together, but also in providing movement. However, some of the joints of the body, for example the sutures of the skull, do not allow for much movement. Instead, they form firm structures that help protect organs. Before looking at the different types of joints it helps to have an understanding of the following vocabulary:

- Arthrology – the study of joints.
- Kinesiology – the study of the motion of the body.
- **Cartilage** – a resilient, strong connective tissue that is less hard and more flexible than bone.
- **Ligament** – a tough band of connective tissue that attaches bones to bones.
- Tendon – a tough band of connective tissue that attaches muscles to bones.

Joints can be classified according to either their structure or their function.

Functional classification of joints

If you consider the degree of movement the joints permit, then you are looking at its functionality and this is classified as follows:

- **Synarthroses** – these are immovable joints and are termed synarthrotic joints. For example, the coronal suture between the frontal and parietal bones is composed of dense fibrous connective tissue that holds the bones together and does not allow movement. Instead, these bones help form the cranium which protects the brain.
- **Amphiarthroses** – these are slightly movable joints that permit a minimal amount of flexibility and movement. They are termed amphiarthrotic joints. For example, the pubic symphysis is an amphiarthrotic joint that joins the pubic bones. It allows for slight movement.
- **Diarthroses** – these are freely movable joints that permit a number of different movements. Diarthroses enable movement because of their structure – they all have a space between the articulating bones. This space is called a *synovial cavity* (in a moment we will take an in depth look at diarthroses). For example, the shoulder joint is a diarthrotic joint formed by the humerus and scapula. It allows for a variety of movements, including flexion, extension, abduction, adduction, medial rotation, lateral rotation and circumduction (these movements will be described shortly).

In general, the joints of the axial skeleton are synarthrotic (immovable) or amphiarthrotic (slightly movable) and their main function is to protect the internal organs and ensure firm attachments between bones. On the other hand, the appendicular skeleton has more diarthrotic (freely movable) joints because mobility and freedom of movement is important in the limbs.

Structural classification of joints

We have looked at how joints can be classified according to their functionality (the degree of movement they permit), now let us look at how they are classified according to their structure. This takes into account the type of connective tissue that binds the joints together and whether or not there is a synovial cavity between the joints. Structurally, joints can be classified as:

- Fibrous joints
- Cartilaginous joints
- Synovial joints

Fibrous joints

In fibrous joints bone ends are held together by fibrous (collagenous) connective tissue and there is no synovial cavity between them. Thus, they are strong joints that do not permit movement. In general, they are synarthrotic. For example, sutures are fibrous joints.

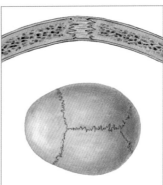

Fig 4.22 Fibrous joint, as shown by the sutures of the skull.

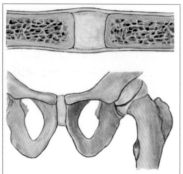

Fig 4.23 Cartilaginous joint, as shown by the pubic bone symphysis

Cartilaginous joints

Bone ends are held together by cartilage and they do not have a synovial cavity between them. Thus, they are also strong joints that permit only minimal movement. Most cartilaginous joints are amphiarthrotic, although some can be synarthrotic. For example, the pubic symphysis is a cartilaginous joint.

Synovial joints

Bone ends are separated by a synovial cavity. This allows for a great deal of movement and all synovial joints are diarthrotic. For example, the glenohumeral (shoulder) joint is a synovial joint.

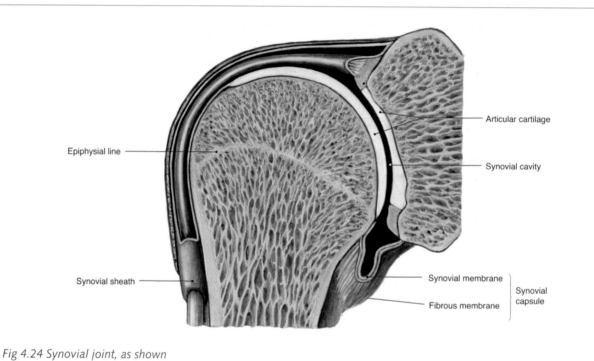

Epiphysial line

Synovial sheath

Articular cartilage

Synovial cavity

Synovial membrane

Fibrous membrane

Synovial capsule

Fig 4.24 Synovial joint, as shown by the shoulder joint

Synovial joints

Synovial joints predominate in the limbs of the body and allow for a great variety of movement. There are many different types of synovial joints, but they all share some common features. Synovial joints have:

- **A synovial cavity** – this is a space that separates the articulating bones. This space is the key to movement within the joints.
- **Articular cartilage** – hyaline cartilage covers the surfaces of the articulating bones in a synovial joint. This cartilage reduces friction between the bones and helps absorb shock.
- **Synovial capsule (articular capsule)** – the entire joint is enclosed in an articular capsule made of:

- An outer fibrous capsule of dense irregular connective tissue that is flexible and has great tensile strength.
- An inner synovial membrane made of areolar connective tissue with elastic fibres and adipose tissue. This membrane secretes synovial fluid. Synovial fluid fills the synovial cavity and:
  • Lubricates the joint, thereby reducing friction.
  • Supplies nutrients to the articular cartilage (remember that cartilage is avascular).
  • Removes waste from the articular cartilage.
  • Contains phagocytes that remove microbes and debris that have resulted from the general wear and tear of the joint.
- **Reinforcing ligaments** – the fibrous capsule is reinforced with ligaments.

In addition to the above features, some synovial joints also have:

- **Articular discs (menisci)** – these are pads of fibrocartilage that lie between the articular surfaces of the bones. They help to maintain the stability of the joint and also direct the flow of synovial fluid to areas of greatest friction. Menisci are found in the knee joint.
- **Bursae** – these are sac like structures made of connective tissue, lined with a synovial membrane and filled with synovial fluid. They cushion the movement of one structure over another and, in addition to being found inside some articular capsules, they can also be located where skin rubs over bone, between tendons and bones, between muscles and bones and between ligaments and bones.

Synovial joints permit a range of different movements some of which are described in the charts on the following pages.

Anatomy and physiology in perspective
The inflammation of a synovial joint is called *synovitis* and it usually accompanies disorders such as arthritis. Symptoms of synovitis include pain, tenderness and swelling.

Did you know?
Synovial fluid is similar in both consistency and appearance to uncooked egg white.

MOVEMENTS AT SYNOVIAL JOINTS – GENERAL

Movement	Definition
Flexion	This is the bending of a joint in which the angle between articulating bones decreases.
Extension	This usually restores a body part to its anatomical position after it has been flexed and is the straightening of a joint in which the angle between the articulating bones increases.
Hyperextension	This is when a body part extends beyond its anatomical position.

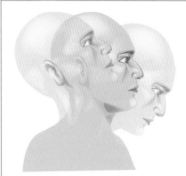

In the classroom…

Here is a simple exercise to demonstrate the movements of flexion, extension and hyperextension. Nod your head. When your chin touches your throat, you are *flexing* your neck. When you return your head to its normal upright position, you are *extending* your neck. When you push your head backwards so that your chin is facing up towards the ceiling, you are *hyperextending* your neck.

Fig 4.25 Flexion, extension and hyperextension

MOVEMENTS AT SYNOVIAL JOINTS – GENERAL

Movement	Definition
Abduction	This is a movement away from the midline of the body.
Adduction	This is the opposite of abduction and is a movement towards the midline of the body.
Circumduction	This is a circular movement of the distal end of a body part and it involves a succession of flexion-extension and abduction-adduction.

In the classroom….

These common arm exercises demonstrate abduction, adduction and circumduction. Do some star-jumps. As you swing your arms up and away from your body, you are *abducting* them. Likewise, you are *abducting* your legs as you jump them outwards. As you bring your arms back down towards the sides of your body, you are *adducting* them. You are also *adducting* your legs as you jump them together again. Swing your arms in a large circular movement as if you are warming up your shoulders. This circular swinging movement is *circumduction*.

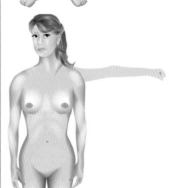

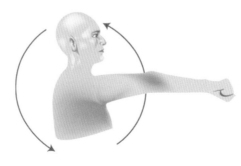

Study tip
Remember: If somebody is abducted they are kidnapped or taken away. Abduction is a movement away.

When you ADDuct a body part, you are ADDing it back to your body.

Fig 4.26 Abduction & adduction   *Fig 4.27 Circumduction*

MOVEMENTS AT SYNOVIAL JOINTS – GENERAL

Movement	Definition
Rotation	This is the movement of a bone in a single plane around its longitudinal axis.
Medial or internal rotation	Involves the movement of the anterior surface of a bone towards the midline.
Lateral or external rotation	Involves the movement of the anterior surface of a bone away from the midine.

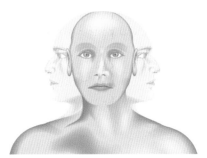

In the classroom...

Turn your head from side to side as if you are saying 'no'. You are *rotating* your head. Now, stand in the anatomical position with your palms facing forwards and turn your palms in towards your thighs so that they face backwards – you have *medially rotated* your forearms. Turn your palms from facing backwards to their original position in which they were facing forwards – you have *laterally rotated* your forearms.

Fig 4.28 Rotation

MOVEMENTS AT SYNOVIAL JOINTS – FOREARM

Movement	Definition
Pronation	This involves turning your palm posteriorly or inferiorly.
Supination	This involves turning your palm anteriorly or superiorly.

In the classroom...

Stand with your arms up in front of you with your palms facing the floor. Now turn your palms up towards the ceiling – you are *supinating* your forearm. Now turn them back down towards the floor – you are *pronating* your forearm.

Fig 4.29 Pronation and supination

MOVEMENTS AT SYNOVIAL JOINTS – FOOT

Movement	Definition
Inversion	This is turning the sole of the foot inwards.
Eversion	This is the opposite of inversion and is turning the sole of the foot outwards.
Dorsiflexion	This is the pulling of the foot upwards towards the shin, in the direction of the dorsum.
Plantar flexion	This is the opposite of dorsiflexion and is the pointing of the foot downwards, in the direction of the plantar surface.

Fig 4.30 Inversion and eversion

Fig 4.31 Plantarflex and dorsiflex

In the classroom…

Here are a few foot exercises you can do to demonstrate inversion, eversion, dorsiflexion and plantar flexion.

Sit on the floor with your legs out straight. Roll your feet inwards so that your soles are facing one another – you are *inverting* them.

Roll your feet away from one another so that your soles are facing away from each other – you are *everting* them.

Still sitting on the floor with your legs out straight, have your toes pointing up towards the ceiling. Now point your toes forwards as if you are trying to touch the floor with them – you are *plantar flexing*.

Do the opposite movement, trying to pull your toes upwards and back towards your shins – you are *dorsiflexing* them.

There are a number of different types of synovial joints and they are classified according to the shapes of their articulating surfaces. They are listed in the table opposite.

Name of joint	Shapes of articulating surfaces	Movements permitted	Examples	Diagram
Gliding (Plane)	Flat surfaces meet.	Side to side Back and forth Note: no angular or rotary motions are permitted.	Patellofemoral joint at knee. Intercarpal joints. Intertarsal joints. Sacro-iliac joint.	
Hinge	A convex surface fits into a concave one.	Flexion Extension Note: movements are in a single plane only.	Elbow joint. Tibiofemoral joint at knee. Ankle joint. Interphalangeal joints.	
Pivot	A rounded/pointed surface fits into a ring.	Rotation	Atlas and axis. Ulna and radius.	
Condyloid (Ellipsoid)	A condyle is a rounded/oval protuberance at the end of a bone and it fits into an elliptical cavity.	Back and forth Flexion Extension Abduction Adduction Circumduction	Wrist joint. Metacarpophalangeal joints.	
Saddle	A surface shaped like the legs of a rider fits into a saddle-shaped surface.	Side to side, Back and forth, Flexion, Extension, Abduction, Adduction, Circumduction, Opposition of thumbs (where the tip of the thumb crosses the palm and meets the tip of a finger)	Thumb joint.	
Ball and socket (Spheroidal)	A ball fits into a cup-shaped socket.	Flexion Extension Abduction Adduction Rotation Circumduction	Shoulder joint. Hip joint.	

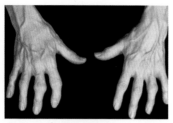

Osteoarthritis

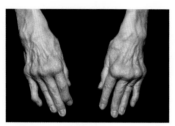

Rheumatoid arthritis

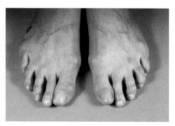

Bunions

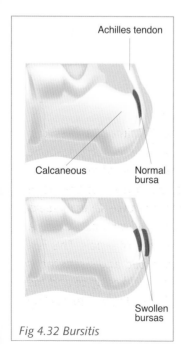

Achilles tendon

Calcaneous

Normal bursa

Swollen bursas

Fig 4.32 Bursitis

Common Pathologies Of The Skeletal System

Pathologies of the skeletal system

Disorders of the skeletal system generally involve pain, stiffness, inflammation and weakness or loss of motion. They can be caused by a number of things, but elderly people are more prone to disorders of this system because of the process of demineralisation that accompanies ageing.

Arthritis

Arthritis is the inflammation of a joint. There are two main types of arthritis:

- **Osteoarthritis** (wear-and-tear arthritis) – Osteoarthritis is a degenerative joint disease caused by ageing, irritation of the joints and general wear and tear. It is a progressive disorder of movable joints, particularly weight-bearing joints such as the knees and hips and is characterised by the deterioration of articular cartilage and the formation of spurs in the joint cavity. Symptoms include pain, swelling and a limited range of movement. Osteoarthritis is common in the elderly.
- **Rheumatoid arthritis** – Rheumatoid arthritis is an autoimmune disease in which the immune system attacks its own tissues, in this case its own cartilage and joint lining. It is a chronic form of arthritis in which the synovial membrane of the joint becomes inflamed and, if left untreated, thickens and synovial fluid accumulates. The resulting pressure causes pain and tenderness. The membrane then produces an abnormal granulation tissue that adheres to the surface of the articular cartilage and sometimes erodes the cartilage completely. The exposed bone ends are then joined by fibrous tissue, which ossifies and renders the joint immovable. Rheumatoid arthritis is characterised by inflammation of the joint, swelling, pain and loss of function. It is thought to be hereditary and can affect any age group. It normally begins in the smaller joints of the hands and feet before developing in other joints and it is always bilateral (for example, it will attack the joints of both the left and right hands at the same time).

Bunions

A bunion is a painful, swelling of the joint between the big toe and the first metatarsal. A bursa, and bursitis, often develops at this joint and the big toe can become laterally displaced. This is known as hallux valgus. It is thought that bunions can be hereditary or caused by ill-fitting shoes.

Bursitis

Bursitis is the inflammation of a bursa. It is characterised by inflammation, pain and limited movement and is usually caused by overuse or irritation from unusual use. Bursitis may also be caused by injury, gout, arthritis or some infections.

Dislocation (luxation)

A dislocation is the displacement of a bone from its normal position in a joint. The bones in the joint lose contact with one another and pain and a loss of motion usually accompany this condition.

Fractures

A break in a bone is called a fracture. Fractures can vary in size and severity, but they are usually accompanied by damage to the surrounding tissue. Symptoms of fractures vary but can include pain (especially when force is applied to the area), swelling, tenderness, loss of function and bruising. There are many different types of fractures, including a:

- **Simple or closed fracture** – this is a clean break of the bone with little damage to the surrounding tissue and no break in the overlying skin.
- **Compound or open fracture** – in this fracture the bone end pierces the skin and the wound is susceptible to contamination by dirt, debris and bacteria.
- **Comminuted fracture** – in this fracture the bone is broken into more than two pieces.
- **Impacted fracture** – this involves the bone ends being driven into each other.
- **Complicated fracture** – in this fracture the bone damages surrounding tissues and/or organs.
- **Greenstick fracture** – this fracture only occurs in children and involves an incomplete break (a crack) in the bone and the bone bends.
- **Stress fracture** – this fracture is caused by the stress of a repeated activity such as walking with a heavy pack.

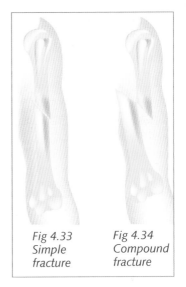

Fig 4.33 Simple fracture

Fig 4.34 Compound fracture

Fig 4.37 Comminuted fracture

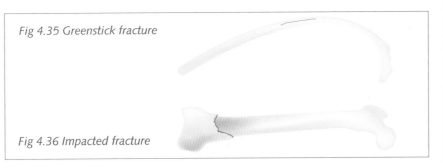

Fig 4.35 Greenstick fracture

Fig 4.36 Impacted fracture

Anatomy and physiology in perspective

How do bones heal?

Unlike most tissues which form scar tissue if injured, bones can completely heal themselves without scarring. This is because bone tissue is constantly replacing itself through the process of remodelling. Immediately after being fractured, the damaged area of the bone becomes swollen. This inflammatory response may take several weeks but is vital to clearing the area of any damaged or dead cells. A fibrocartilaginous callus is then formed in the area of repair. This is a soft, rubbery tissue that slowly becomes mineralised and strengthens into a bony callus. Finally, remodelling of the bone takes place and the external callus is slowly replaced by stronger bone. The entire healing process can take many months.

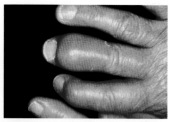

Gout

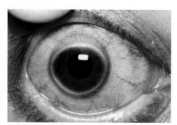

Osteogenesis imperfecta

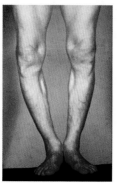

Rickets

Gout

Gout is the build-up of uric acid and its salts in the blood and joints. Crystals accumulate in, irritate and erode the cartilage of joints. Eventually, the bones can fuse, leading to an immovable joint. Symptoms include inflammation, swelling, pain, tenderness and a loss of mobility. Gout occurs primarily in middle-aged and older males and is suspected to be caused by diet, an abnormal gene or environmental factors such as stress.

Osteogenesis imperfecta (brittle bone disease)

Osteogenesis imperfecta is a genetic disease in which the bones are abnormally brittle. Symptoms can include frequent fracturing of bones, bone deformity, discoloration of the sclera of the eyes, translucent skin, possible deafness and thin dental enamel of the teeth.

Osteomalacia and rickets

Osteomalacia is the softening of the bones and it is caused by the progressive demineralisation of the bones due to a deficiency of vitamin D. Osteomalacia in children is known as rickets.

Osteoporosis

Osteoporosis is a progressive disease in which bones lose their density and become brittle and prone to fractures. It is common in the elderly and in post-menopausal women.

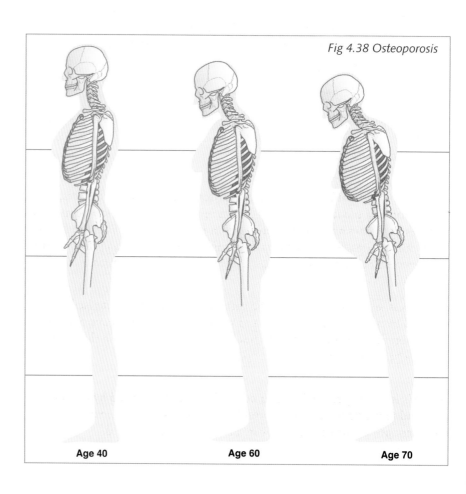

Fig 4.38 Osteoporosis

Age 40   **Age 60**   **Age 70**

Risk factors of osteoporosis include:

- Women whose bodies produce inadequate levels of oestrogen (menopause, underweight/eating disorders, athletes, amenorrhea, nursing mothers).
- Prolonged use of certain drugs (alcohol, some diuretics, cortisone, tetracycline).
- Smoking.
- Calcium deficiency or malabsorption.
- Vitamin D deficiency.
- Lack of exercise/sedentary lifestyle.
- Family history of osteoporosis.

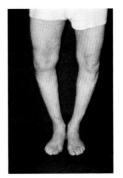

Osteoporosis

Paget's Disease

Paget's disease is a chronic bone disease in which the rate at which bone is broken down and rebuilt increases to such an extent that the infected areas are abnormally enlarged and weak. Paget's disease has few symptoms although pain, bone enlargement and bone deformity may occur. The cause of Paget's disease is unknown.

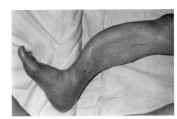

Paget's Disease

Sprain

A sprain is the injury to a ligament in a joint and it is characterised by considerable swelling and pain. The blood vessels, muscles, nerves and tendons associated with the joint can also be injured.

Torn cartilage

Torn cartilage is the tearing of the menisci in the knee joint and it is quite common in athletes. Symptoms of torn cartilage include pain, tenderness, swelling and limited motion.

Common pathologies of the spine

Some pathologies of the skeletal system specifically affect the vertebrae and discs of the spine. These are generally characterised by back ache, spinal stiffness and sometimes nerve disorders.

Cervical spondylosis

Cervical spondylosis involves the degeneration of the intervertebral discs and the vertebrae of the neck. It usually affects middle-aged to elderly people and it causes neck pain and possible pressure on the spinal cord. This can lead to spinal cord compression which can result in changes in walking such as jerky leg movements and unsteadiness, as well as weakness in the arms.

Kyphosis

Kyphosis is an exaggerated thoracic curve that is often characterised by a hunched back, rounded shoulders and mild, persistent back pain. Scoliosis can sometimes develop from kyphosis. Kyphosis is common in the elderly or it can be a result of rickets, osteoporosis or poor posture.

Lordosis

Lordosis is an exaggerated lumbar curve that is characterised by a sway back and lower back ache. It can result from excess weight around the abdomen, poor posture, rickets or it can be pregnancy-related.

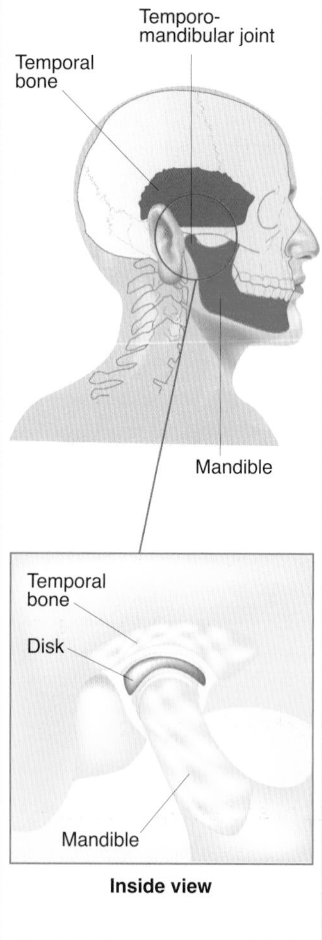

Fig 4.39 Temporo-mandibular joint

Did you know?

In a lifetime, an average person walks approximately five times the circumference of the globe.

Prolapsed (slipped) intervertebral disc – PID

A prolapsed intervertebral disc is also known as a herniated or slipped disc. It occurs when the disc's outer ring of fibrocartilage ruptures and its inner nucleus pulposus protrudes through this rupture and compresses a spinal nerve. A PID usually occurs if the ligaments surrounding the discs are injured or weakened and the discs are put under excess pressure. It can cause pain, numbness and weakness along the pathway of the spinal nerve that it compresses.

Scoliosis

Scoliosis is a lateral curvature of the spine. It has few symptoms, but can cause backache after sitting or standing for a long time. It can be hereditary or caused by conditions such as poor posture, having one leg shorter than the other, paralysis of the muscles on one side of the body or chronic sciatica.

Spinal stenosis

Spinal stenosis is the narrowing of the spinal canal and it is a cause of lower back pain and sciatica in elderly people. Spinal stenosis usually results from other diseases such as Paget's disease or osteoarthritis.

Spondylitis

Spondylitis is inflammation of the joints of the spine and its primary symptoms are pain and spinal stiffness. Different forms of spondylitis exist, the main one being *ankylosing spondylitis*. This is inflammation of the spine and large joints such as the sacroiliac joint. It develops mainly in men between the ages of 20–40 years and its cause is not known although it is thought that genetics plays a role. Ankylosing spondylitis has symptom-free periods followed by flare-ups that differ in severity. Common symptoms include back pain, stiffness and muscular spasms. More severe symptoms can include loss of appetite, weight loss, fatigue, anaemia and postural changes.

Temporo-mandibular joint disorder (TMJ)

Because the temporo-mandibular joint is one of the most complicated joints in the body, it is prone to a number of different disorders which are all classified as temporo-mandibular joint disorder or syndrome (TMJ disorder/syndrome). TMJ disorder includes symptoms such as muscle pain and tightness around the jaw, tenderness in the muscles that move the jaw, clicking of the jaw and limitation of the movement or opening of the jaw. Associated symptoms can be headaches, pain or stiffness in the neck that radiates into the arms, dizziness, earache and disrupted sleep. TMJ disorder can be caused by muscular tension, anatomic problems within the joints, arthritis or by psychological stress.

Whiplash injury

Caused by the sudden jerking back of the head and the neck, for example, in car accidents. It is characterised by pain and stiffness resulting from damage to the ligaments, vertebrae and sometimes spinal cord of the neck.

Common disorders of the feet

Clubfoot (talipes equinovarus)

Clubfoot, or talipes equinovarus, is a birth defect in which the foot is twisted out of shape. Two types of clubfoot exist: positional and true. Positional clubfoot is caused when the foot has been held in an unusual position in utero and it can often be corrected through physical therapy after birth. True clubfoot, on the other hand, is usually only corrected through surgery because either the bones of the leg or foot or the muscles of the calf are structurally underdeveloped.

Hammer toes

Hammer toes are painful, rigid toes that are fixed in a contracted position and cannot be straightened. Usually the second, third and fourth toes are affected. Hammer toes often accompany other foot disorders such as bunions or high arches and can also lead to corns and nail problems. The most common cause of hammer toes is poorly fitted shoes.

Heel spur

Heel spurs are bony growths under the heel of the foot and are usually caused by excessive pulling on the heel bone by tendons. They are not always painful, but can cause pain and difficulty in standing or walking, especially if the tissues surrounding the spur are inflamed.

High arches (pes cavus)

Pes cavus is the opposite of pes planus and is characterised by unusually high arches in the feet. These high arches result in stiffness of the foot and limited movement and can cause pain, calluses and claw foot. High arches are generally inherited or can be caused by chronic illness.

Pes planus (flat foot or dropped arches)

Pes planus, commonly called flat foot, is a condition in which the arches of the feet drop and the foot spreads. Because the arches support the foot and absorb shock when the body is in motion, pes planus often results in fatigue, pain and backache. Pes planus is sometimes hereditary but can also be the result of joint weakness, nutritional deficiencies in children or chronic illness.

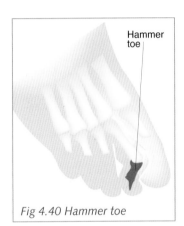

Fig 4.40 Hammer toe

NEW WORDS	
Arthrology	the study of joints.
Articulation	the point of contact between two bones. Commonly called a joint.
Cartilage	a resilient, strong connective tissue.
Demineralisation	the process through which minerals such as calcium and phosphorous are lost from the bones.
Haemopoiesis	the production of blood cells and platelets.
Kinesiology	the study of the motion of the body.
Ligament	a tough band of connective tissue that attaches bones to bones.
Osseous tissue	bone tissue.
Ossification	the process of bone formation.
Osteology	the study of the structure and function of bones.
Process	a bony projection or prominence.
Remodelling	the process through which new bone tissue replaces old, worn out or injured bone tissue.
Tendon	a tough band of connective tissue that attaches muscles to bones.

Study Outline

Functions of the skeletal system

1. Functions of the skeletal system include protection, support, shape, homeostasis of minerals, movement, energy storage, site of blood cell production.

> People's Skeletal Systems Have Many Essential Bones.
>
> Protection, Support, Shape, Homeostasis of minerals, Movement, Energy storage, Blood cell production

2. Blood cells are produced in red bone marrow.
3. Lipids are stored in yellow bone marrow.

Anatomy of bones

Bone tissue

1. There are two types of bone tissue: compact and spongy.
2. Compact bone tissue is made up of osteons which consist of concentric rings of lamellae through the centre of which run Haversian canals in which nerves, blood and lymph vessels are found.
3. Compact bone functions in protection and support and is the external layer of all bones.
4. Spongy bone is a light tissue made up of lamellae arranged into trabeculae.
5. Spongy tissue contains red bone marrow in which blood cells and platelets are produced.

Bone formation and remodelling

1. Bones are formed through the process of ossification.
2. Bones are continually replaced through the process of remodelling.

Types of bone

1. Bones can be classified as short, long, irregular, flat or sesamoid.

> Skeletons Linger Inside Fluffy Shirts.
>
> Short, Long, Irregular, Flat, Sesamoid.

Structure of a long bone

1. Long bones have a long, central shaft called a diaphysis which is covered by a membrane called the periosteum.
2. The diaphysis contains the medullary/marrow cavity which is lined by a membrane called the endosteum. This cavity contains fatty yellow bone marrow.
3. Each end of the diaphysis is called an epiphysis and is covered by articular cartilage. The epiphysis is made of spongy bone tissue which contains red bone marrow.

4. The region where the diaphysis joins the epiphysis is the metaphysis and in a growing bone it houses the epiphyseal plate where growth in length takes place.

Organisation of the skeleton

1. The skeleton is divided into the axial and appendicular skeletons.
2. The axial skeleton consists of the bones along the centre/axis of the body. These are the skull, hyoid, ribs, sternum and the vertebrae of the spine.
3. The appendicular skeleton consists of the appendages. Namely the legs, arms and their girdles.
4. The 8 bones of the cranium are: 1 frontal, 2 parietals, 2 temporals, 1 occipital, 1 ethmoid, 1 sphenoid.

> To help you remember the bones of the cranium, take a moment to picture yourself wearing a cap of fluffy pink toes. Take some time to remember this image, imagine you are looking in the mirror and seeing lots of fluffy pink toes coming out of your head. Then if you ever need to remember the bones of the cranium, simply put your fluffy pink toed cap on:
>
> ### Fluffy Pink TOES
>
> Frontal, Parietals, Temporals, Occipital, Ethmoid, Sphenoid

5. The 14 bones of the face are: 2 nasals, 2 maxillae, 2 zygomatics, 1 mandible, 2 lacrimals, 2 palatines, 2 inferior nasal conchae, 1 vomer.

> Most Interesting Men Prefer Naughty, Zany, Voluptuous Ladies.
>
> Maxillae, Inferior nasal conchae, Mandible, Palatines, Nasals, Zygomatics, Vomer, Lacrimals.

6. The hyoid bone is found in the neck and does not articulate with any other bone.
7. The spine is composed of 33 vertebrae: 7 cervical, 12 thoracic, 5 lumbar, 5 sacral, 4 coccygeal.
8. A vertebra is composed of a thick, disc-shaped body and a vertebral arch formed by processes. The space formed by the body and arches is the vertebral foramen through which runs the spinal cord. Transverse processes project from the sides of the arch and a spinous process projects posteriorly. These processes serve as sites of attachment for muscles. Superior and inferior articular processes form joints with the vertebrae above and below.
9. Intervertebral discs are found between the vertebrae. They are composed of the outer annulus fibrosus and the inner nucleus pulposus. These discs provide cushioning and absorb shock.
10. There are four natural curves to the spine: cervical, thoracic, lumbar and sacral.
11. The thorax is made up of the sternum, 12 pairs of ribs and the thoracic vertebrae. The sternum itself consists of the manubrium, body and xiphoid process.

12. The shoulder girdle consists of the clavicle and scapula. The shoulder joint is a diarthrotic, synovial ball-and-socket joint.

13. The arm bone is the humerus. The bones in the forearm are the ulna and radius. The bones of the hand are the 8 carpals (trapezium, trapezoid, capitate, hamate, scaphoid, lunate, triquetrum, pisiform), 5 metacarpals and 14 phalanges.

> Hands Take Crazy Things To Silly Lunch Parties.
>
> Hamate, Trapezium, Capitate, Trapezoid, Triquetrum, Scaphoid, Lunate, Pisiform.

14. The pelvic girdle consists of the ilium, ischium and pubis. The hip joint is a diarthrotic, synovial ball-and-socket joint.

15. The thigh bone is the femur. The knee bone is the patella. The bones of the leg are the tibia and fibula. The bones of the foot are the 7 tarsals (talus, calcaneus, cuboid, navicular, 3 cuneiforms), 5 metatarsals and 14 phalanges.

16. The foot has three arches that distribute the weight of the body over the entire foot. They are the medial and lateral longitudinal arches and the transverse arch.

Joints

Joints, or articulations, are the points of contact between bones. They hold bones together, provide movement and protect vital organs.

Classification of joints

1. Joints can be classified according to the degree of movement they permit. This is called functional classification and includes: synarthroses (immovable joints), amphiarthroses (slightly movable joints) and diarthroses (freely movable joints).

2. Joints can also be classified according to what type of connective tissue binds them together and whether or not there is a synovial cavity between them. This is called structural classification and includes: fibrous joints, cartilaginous joints and synovial joints.

Synovial joints

1. Synovial joints are joints that are lined by articular cartilage, have a synovial cavity between them and are enclosed in a synovial capsule. This capsule is lined with a membrane that secretes synovial fluid which lubricates joints and supplies nutrients to, and removes waste from, the articular cartilage. Synovial joints allow for a great deal of movement.

2. Movements at synovial joints can include flexion, extension, hyperextension, abduction, adduction, circumduction and rotation.

3. Movements of the forearm include pronation and supination.

4. Movements of the foot include inversion, eversion, dorsiflexion and plantarflexion.

5. Synovial joints are classified as: gliding, hinge, pivot, condyloid, saddle, ball-and-socket.

Revision

1. Name seven functions of the skeletal system.
2. Identify where in the body blood cells are produced.
3. Describe the structure and function of compact bone.
4. Describe the structure and function of spongy bone.
4. Explain what ossification is.
5. Explain what remodelling is.
6. Identify five types of bones. For each type of bone give one example.
7. Name the central shaft of a long bone.
8. Name the ends of a long bone.
9. Explain the organisation of the skeleton.
10. Name the bones of the cranium.
11. Name the bones of the face.
12. Name the bone found in the neck that does not articulate with any other bones.
13. Describe the spine, giving the names of the different regions and the numbers of how many vertebrae are found in each of these regions.
14. Name the space in a vertebra through which the spinal cord runs.
15. Describe the structure and function of an intervertebral disc.
16. Identify the bones of the thorax.
17. Explain the structure of the shoulder girdle.
18. Describe the shoulder joint and name the movements it allows.
19. Name the bones of the upper limb.
20. Name the carpals.
21. Describe the hip joint and name the movements it allows.
22. Describe the knee joint and name the movements it allows.
23. Name the bones of the lower limb.
24. Name the tarsals.
25. Explain the three arches of the foot.
26. Explain the meaning of the following words: synarthrotic, amphiarthrotic, diarthrotic.
27. Describe the following types of joints: fibrous, cartilaginous, synovial.
28. Describe a synovial joint.
29. Explain the functions of synovial fluid.

Multiple choice questions

1. **An example of a saddle joint is:**
   a. The shoulder joint.
   b. The hip joint.
   c. The thumb joint.
   d. The elbow joint.

2. **Ossification is:**
   a. The process by which bone is formed.
   b. The process by which blood cells are formed.
   c. The process in which lipids are stored.
   d. The process by which cartilage is formed.

3. **The diaphysis is:**
   a. The area at both ends of a long bone.
   b. The main shaft of a long bone.
   c. The lining of the ends of a long bone.
   d. The lining of the main shaft of a long bone.

4. **How many lumbar vertebrae are there?**
   a. 4
   b. 5
   c. 6
   d. 7.

5. **A hinge joint allows which of the following movements?**
   a. Side-to-side
   b. Circumduction
   c. Supination
   d. Extension.

6. **What type of joint do the atlas and axis form?**
   a. Pivot
   b. Saddle
   c. Ball-and-socket
   d. Hinge.

7. **Which of the following statements is correct?**
   a. Functions of the skeletal system include protection, movement and excretion.
   b. Functions of the skeletal system include mineral homeostasis, vitamin D production and movement.
   c. Functions of the skeletal system include support, protection and movement.
   d. Functions of the skeletal system include movement and support only.

8. **Minerals present in bone include:**
   a. Calcium and haemoglobin.
   b. Calcium only.
   c. Calcium and nickel.
   d. Calcium and phosphorous.

9. **Articular cartilage is the cartilage found:**
   a. Covering the ends of articulating bones.
   b. Covering the bodies of articulating bones.
   c. Covering synovial capsules.
   d. Covering menisci.

10. **Which of the following movements is the opposite of inversion?**
    a. Pronation
    b. Eversion
    c. Supination
    d. Opposition.

5 The Muscular System

Introduction

All your body's movements are created by tiny filaments sliding to overlap one another. As they overlap, they cause a muscle to shorten and this creates movement – either of the bones and joints, internal tracts or even the heart.

The study of muscles is called myology and in this chapter you will discover how your muscles work and also learn about some of the important muscles that help you move.

Student objectives

By the end of this chapter you will be able to:

- Describe the functions of the muscular system.
- Describe the different muscle types such as skeletal, cardiac and smooth.
- Identify the structure of a skeletal muscle.
- Describe how a muscle contracts.
- Identify the different types of muscular contraction and muscle fibres.
- Identify and name the skeletal muscles.
- Identify the common pathologies of the muscular system.

Did you know?
Muscles can only shorten.
This means that every
movement you make is a
pull, not a push. Even
when you are pushing
something like a shopping
trolley, your muscles are
actually still pulling on
bones, not pushing.

Functions Of The Muscular System

Muscles have the unique ability of shortening themselves. This is known as contraction and is the essential function of all muscles. Through contracting, muscles can produce movement, maintain posture, move substances within the body, regulate organ volume and produce heat.

Movement of the skeleton (locomotion)

Skeletal muscles are a type of muscle that are mostly attached to bones by strong cords of dense connective tissue called tendons. When these muscles contract, they move the bones at the joints of the body and this produces movement of the skeleton. Thus, we can run, jump and even somersault. Some skeletal muscles are not attached to bones. Instead, they are attached to skin. These are the muscles of the face and they enable us to smile, frown and reveal our inner emotions.

Maintenance of posture

When we are awake, certain skeletal muscles are always partially contracted to keep our bodies in an upright position. For example, the muscles of the neck maintain a sustained partial contraction to keep our heads upright. Muscle tendons also surround, protect and stabilise joints, thus stabilising body positions.

Movement of substances and the regulation of organ volume

In addition to skeletal muscle, there are two other types of muscle: smooth and cardiac (more on these in a minute). All three types of muscle function in moving substances within the body and in regulating organ control. For example:

- **Skeletal muscle** – Skeletal muscle contracts to help return venous blood to the heart and to move lymph through the lymphatic vessels.
- **Cardiac muscle** – The heart is made up of cardiac muscle tissue which contracts to pump blood around the body. Cardiac muscle also helps regulate blood pressure.
- **Smooth (visceral) muscle** – Smooth muscle lining hollow tracts of the body contracts to move food through the gastrointestinal tract; moves urine through the urinary tract; and a baby through the birthing canal.

Heat production

Skeletal muscles release a lot of energy as they contract. This energy takes the form of heat which is considered a by-product of muscle contraction. The heat generated by muscular contraction is used to maintain our normal body temperature. If the body needs to increase its temperature, skeletal muscles contract involuntarily (shiver) to increase heat production. The generation of heat in the body is called thermogenesis.

In the classroom

Take a moment to think back to the functions of the skin and remember how smooth muscles also play a role in helping to maintain normal body temperatures by vasodilation and vasoconstriction.

Muscle Tissue

Types of muscle tissue

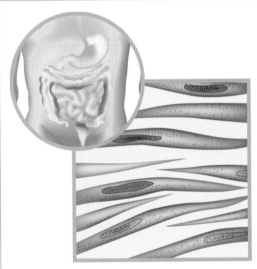

Smooth muscle

Skeletal muscle

Muscle types in a nutshell

Smooth (visceral) muscle is non-striated, involuntary and moves substances within the body and regulates organ volume.

Skeletal muscle is striated, voluntary and moves the skeleton, maintains posture and generates heat.

Cardiac muscle is striated, involuntary and pumps blood.

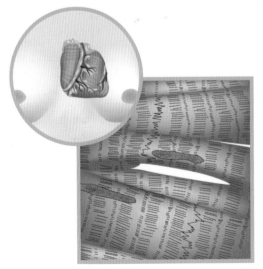

Cardiac muscle

Fig 5.1 Types of muscle tissue

There are three different types of muscle tissue that differ in their anatomy, location, function and what they are controlled by.

- **Anatomy** (striated/non-striated) – Muscle tissue is either striated or non-striated. When looked at under a microscope, striated tissue is made up of light and dark bands.
- **Location** – The type of muscle tissue often depends on where it is in the body, or what it is attached to.
- **Function** – Different muscle tissues have very specific functions. For example, cardiac muscle tissue functions only in contracting the heart.
- **Control** (voluntary/involuntary) – Muscle tissue is under voluntary or involuntary control. Voluntary control means it is under conscious control while involuntary control means it is not under conscious control. Most involuntary muscles are controlled by neurotransmitters and hormones. Some also contain autorhythmic cells which are self-excitable cells.

The table below lists the three types of muscle tissue.

TYPE	ANATOMY	LOCATION	FUNCTION	CONTROL
Skeletal muscle	Muscle fibres: • are striated. • are generally long and cylindrical-shaped (although some are circular). • have multiple nuclei and many mitochondria. • are strengthened and reinforced by connective tissue. • have tendons.	Attached by tendons to bones, skin or other muscles.	• Movement of the skeleton. • Movement of lymph and venous blood. • Maintenance of posture. • Heat production.	Voluntary.
Cardiac muscle	Muscle fibres: • are striated. • are arranged in spiral-shaped bundles of branching cells.	Forms most of the heart.	• Pumps blood around the body. • Helps regulate blood pressure.	Involuntary. Contracts at a steady rate set by a 'pace-maker' that is adjusted by neuro-transmitters and hormones.
Smooth muscle (Visceral muscle)	Muscle fibres: • are non-striated. This is why they are called smooth. • are spindle-shaped. • have a single nucleus. • are arranged in sheets or layers that alternately contract and relax to change the size or shape of structures.	Forms the walls of hollow internal structures such as blood vessels, the gastrointestinal tract, and the bladder.	• Move substances through tracts. • Regulate organ volume.	Involuntary. Contracts auto-rhythmically and is controlled by neurotransmitters and hormones.

Skeletal Muscle

Now that you have a basic understanding of the functions and types of muscle tissue, we will take a look at skeletal muscles and see how they work together with the skeleton to produce movement. But before we go any further, here is some important vocabulary that describes the characteristics of muscle tissue:

- Excitability (irritability) – this is the ability of muscle cells (and also nerve cells) to respond to stimuli.
- Conductivity – the ability of muscle cells to move action potentials along their plasma membranes.
- Action potentials – an electrical change that occurs on the membrane of a muscle fibre in response to a nervous impulse.
- Contractility – the ability of muscles to contract and shorten.
- Extensibility – the ability of muscles to extend and lengthen.
- Elasticity – the ability of muscles to return to their original shape after contracting or extending.

Structure of a skeletal muscle

It is often easier to understand how something works if you know what it is made of. So, before we look at exactly how a muscle contracts, let us slowly break a muscle down into its smallest components.

Connective tissue: the outer protection of a muscle

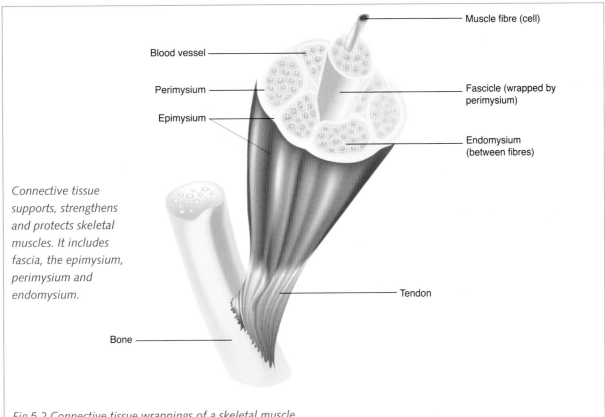

Connective tissue supports, strengthens and protects skeletal muscles. It includes fascia, the epimysium, perimysium and endomysium.

Muscle fibre (cell)

Blood vessel

Perimysium

Epimysium

Fascicle (wrapped by perimysium)

Endomysium (between fibres)

Tendon

Bone

Fig 5.2 Connective tissue wrappings of a skeletal muscle

Study tip
When learning about
muscles remember that:
- 'myo…' = muscle
- 'sarco…' = flesh

Lining the walls of the body, and holding the muscles and limbs together, is a dense, irregular connective tissue called fascia (fascia also surrounds and protects organs of the body). Fascia:

- Separates muscles into different functional groups and fills spaces between muscles, thus allowing free movement of the muscles.
- Supports the nerves, blood and lymphatic vessels that serve the muscles.

Beneath the fascia lie muscles. Looking at a muscle under a microscope, you will see hundreds of muscle fibres surrounded and held together by connective tissue. This connective tissue surrounds, protects and reinforces the fibres because, although they are capable of producing great power, they are still quite fragile cells that can be damaged. Three types of dense, irregular connective tissue protect and strengthen a muscle:

- **The epimysium** – this is the outermost layer that encircles the whole muscle.
- **The perimysium** – this surrounds bundles of 10–100 muscle fibres. These bundles are called *fascicles*.
- **The endomysium** – this surrounds each individual muscle fibre within the fascicle. It contains many blood capillaries so that each muscle fibre has a good supply of blood that brings oxygen and nutrients to the muscles and removes their waste products.

Together the epimysium, perimysium and endomysium extend beyond the muscle fibres and become tendons or aponeuroses. Tendons are cylindrical cords of connective tissue that attach muscles to the periosteum of bones. Aponeuroses are flat, sheet-like tendons that attach muscles to the periosteum of bones or to the skin.

Muscle fibres: the inner cells of a skeletal muscle

Each muscle fibre within the fascicle is a single cell that has similar properties to the generalised animal cell you learnt about in Chapter 2. However, there are slight variations (mainly in the vocabulary):

- A muscle cell is called a *muscle fibre* or *myofibre*.
- The fibre is a long, cylindrical shape.
- The plasma membrane is called the *sarcolemma*.
- The cytoplasm is called the *sarcoplasm*.
- There are many nuclei in a muscle fibre and they are located at the periphery of the fibre, out of the way of the contractile elements (described shortly).
- There are many mitochondria that lie in rows throughout the fibre and are close to the muscle proteins that use adenosine triphosphate (ATP) during contraction.
- Long thread-like organelles nearly fill the sarcoplasm. These are called *myofibrils* and are the contractile elements of skeletal muscles. They are made of filaments called *myofilaments* that are arranged into compartments called *sarcomeres*. Sarcomeres are the basic functional units of skeletal muscles and we will study them in detail shortly.
- The endoplasmic reticulum of muscle fibres is called the *sarcoplasmic reticulum* and it stores calcium which is necessary for muscular contraction.

In the classroom
Take a few minutes to review the structure and functions of a cell discussed in Chapter 2. This will help you to understand the structure of a skeletal fibre which is essentially a type of cell.

Did you know?
Some muscle fibres are as long as 30cm.

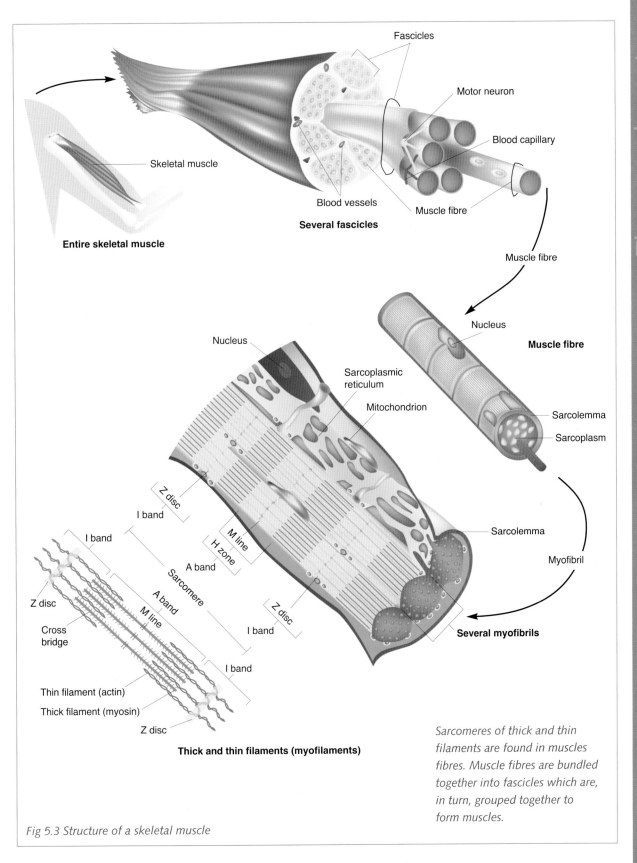

Entire skeletal muscle

Skeletal muscle

Fascicles

Motor neuron

Blood capillary

Several fascicles

Blood vessels

Muscle fibre

Muscle fibre

Nucleus

Muscle fibre

Sarcolemma

Sarcoplasm

Nucleus

Sarcoplasmic reticulum

Mitochondrion

Sarcolemma

Myofibril

Several myofibrils

Z disc

I band

I band

Z disc

Cross bridge

M line

H zone

A band

Sarcomere

A band

M line

Z disc

I band

I band

Thin filament (actin)

Thick filament (myosin)

Z disc

Thick and thin filaments (myofilaments)

Sarcomeres of thick and thin filaments are found in muscles fibres. Muscle fibres are bundled together into fascicles which are, in turn, grouped together to form muscles.

Fig 5.3 Structure of a skeletal muscle

147

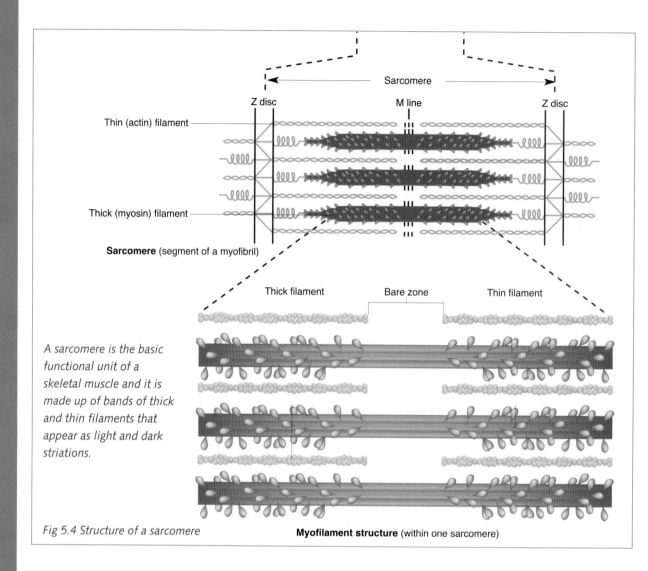

A sarcomere is the basic functional unit of a skeletal muscle and it is made up of bands of thick and thin filaments that appear as light and dark striations.

Fig 5.4 Structure of a sarcomere

Myofilament structure (within one sarcomere)

Sarcomeres: the basic functional units of a skeletal muscle

We have now broken down a muscle by removing its outer covering of protective connective tissue and breaking down its cell, or fibre, into its different components. As mentioned above, one of the components of a muscle fibre is a myofibril which is the contractile element of a skeletal muscle. Myofibrils are made up of myofilaments which are arranged in compartments called sarcomeres. We will now look at a sarcomere.

A sarcomere is the basic functional unit of a striated muscle fibre. In other words, it contains the filaments that move to overlap one another and cause a muscle to shorten.

Sarcomeres contain three types of filaments:

- **Thick filaments** – Thick filaments contain molecules of the protein *myosin* which are shaped like two golf clubs twisted together:
  - The handles of the clubs form the tails of the molecules and point towards the centre of the sarcomere.

- The heads of the clubs form the heads of the molecules and are called *myosin heads* or *cross-bridges*. These extend towards the thin filaments.
    - The molecules lie parallel to one another.
- **Thin filaments** – Thin filaments contain the proteins *actin*, *tropomyosin* and *troponin*. These molecules have irregular shapes but appear together as a chain of twisted molecules. On each molecule is a *myosin-binding site* where myosin heads can attach to bring about contraction. When a muscle is relaxed, these sites are blocked by a tropomyosin-troponin complex so that myosin cannot bind to them.
- **Elastic filaments** – Elastic filaments contain the protein *titin (connectin)* which helps stabilise the position of the thick filaments.

Sarcomeres are made up of two different regions or bands that give skeletal muscles their striated appearance:

- **A-Band** – The A-band is a darker area composed mainly of thick filaments with only a few thin filaments that have overlapped into the area. In the centre of the A-band is a narrow *H-zone* that contains thick filaments only. It is divided by an *M-line* of protein molecules that hold the thick filaments together.
- **I-Band** – The I-band is a lighter area composed of thin filaments only.

Sarcomeres are separated from one another by *Z-discs (Z-lines)* that are narrow regions of dense material.

Mechanisms of Muscle Contraction

How do muscles contract?

Muscles shorten when the thick and thin filaments in the sarcomere slide past one another. This is known as the *sliding filament mechanism*. Let's take a look at what starts this sliding and how exactly it works:

- **Relaxed muscle** – when a muscle is relaxed the myosin-binding sites on each actin molecule are covered by a tropomyosin-troponin complex. Thus, the myosin heads cannot bind with the actin molecules. ATP is attached to ATP-binding sites on the myosin heads and it is split into ADP and P (a phosphate group). This means that energy has been transferred from the ATP to the myosin heads. They are in an energised or activated state ready to bind to the myosin-binding sites on the actin molecules as soon as the spaces become available.
- **A nerve impulse starts the process** – a nerve impulse from the central nervous system triggers the release of a neurotransmitter called *acetylcholine* (please refer to Chapter 6 for further details). This neurotransmitter in turn triggers a *muscle action potential* which is an electrical change that occurs on the membrane of the muscle fibre. This muscle action potential travels along the sarcolemma and causes the release of calcium which is stored in the sarcoplasmic reticulum of the fibre.

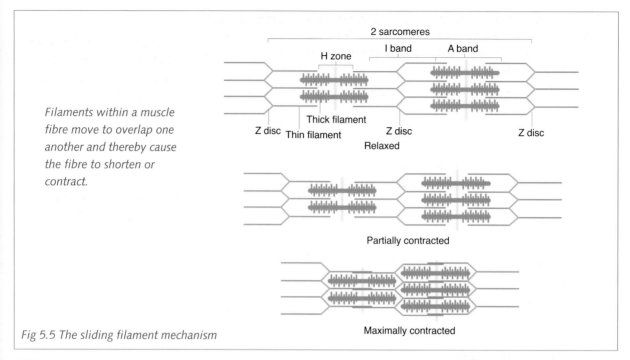

Filaments within a muscle fibre move to overlap one another and thereby cause the fibre to shorten or contract.

Fig 5.5 The sliding filament mechanism

Study tip
Just a reminder: ATP stands for adenosine triphosphate and it is the main energy-transferring molecule in the body.

Muscle contraction in a nutshell
Contraction is all about the myosin heads of thick filaments binding to sites on thin actin filaments and pulling these filaments inwards. A nervous impulse, ATP and calcium are needed.

- **Calcium frees up the myosin-binding sites** – calcium binds with the troponin on the myosin-binding site and causes the tropomyosin-troponin complex to move away from the sites. This frees up the sites so that the myosin can bind with the actin.
- **Myosin binds to actin in a power stroke** – a 'power-stroke' occurs when the myosin heads bind to the myosin-binding sites on the actin and change their shape by swivelling their heads towards the centre of the sarcomere. It is this change in shape that draws the thin filaments past the thick filaments. The thin filaments slide inwards towards the H-zone. The thick filaments remain in the same place. This inward sliding causes the Z-discs to come towards each other and the entire sarcomere shortens even though the actual lengths of the thin and thick filaments do not change. The sarcomere shortens, the muscle fibres shorten, the whole muscle shortens and pulls on a bone – thus movement is generated.
- **Myosin heads detach from actin** – after the power stroke, ATP binds to the myosin heads at ATP-binding sites. This causes the heads to detach themselves from the actin. The ATP on the myosin heads is then split by an enzyme and it transfers its energy to the myosin heads so they are once again in an activated state and ready to combine with another site further along the thin filament.
- **The cycle begins again** – the above cycle begins again and continues as long as ATP and calcium are present.

How do muscles relax?
Two processes take place to stop muscle contraction:

- **Acetylcholine is broken down** – an enzyme breaks down acetylcholine and this stops any further muscle action potentials. Therefore, calcium is no longer released.

- **Calcium levels drop** – calcium levels drop and there is no longer enough calcium available to bind with the troponin. Thus, the tropomyosin-troponin complex moves back over the myosin-binding sites on the actin and so prevents the myosin heads from binding with the actin.

The thin filaments, therefore, slip back into their relaxed position and no further contraction takes place.

Did you know?
During maximum muscle contraction, the distance between the Z-discs can be halved.

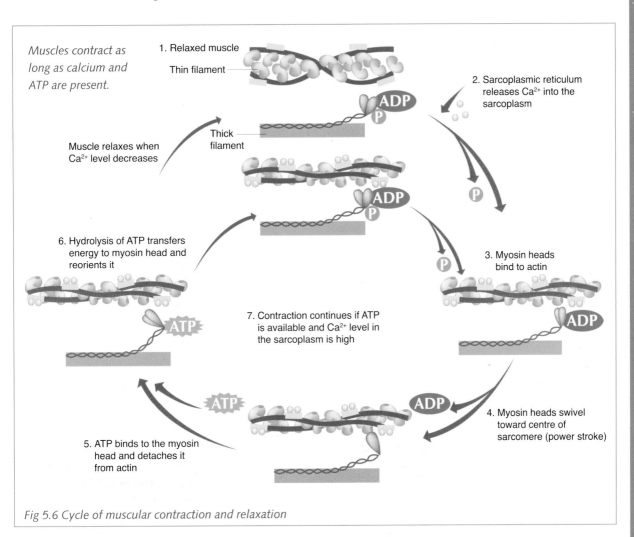

Fig 5.6 Cycle of muscular contraction and relaxation

Muscles contract as long as calcium and ATP are present.

1. Relaxed muscle
Thin filament
2. Sarcoplasmic reticulum releases Ca²⁺ into the sarcoplasm
Muscle relaxes when Ca²⁺ level decreases
Thick filament
6. Hydrolysis of ATP transfers energy to myosin head and reorients it
3. Myosin heads bind to actin
7. Contraction continues if ATP is available and Ca²⁺ level in the sarcoplasm is high
4. Myosin heads swivel toward centre of sarcomere (power stroke)
5. ATP binds to the myosin head and detaches it from actin

Infobox

Anatomy and physiology in perspective
Have you ever tried to lift or push a really heavy object and noticed that no matter how much tension is in your muscles, you are not generating any movement? This is because muscular contraction is taking place, but no shortening of the muscles is occurring. Why not? This is because the myosin heads are swivelling and generating force, but the thin filaments are not sliding inwards. This is called an isometric contraction (more on this later).

Anatomy and physiology in perspective

To understand the workings of the sliding filament mechanism, picture yourself jogging on a treadmill. As one foot hits the belt it pushes it backwards, then the other foot hits the belt and pushes it backwards. These strokes are repeated again and again until the belt is moving smoothly. All the time, however, you are staying in the same place. Now, imagine you are the thick filament, your feet are the myosin heads and the belt of the treadmill is the thin filament. As each myosin head (foot) connects with the thin filament (belt), it moves it backwards towards the H-zone. Each repeated movement moves the thin filament more smoothly. Meanwhile, the thick filament (you) is staying in the same place. Don't forget... your legs, the myosin heads, need a constant supply of energy to keep going.[i]

Did you know?

When a person dies, calcium leaks out of the sarcoplasmic reticulum and binds with troponin, causing the tropomyosin-troponin complex to move off the myosin-binding sites. The myosin heads are, therefore, able to bind with the actin and contraction occurs. However, because the person is dead, there is no ATP available to activate the release of the myosin heads from the actin and so the muscles are in a permanent state of contraction. This is called *rigor mortis*. It lasts about 24 hours until the tissues begin to degenerate.

Types of muscular contraction

Muscle tone

Even when we think our muscles are completely relaxed, a few of the muscles fibres are still involuntarily contracted. This constant contraction gives muscles their firmness or tension. This is called *muscle tone* or *tonus*. Tone is essential for maintaining posture and keeping our bodies in an upright position. Here are some more terms associated with muscle tone:

- Atony – a lack of muscle tone.
- Atrophy – the wasting away of muscles. Muscles decrease in size.
- Hypotonia – a loss of muscle tone. Muscles appear loose and flattened. They are described as hypotonic.
- Hypertonia – an increase in muscle tone. Muscles appear stiff or rigid. They are described as hypertonic.

Isotonic and isometric contractions

Most physical activities include two types of contraction: isotonic and isometric contractions.

- **Isotonic contractions** – Isotonic contractions are regular contractions in which muscles shorten and create movement while the tension in the muscle remains constant. They improve muscle strength and joint mobility and come in two forms:
  - **Concentric contractions** – these are always towards the centre and are contractions in which the muscle shortens and generates a movement that shortens the angle at a joint.
  - **Eccentric contractions** – these are always away from the centre and are contractions in which muscles lengthen.

- **Isometric contractions** – In isometric contractions, the muscle contracts but it does not shorten and no movement is generated. This type of contraction stabilises some joints while others are moved. It also improves muscle tone.

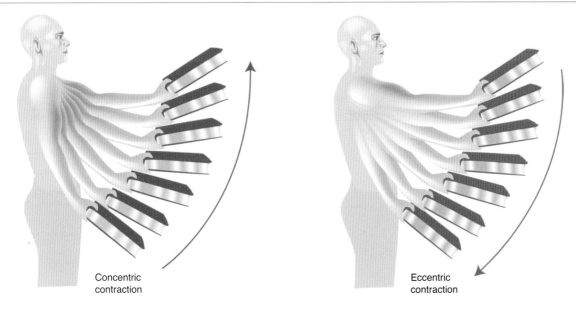

Concentric
contraction

Eccentric
contraction

In the classroom...
Pick a book up off your desk and bring it up towards your face. The muscle in your arm is shortening and reducing the angle at your elbow joint. It is performing a concentric contraction. Now replace the book on the table. The muscle in your arm is still contracting even though it is lengthening. It is performing an eccentric contraction.

Fig 5.7 Isotonic contraction

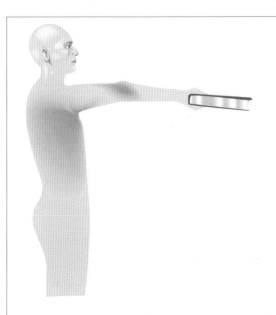

In the classroom...
Hold your book in a steady position in front of you and ensure there is no movement. The muscles in your shoulder and arm are contracting but not allowing any movement. This is isometric contraction.

Fig 5.8 Isometric contraction

Muscle metabolism

Muscles need energy in the form of ATP to contract. They obtain this energy in a number of different ways, including:

- **The phosphagen system** – Muscles store a small amount of ATP in their fibres and are able to convert it into energy through what is called the phosphagen system. However, this is only enough energy to last for around 15 seconds of maximal muscular activity.
- Glycolysis – Once the ATP inside the fibres has been used up, muscles break down glucose through a process called glycolysis. This process does not need oxygen and is called an anaerobic process. Through glycolysis, glucose is broken down into *pyruvic acid* (*pyruvate*) and ATP. The ATP is used by the muscles while the pyruvic acid enters the mitochondria of the muscle fibre where it needs oxygen to be broken down completely. If there is not enough oxygen present to completely break down the pyruvic acid, then it is converted into lactic acid. Lactic acid can diffuse out of the skeletal muscles into the blood and be used by the heart muscle fibres, kidney and liver cells to produce ATP. However, some lactic acid does accumulate in the muscle tissues and blood and can cause muscle fatigue and soreness.
- Aerobic respiration – If there is enough oxygen present, then the pyruvic acid is completely oxidised into carbon dioxide, water, ATP and heat. This process is called *cellular respiration* or *biological oxidation*. Aerobic respiration provides energy for activities of longer than 10 minutes, as long as there is an adequate supply of oxygen and nutrients.

Where do the muscles get their glucose and oxygen from?

Glucose – Carbohydrates are broken down into glucose and stored in the body as *glycogen*. Glycogen can be stored in both the liver and the muscles.

Oxygen – Oxygen is supplied to muscle fibres in two forms. Firstly, it is stored in the muscle cells themselves in the form of myoglobin and secondly, it is stored in the blood in the form of *haemoglobin*.

Why do muscles sometimes stop working?

If a muscle, or group of muscles, is overstimulated it can become progressively weaker until it no longer responds to any stimulus and cannot maintain its contractions. This is called muscle fatigue. Muscle fatigue can occur for a number of reasons, including:

- An insufficient supply of oxygen or glycogen.
- A build up of lactic acid.
- Failure of action potentials to release adequate acetylcholine.

Did you know?
After your first year of life, the growth of your muscles is due to the enlargement of the existing cells and not an increase in the number of cells.

Anatomy and physiology in perspective

Marathon runners can sometimes collapse when their muscles suddenly stop working. This is an example of true muscle fatigue. Their muscles have literally stopped contracting due to a lack of oxygen and glucose and a build up of lactic acid.

Runners (or people who do other forms of strenuous exercise) are also usually seen to breathe heavily for a long time after they have stopped exercising. This is because they need to recover oxygen to restore the body to its resting condition by making ATP reserves, eliminating accumulated lactic acid, repairing tissues and cooling the body down. This heavy breathing is called *recovery oxygen consumption* or recovering the *oxygen debt*.

Types of skeletal muscle fibres

Not all muscle fibres are identical and they differ not only in their appearance (colour and size), but also in the way in which they produce ATP, how quickly they contract and how quickly they fatigue.

There are three different types of muscle fibres: slow oxidative, fast oxidative and fast glycolytic fibres. Most muscles are composed of a combination of all three types. The proportion of the types of fibres is dependent on the usual activity of the muscle. For example, postural muscles that do not need to contract quickly but that do need to maintain a constant, fatigue-resistant contraction contain a large proportion of slow oxidative fibres that are very resistant to fatigue. The table on the following page explains the different types of fibres.

Did you know?
The colour of muscle fibres varies according to the amount of *myoglobin* found in them. Like haemoglobin, myoglobin is a red-coloured, iron-containing protein that binds to oxygen and so provides oxygen to muscle fibres.
Red muscle fibres – These have a high myoglobin content and also have more mitochondria and more blood capillaries than white muscle fibres.
White muscle fibres – These have a low myoglobin content and also have fewer mitochondria and blood capillaries than red muscle fibres.

MUSCLE FIBRE TYPES		
Features	**Type I Slow-twitch**	**Type II Fast-twitch**
	Type I Slow oxidative	Type IIA Fast oxidative	Type IIB Fast glycolytic
Colour	Red	Red to pink	White
Diametre of fibre	Smallest	Medium	Largest
Oxygen supply	Contains large amounts of myoglobin (therefore red), many mitochondria and many blood capillaries.	Contains large amounts of myoglobin (therefore red to pink), many mitochondria and many blood capillaries.	Has a low myoglobin content (therefore white), few mitochondria and few blood capillaries.
ATP production	Generates ATP by aerobic processes (therefore called oxidative fibres).	Generates ATP by aerobic processes (therefore called oxidative fibres).	Generates ATP by anaerobic processes (glycolysis). Therefore cannot supply the muscle continuously with ATP.
Contraction velocity	Split ATP slowly and therefore have a slow contraction velocity.	Split ATP quickly and therefore have a fast contraction velocity.	Due to its large diameter, it splits ATP very fast and therefore has a strong, rapid contraction velocity.
Fatigue resistance	Very resistant to fatigue	Resistant to fatigue, but not as much as Type 1.	Fatigues easily.
Activities	Maintaining posture and endurance activities.	Walking and running.	Fast movements such as throwing a ball.

Study tip
Before learning your muscles, review the chart on Movements at Synovial Joints in Chapter 4. This will help you to understand and visualise the different actions of the muscles.

Skeletal Muscles and Movement

How skeletal muscles produce movement

Before looking at the different skeletal muscles that make up the muscular system, it is necessary to understand how they produce movement. A muscle is usually attached to two bones that form a joint and when the muscle contracts, it pulls the movable bone towards the stationary bone. All muscles have at least two attachments:

- Origin – The point where the muscle attaches to the stationary bone is called the origin.
- Insertion – The point where the muscle attaches to the moving bone is called the insertion. During contraction, the insertion usually moves toward the origin.

Skeletal muscles are attached to two bones that form a joint. When the muscle contracts it pulls one bone towards the other to produce movement.

(Origin)

Fascia

Head

Line of force of the muscle

Belly

Axis of rotation of the joint

Tendon (Attachment)

Virtual lever arm of the muscle

Fig 5.9 Relationship of skeletal muscles to bones

In the classroom

Flex your forearm at your elbow joint. Two specific muscles are at work here: your biceps brachii is contracting. It is the prime mover. In opposition to this, your triceps brachii muscle is lengthening. It is the antagonist.

Now extend your forearm at your elbow joint. Once again, two muscles are working here but this time it is your triceps brachii that is contracting and that is now the prime mover. The biceps brachii is lengthening and is the antagonist.

Muscles only shorten when they contract and thus can only pull, they can never push. Therefore, most body movements are the result of two or more muscles acting together or against each other so that whatever one muscle does, another muscle can undo. For example, one muscle contracts to bend your forearm and another muscle contracts to pull it straight again.

This means that muscles at joints are usually arranged in opposing pairs of flexors-extensors, abductors-adductors etcetera. These pairs are also accompanied by other muscles to ensure that movements are smooth and efficient.

- The muscle responsible for causing a particular movement is called the prime mover or agonist.
- The muscle that opposes this movement is called the antagonist. It relaxes and lengthens in a controlled way to ensure the movement is performed smoothly by the prime mover.
- Additional muscles at the joint ensure a steady movement and help the prime mover function effectively. These muscles are called synergists and they are usually found alongside the prime mover.
- Specialised synergists stabilise the bone of the prime mover's origin so that it can act efficiently. These muscles are called fixators or stabilisers.

Anatomy and physiology in perspective – The effects of ageing on muscle tissue

As we age we progressively lose skeletal muscle mass and it is replaced mainly by fat. In addition, slow oxidative fibres increase while the other types decrease. It is unknown as to whether these processes are due to a decrease in physical activity or simply part of the ageing process.

Muscles of the body

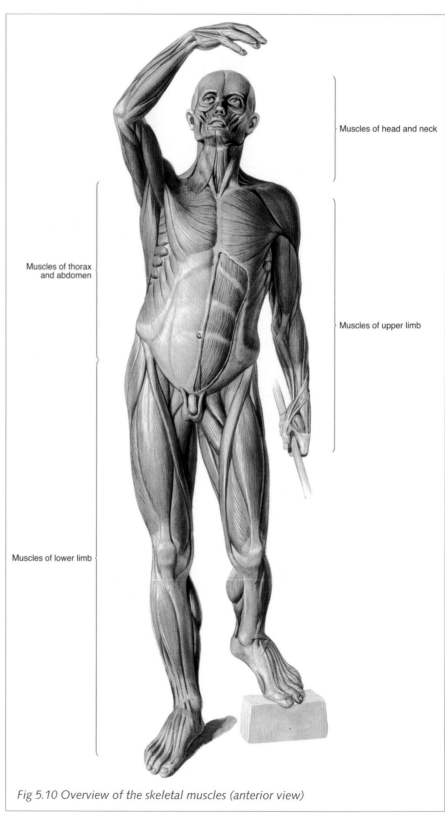

Muscles of head and neck

Muscles of thorax
and abdomen

Muscles of upper limb

Muscles of lower limb

Fig 5.10 Overview of the skeletal muscles (anterior view)

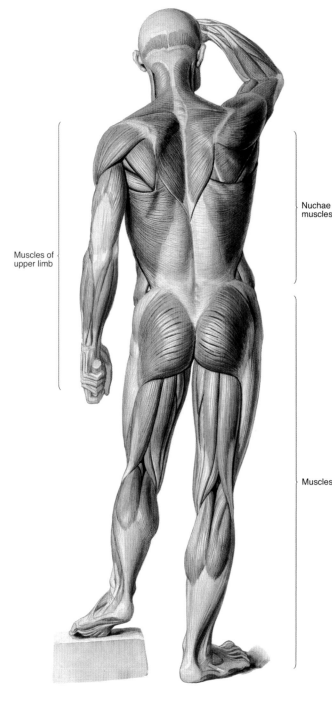

Muscles of upper limb

Nuchae muscles and muscles of back

Muscles of lower limb

Fig 5.11 Overview of the skeletal muscles (posterior view)

Study tip

Muscle names often reflect a muscle's characteristics. So, when you are trying to learn them be aware of how their names describe their:

- **Shape** – the trapezius is shaped like a trapezium.
- **Direction** – the obliques run diagonal to the midline.
- **Position** – the tibialis is found next to the tibia bone.
- **Movement** – the flexor digitorum longus flexes the toes.
- **Number of origins** – the biceps brachii has two heads (bi = 2, ceps = heads).
- **Attachments** – the carpi radialis is attached to the carpum (carpum = wrist).
- **Size** – the gluteus maximus is the largest of the gluteal muscles.
- **Origin and insertion** – the sternocleidomastoid originates on the sternum and clavicle and inserts on the mastoid process.

Muscles of the scalp and face

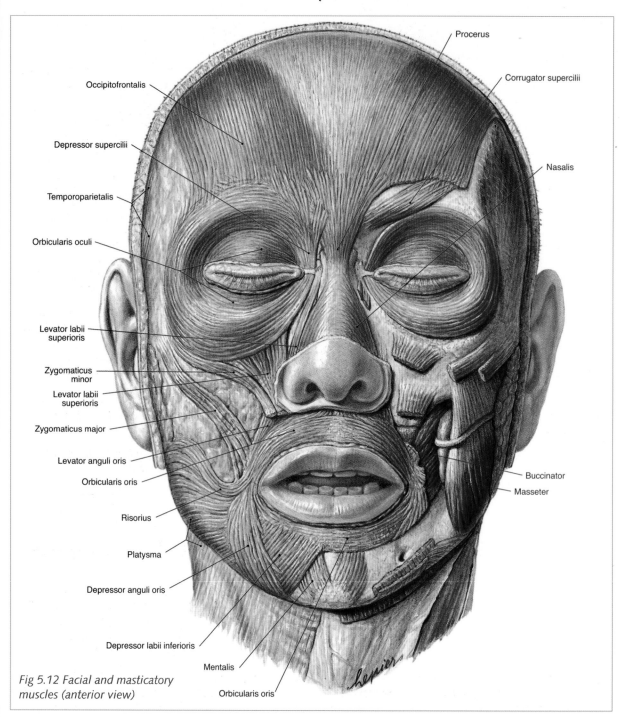

Procerus

Corrugator supercilii

Occipitofrontalis

Nasalis

Depressor supercilii

Temporoparietalis

Orbicularis oculi

Levator labii
superioris

Zygomaticus
minor

Levator labii
superioris

Zygomaticus major

Levator anguli oris

Buccinator

Orbicularis oris

Masseter

Risorius

Platysma

Depressor anguli oris

Depressor labii inferioris

Mentalis

Orbicularis oris

*Fig 5.12 Facial and masticatory
muscles (anterior view)*

Note...
The tables on the following pages classify the muscles firstly according to
their location and secondly according to their function. For example, the
teres minor muscle moves the arm and is located on the back. Therefore, it
is found in the table entitled *Muscles of the back* and under the section
Muscles of the back that move the arm.

MUSCLES OF THE SCALP AND FACE				
Name	Meaning of name	Origin	Insertion	Basic actions
Facial muscles of expression The facial muscles of expression are a group of muscles that usually originate in the fascia or bones of the skull and insert into the skin. They move the skin and enable us to express emotions such as surprise, fear, happiness and sadness.				
The frontalis and occipitalis discussed below are often labelled together as the occipitofrontalis or epicranius. This is because they are united by the galea aponeurotica and work together to create facial expressions such as raising the eyebrows and wrinkling the forehead.				
Frontalis	*Front* = forehead	Galea aponeurotica (cranial aponeuroses between the frontalis and occipitalis)	Fascia and skin above the eyes and nose	Pulls scalp forwards when raising eyebrows or wrinkling forehead.
Occipitalis	*Occipito* = base of the skull	Occipital bone and mastoid process	Galea aponeurotica	Pulls scalp backwards when raising eyebrows or wrinkling forehead.
The orbicularis oculi and levator palpebrae function in moving the eyelids and protecting the eye. For example, the orbicularis oculi can cause the eyelids to close involuntarily or blink if an object is brought too close to the eye.				
Orbicularis oculi	*Orb* = circular *Oculus* = eye	Medial wall of orbit	Circular path around orbit	Closing eye.
Levator palpebrae superioris	*Palpebrae* = eyelids	Roof of orbit	Skin of upper eyelid	Opening eye.
Frowning, wrinkling your nose or flaring your nostrils involves the muscles found in the centre of your face and around your nose. These include the corrugator supercilii, procerus and nasalis.				
Corrugator supercilii	*Corrugo* = wrinkle *Supercilium* = eyebrow	Frontal bone	Skin of eyebrow	Frowning.
Procerus	Long, slender	Nasal bone and cartilage	Skin between the eyebrows	Wrinkling nose.
Nasalis	*Nasus* = nose	Middle of maxilla and nose	Bridge and tip of nose	Compresses and dilates the nostrils eg flaring nostrils when breathing in strongly.
The orbicularis oris is a sphincter muscle that encircles the mouth and enables you to close or purse your lips and move them when speaking. Also important because it serves as the insertion for some of the other facial muscles that are responsible for moving the lips. Muscles that move the lips include the buccinator, levator labii superioris, zygomaticus, risorius, depressor anguli oris, depressor labii inferioris and the mentalis.				
Orbicularis oris	*Orb* = circular *Or* = mouth	Muscle fibres surrounding the opening of the mouth	Skin at the corner of the mouth	Moving lips during speech.

Name	Meaning of name	Origin	Insertion	Basic actions
Buccinator	*Bucc* = cheek	Maxilla and mandible	Muscles of the lips	Moves cheeks when sucking or blowing.
Levator labii superioris	*Labii* = lip	Maxilla and zygomatic bone	Upper lip and corner of mouth	Elevates upper lip, eg snarling or kissing.
Zygomaticus	*Zygomatic* = cheek bone	Zygomatic bone	Corner of the mouth	Draws lips upwards and outwards as in smiling.
Risorius	*Risor* = laughter	Fascia over parotid salivary gland	Corner of the mouth	Draws the mouth sideways and outwards.
Levator anguli oris	*Angulus* = angle *Oris* = mouth	Maxilla	Corner of mouth	Draws mouth upwards, eg smiling.
Depressor anguli oris (triangularis)	*Angulus* = angle *Oris* = mouth	Mandible	Corner of the mouth	Pulls mouth downwards as in sadness.
Depressor labii inferioris	*Labii* = lip	Mandible	Lower lip	Pulls lower lip downwards as in sadness.
Mentalis	*Mentum* = chin	Mandible	Chin	Protrudes lower lip and pulls up chin, eg pouting. Also causes wrinkling of the chin.

The platysma is included in the facial muscles of expression because it moves the lower lip. It does, however, also move the lower jaw and draw the skin of the chest upwards. It is a large, flat muscle that often stands out after a runner has finished a hard race.

Platysma	*Platy* = broad, flat	Fascia overlying the muscles of the chest	Mandible and fascia and muscles of chin and jaw	Depresses mandible and moves lower lip as in pouting and draws skin of the chest upwards.

Facial muscles of mastication
The muscles of mastication move the mandible (lower jaw) and are involved in biting and chewing.

Masseter	*Maseter* = chewer	Maxilla and zygomatic bone	Mandible	Moves mandible when chewing.
Temporalis	*Tempora* = temples	Temporal and frontal bones	Mandible	Moves mandible when chewing.
Pterygoideus medialis (medial pterygoid)	*Pterygoid* = like a wing (the sphenoid bone is shaped like a butterfly)	Sphenoid and maxilla	Mandible	Closes jaw and moves mandible when chewing.
Pterygoideus lateralis (lateral pterygoid)	*Pterygoid* = like a wing	Sphenoid	Mandible and temporo-mandibular joint	Opens jaw and moves mandible when chewing.

Muscles of the neck

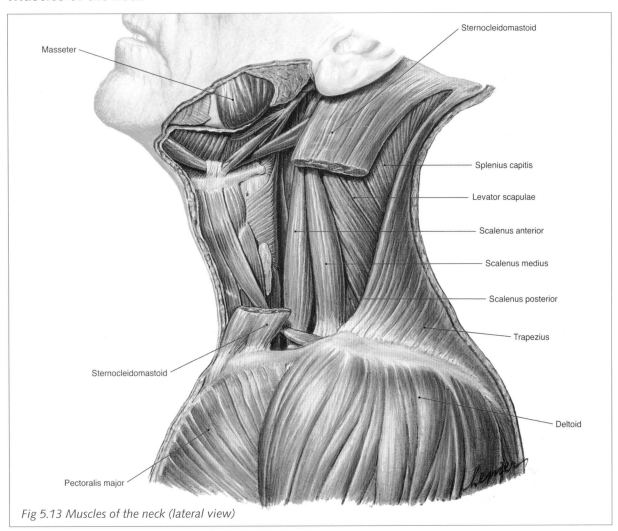

Fig 5.13 Muscles of the neck (lateral view)

Muscles of the neck

There are a number of muscles found in the neck, the main ones being the sternocleidomastoid, semispinalis capitis, splenius capitis, longissimus capitis (caput = head) and the scalenes. Here we will only look at the sternocleidomastoid. The sternocleidomastoid is shaped like a strap and plays an important role in moving the head. It is often affected in whiplash injury and can also cause headaches and neck pain.

MUSCLES OF THE NECK				
Name	**Meaning of name**	**Origin**	**Insertion**	**Basic actions**
Sternocleido-mastoid	Refers to origin (sternum and clavicle) and insertion (mastoid process)	Sternum, clavicle	Mastoid process	Action of two sides at the same time: flexes the neck. Action of one side: laterally flexes the neck to the same side and rotates it to the opposite side.

Muscles of the thorax and abdomen

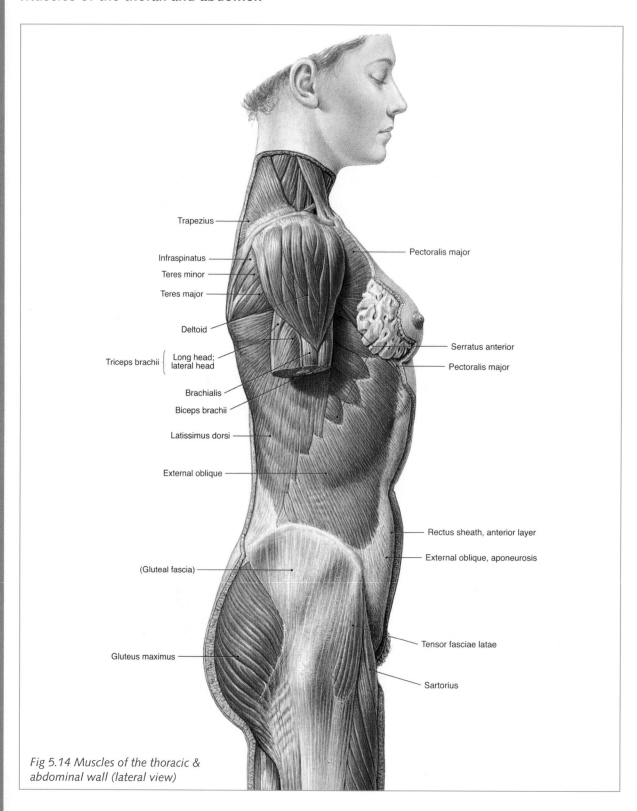

Trapezius

Infraspinatus

Teres minor

Teres major

Deltoid

Triceps brachii { Long head; lateral head

Brachialis

Biceps brachii

Latissimus dorsi

External oblique

(Gluteal fascia)

Gluteus maximus

Pectoralis major

Serratus anterior

Pectoralis major

Rectus sheath, anterior layer

External oblique, aponeurosis

Tensor fasciae latae

Sartorius

Fig 5.14 Muscles of the thoracic & abdominal wall (lateral view)

Muscles of the thorax and abdomen

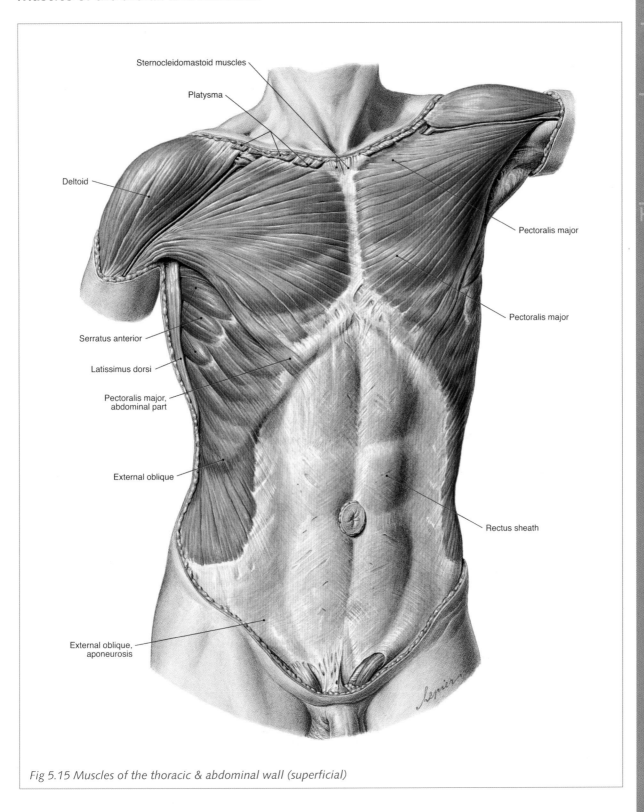

Sternocleidomastoid muscles

Platysma

Deltoid

Pectoralis major

Pectoralis major

Serratus anterior

Latissimus dorsi

Pectoralis major,
abdominal part

External oblique

Rectus sheath

External oblique,
aponeurosis

Fig 5.15 Muscles of the thoracic & abdominal wall (superficial)

Muscles of the thorax and abdomen

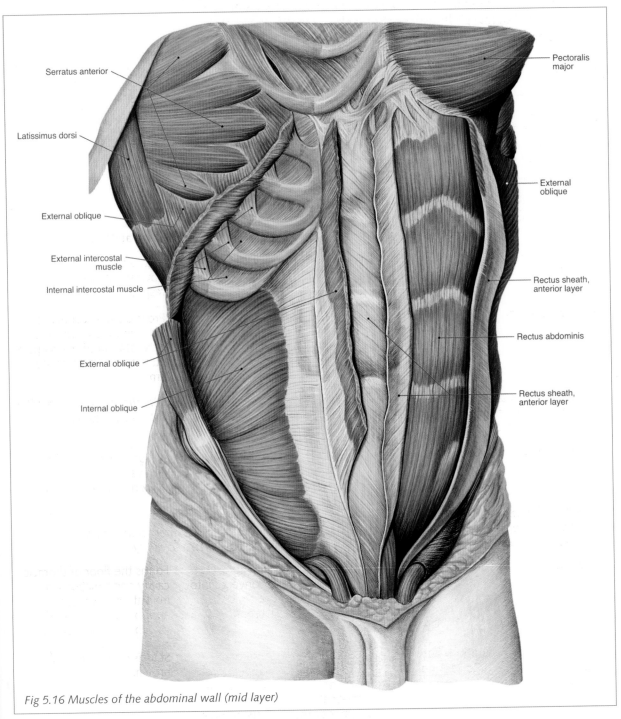

Serratus anterior

Latissimus dorsi

External oblique

External intercostal muscle

Internal intercostal muscle

External oblique

Internal oblique

Pectoralis major

External oblique

Rectus sheath, anterior layer

Rectus abdominis

Rectus sheath, anterior layer

Fig 5.16 Muscles of the abdominal wall (mid layer)

Muscles of the anterior thorax

The muscles of the anterior thorax can be divided into muscles that move the shoulder and arm and muscles used in breathing. The muscles that move the shoulder (pectoral girdle) are divided into anterior and posterior groups.

They function mainly in stabilising the scapula which acts as the point of origin for many of the muscles that move the humerus. Anterior muscles that move the shoulder girdle include the subclavius (not featured here), pectoralis minor and serratus anterior. NB: Certain terms are used when describing movements of the shoulder (and also the mandible). These terms are:

- Protraction – a forward movement of the shoulder on a plane parallel to the ground.
- Retraction – the opposite of protraction. A backward movement on a plane parallel to the ground.
- Elevation – lifting the shoulders upwards.
- Depression – dropping the shoulders downwards.

MUSCLES OF THE THORAX AND ABDOMEN				
Name	Meaning of name	Origin	Insertion	Basic actions
Pectoralis minor	Pectus = breast Minor = small	Ribs 3–5	Scapula	Draws the shoulders downwards and aids respiration.
Serratus anterior	Serratus = saw toothed	First 8–9 ribs	Scapula	Protracts the scapula (pulls it around the chest) and aids rotation of the scapula during abduction of the arm.
The pectoralis major is the large muscle of the chest and, together with the pectoralis minor, it forms the front wall of the armpit. This muscle helps move the arm and when it is very tight it can restrict expansion of the chest.				
Pectoralis major	Pectus = breast	Clavicle, sternum, ribs 2–6	Humerus	Adducts and medially rotates the arm. Aids flexion and respiration.
Muscles of breathing The principle muscles of breathing are the diaphragm and the external and internal intercostals. They are attached to the ribs and contract and relax to alter the size of the thoracic cavity.				
Diaphragm	Dia = across Phragma = wall	Xiphoid process, inferior 6 ribs, lumbar vertebrae	All fibres converge into a central tendon	Forms the floor of thoracic cavity and functions in inhalation. It produces approximately 60% of your breathing capacity
External and Internal intercostals	Inter = between Costa = rib	External – lower border of a rib Internal – upper border of a rib	External – upper border of rib below Internal – lower border of rib above*	Stabilising ribcage and breathing.

* R.Putz & R. Pabst, Sobotta Atlas of Human Anatomy – Muscles, Joints and Nerves, p.19,

Cross-referenced: G. Tortora & S. Grabowski, Principles of Anatomy and Physiology, p.292

Muscles of the abdominal wall

Four pairs of flat, sheet-like muscles make up the abdominal wall. These muscles are the rectus abdominis, external oblique, internal oblique and transversus abdominis. Together with the iliopsoas (discussed later), these muscles are crucial to supporting and stabilising the lower back. A weakness in these core muscles can often lead to injury of the lumbar spine.

MUSCLES OF THE ABDOMINAL WALL				
Name	Meaning of name	Origin	Insertion	Basic actions
Rectus abdominis	*Rectus* = fibres parallel to midline *Abdomino* = abdomen	Pubic crest and pubic symphysis	Ribs 5-7 and xiphoid process	Stabilises pelvis during walking and compresses abdomen during defecation/urination. Action of one side: flexes trunk laterally. Action of both sides together: forward flexion of trunk.
External and Internal obliques	*Oblique* = fibres diagonal to midline	External – lower 8 ribs Internal – iliac crest, inguinal ligament and thoracolumbar fascia	External – iliac crest and linea alba Internal – inferior 3-4 ribs and linea alba	Compress abdomen and support abdominal viscera. Action of one side: flexes trunk laterally and rotates it. Action of both sides together: flexes trunk.
Transversus abdominis	*Transverse* = fibres perpendicular to midline	Iliac crest, inguinal ligament, lumbar fascia, inferior 6 ribs	Xiphoid process, linea alba, pubis	Maintain posture and compress the abdomen as in when forcing expiration, sneezing and coughing.

Muscles of the back

Muscles of the back that move the shoulder

The trapezius, levator scapulae and rhomboideus muscles help to stabilise the scapula which functions as a point of origin for many muscles. These muscles are easily affected by stress and can become hypertonic, causing neck pain, shoulder stiffness and headaches.

MUSCLES OF THE BACK THAT MOVE THE SHOULDER				
Name	Meaning of name	Origin	Insertion	Basic actions
Trapezius	Shaped like a trapezoid	Occipital bone, cervical 7 and all thoracic vertebrae	Clavicle, acromion process, and scapula	The trapezius has three sets of muscle fibres: the upper, middle and lower fibres. **Upper fibres:** *Action of both sides:* extends the neck and elevates the shoulders. *Action of one side:* laterally flexes the neck. **Middle fibres:** Retract the scapula. **Lower fibres:** Help move the scapula around the chest during abduction of the shoulder.
Levator scapulae	Levator = raises	First 4–5 cervical vertebrae	Superior border of scapula	Elevates shoulder and rotates the scapula.
Rhomboideus major and minor	Shaped like a rhomboid (diamond)	Major – thoracic vertebrae 2–5 Minor – Cervical 7 and thoracic 1	Scapula	Retracts the scapula and fixes it to the thoracic wall.

Muscles of the back

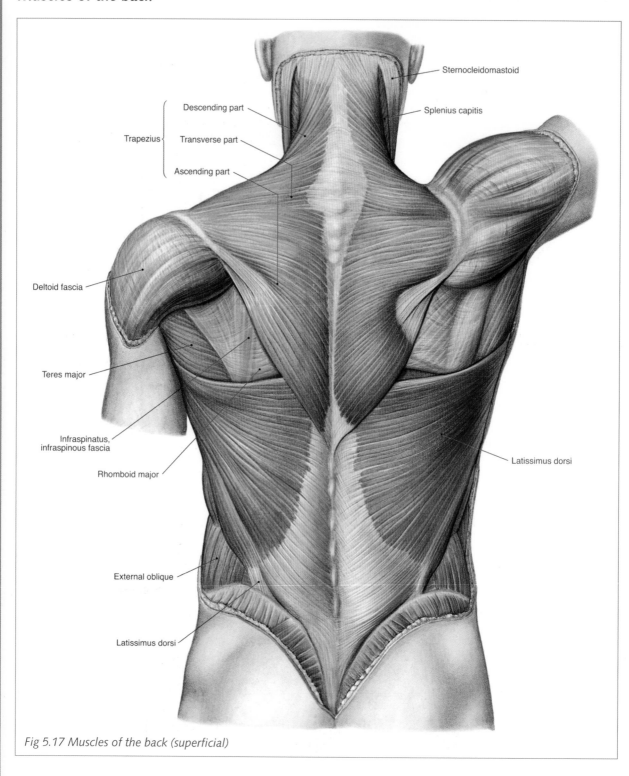

Fig 5.17 Muscles of the back (superficial)

Muscles of the back

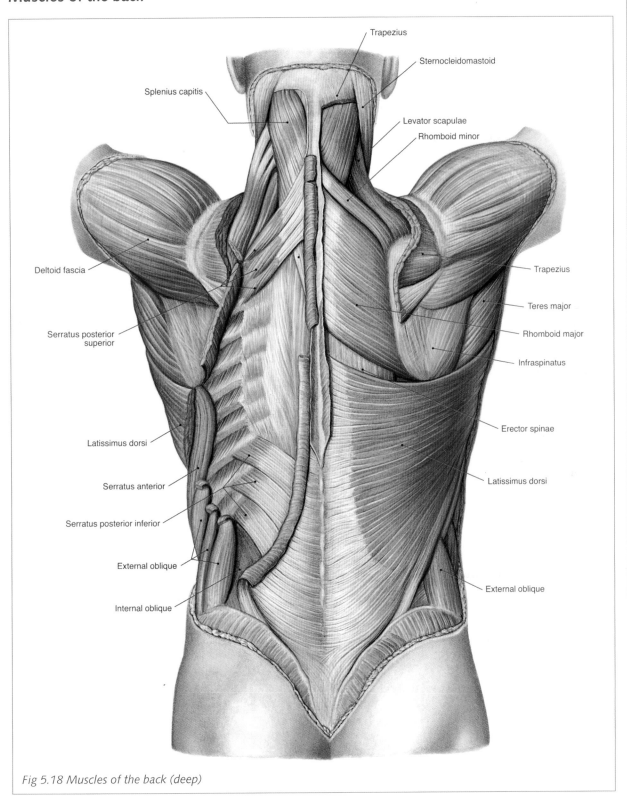

Trapezius

Sternocleidomastoid

Splenius capitis

Levator scapulae

Rhomboid minor

Deltoid fascia

Trapezius

Teres major

Rhomboid major

Serratus posterior superior

Infraspinatus

Erector spinae

Latissimus dorsi

Serratus anterior

Latissimus dorsi

Serratus posterior inferior

External oblique

External oblique

Internal oblique

Fig 5.18 Muscles of the back (deep)

Muscles of the back that move the arm

Muscles located on the back that move the arm can be divided into those that originate on the scapula and those that originate on the axial skeleton. The muscles originating on the scapula include the muscles of the rotator cuff, which strengthen and reinforce the shoulder joint. More muscles originate on the scapula but are located on the arm itself. These are discussed in the section entitled *Muscles of the Arm*.

Rotator cuff

Four muscles work together to hold the head of the humerus in the glenoid cavity of the scapula. These muscles are the subscapularis, supraspinatus, infraspinatus and teres minor muscles. Together, their tendons form the rotator cuff which encircles the ball-and-socket joint of the shoulder and strengthens and reinforces it.

MUSCLES OF THE BACK THAT MOVE THE ARM

Name	Meaning of name	Origin	Insertion	Basic actions
Subscapularis	*Sub* = below *Scapularis* = scapula	Scapula	Humerus	Medially rotates the shoulder joint.
Supraspinatus	*Supra* = above *Spinatus* = spine of scapula	Scapula	Humerus	Helps deltoid abduct the arm and initiates abduction of the shoulder joint (first 15° – 20°).
Infraspinatus	*Infra* = below	Scapula	Humerus	Laterally rotates the shoulder joint.
Teres minor	*Teres* = rounded	Scapula	Humerus	Laterally rotates the shoulder joint.
Teres major	*Teres* = rounded	Scapula	Humerus	Extends, medially rotates and adducts the arm.
The pectoralis major (discussed as a muscle of the chest) and latissimus dorsi originate on the axial skeleton. Together with the subscapularis and teres major, the latissimus dorsi forms the posterior wall of the armpit and moves the humerus.				
Latissimus dorsi	*Latissimus* = widest *Dorsum* = back	Lower 6 thoracic vertebrae, lumbar vertebrae, sacrum, ilium, lower 4 ribs	Humerus	Extends, medially rotates and adducts the arm. When the arms are fixed above the head it will raise the trunk (as in chin-ups).

Muscles that move the vertebral column

Many muscles are responsible for moving the vertebral column and some have already been mentioned elsewhere. In general, the muscles of the vertebral column are complex, have many origins and insertions and overlap one another.

The erector spinae (sacrospinalus) is actually made up of three sets of muscles organised in parallel columns and it is important in maintaining posture. It is often the muscle you injure when you lift a heavy object without bending your knees first.

MUSCLES THAT MOVE THE VERTEBRAL COLUMN				
Name	Meaning of name	Origin	Insertion	Basic actions
Erector spinae/ sacrospinalus (made up of the iliocostalis, longissimus and spinalis muscles)	Refers to its action i.e. holds spine erect	Sacrum, iliac crest, vertebrae and ribs	Ribs, vertebrae and occipital bone	Action of one side: flexes trunk laterally. Action of both sides together: extends the trunk. It is one of the main postural muscles that hold the body upright.
Quadratus lumborum	Quad = four Lumbo = lumbar region	Iliac crest, iliolumbar ligament	Rib 12 and first 4 lumbar vertebrae	Action of one side: flexes the trunk laterally. Action of both sides together: extends the trunk.

Muscles of the arm

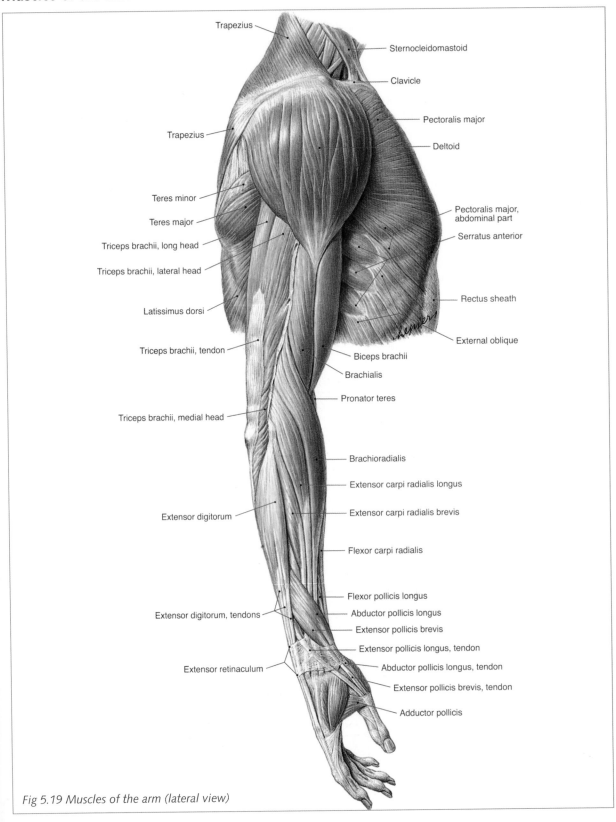

Trapezius

Sternocleidomastoid

Clavicle

Pectoralis major

Trapezius

Deltoid

Teres minor

Pectoralis major, abdominal part

Teres major

Serratus anterior

Triceps brachii, long head

Triceps brachii, lateral head

Latissimus dorsi

Rectus sheath

Triceps brachii, tendon

External oblique

Biceps brachii

Brachialis

Pronator teres

Triceps brachii, medial head

Brachioradialis

Extensor carpi radialis longus

Extensor carpi radialis brevis

Extensor digitorum

Flexor carpi radialis

Flexor pollicis longus

Extensor digitorum, tendons

Abductor pollicis longus

Extensor pollicis brevis

Extensor pollicis longus, tendon

Abductor pollicis longus, tendon

Extensor retinaculum

Extensor pollicis brevis, tendon

Adductor pollicis

Fig 5.19 Muscles of the arm (lateral view)

Muscles of the arm

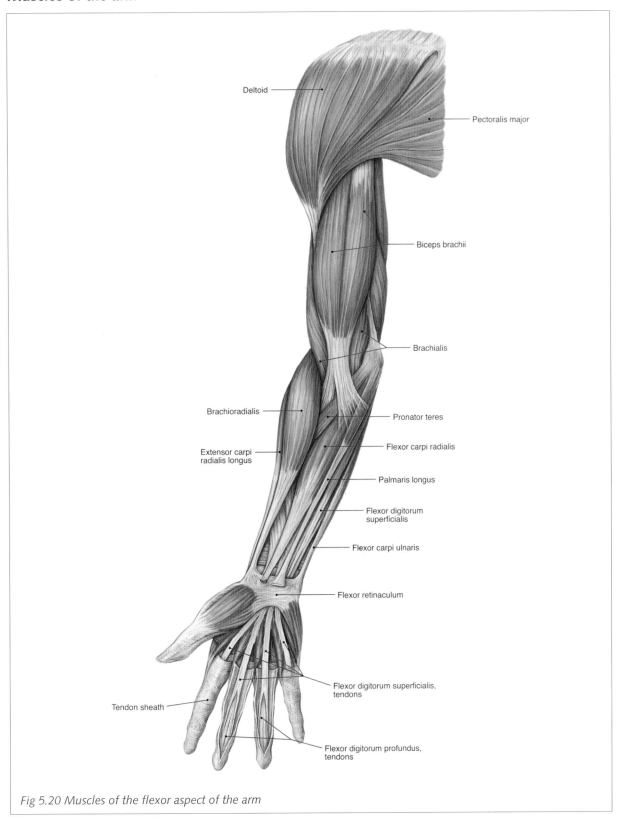

Deltoid

Pectoralis major

Biceps brachii

Brachialis

Brachioradialis

Pronator teres

Extensor carpi radialis longus

Flexor carpi radialis

Palmaris longus

Flexor digitorum superficialis

Flexor carpi ulnaris

Flexor retinaculum

Flexor digitorum superficialis, tendons

Tendon sheath

Flexor digitorum profundus, tendons

Fig 5.20 Muscles of the flexor aspect of the arm

Muscles of the arm

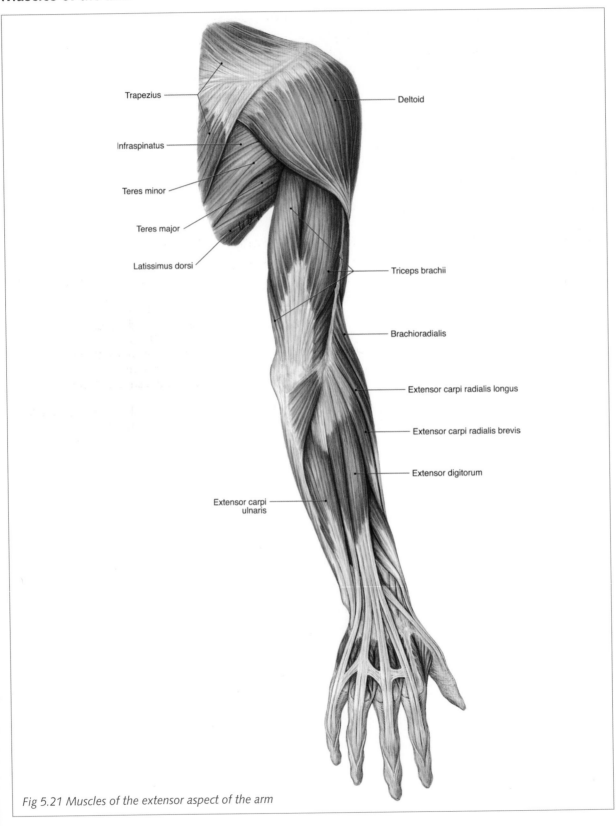

Trapezius

Deltoid

Infraspinatus

Teres minor

Teres major

Latissimus dorsi

Triceps brachii

Brachioradialis

Extensor carpi radialis longus

Extensor carpi radialis brevis

Extensor digitorum

Extensor carpi ulnaris

Fig 5.21 Muscles of the extensor aspect of the arm

Muscles of the arm

The humerus bone is moved by a number of muscles that all originate on the trunk of the body. Only two of them, the pectoralis major and latissimus dorsi discussed above, originate on the axial skeleton. All the other muscles that move the humerus originate on the scapula, including the rotator cuff muscles which have already been discussed.

The shoulder joint is capable of a large variety of movements including flexion, extension, abduction, adduction, medial rotation, lateral rotation and circumduction. It has more freedom of movement than any other joint in the body. The muscles listed in the table below are located on the arm and move the humerus.

MUSCLES OF THE ARM				
Name	Meaning of name	Origin	Insertion	Basic actions
Deltoid	Shaped like a deltoid (triangle)	Clavicle, acromion, scapula	Humerus	The deltoid has three sets of muscle fibres: the anterior, middle and posterior fibres. **Anterior fibres:** Flex the shoulder joint. **Middle fibres:** Abduct the shoulder joint. **Posterior fibres:** Extend the shoulder joint.
Coraco-brachialis	Coraco = coracoid process of scapula Brachion = arm	Scapula	Humerus	Flexes and adducts arm and helps stabilise the humerus.

Muscles located on the upper arm that move the forearm

Some of the muscles located on the upper arm move the forearm. They originate on the scapula or humerus, pass over the elbow joint and insert into the radius and ulna. The elbow joint is a hinge joint and therefore only capable of flexion and extension. Thus, the muscles that move the forearm can be categorised into flexors and extensors. Some of these muscles also allow pronation and supination of the forearm. The muscles listed in the table below are all flexors.

MUSCLES LOCATED ON THE UPPER ARM THAT MOVE THE FOREARM				
Name	Meaning of name	Origin	Insertion	Basic actions
Biceps brachii	Biceps = 2 heads Brachion = arm	Scapula	Radius and bicipital aponeurosis	Flexes the elbow joint, supinates the forearm and aids flexion of the shoulder.
Brachialis	Brachion = arm	Humerus	Ulna	Flexes elbow joint.
Brachioradialis	Describes position	Humerus	Radius	Flexes elbow joint.
The triceps brachii and anconeus muscles are extensors. Although it is not discussed here, the anconeus extends the elbow.				
Triceps brachii	Triceps = 3 heads	Scapula and humerus	Ulna	Extends elbow joint.
The pronator teres and pronator quadratus muscles are pronators. Although it is not discussed here, the pronator quadratus pronates the forearm and hand.				
Pronator teres	Describes action Pronation = to turn palm posteriorly	Humerus and ulna	Radius	Pronates forearm and hand.
The supinator does as its name describes: it supinates the forearm.				
Supinator	Describes action Supination = to turn palm anteriorly	Humerus and ulna	Radius	Supinates forearm and hand.

Muscles that move the wrist

The muscles of the forearm move the wrist and can be divided into anterior and posterior compartments. The tendons of these muscles are held close to the bones by strong fibrous bands called retinacula. The flexor retinaculum (transverse carpal ligament) is found on the palmar surface of the carpal bones while the extensor retinaculum (dorsal carpal ligament) is found over the dorsal surface of the carpal bones.

The anterior compartment muscles originate on the humerus, insert on the carpals, metacarpals and phalanges and are flexors. These muscles include the flexor carpi radialis, palmaris longus, flexor carpi ulnaris, flexor digitorum superficialis, flexor digitorum profundus and the flexor pollicis longus. Not all of these muscles are discussed in the table below.

MUSCLES THAT MOVE THE WRIST

Name	Meaning of name	Origin	Insertion	Basic actions
Flexor carpi radialis	Describes action Carpi = wrist	Humerus	Metacarpals 2-3	Flexes and abducts wrist.
Flexor carpi ulnaris	Describes action	Humerus and ulna	Carpals and metacarpals	Flexes and adducts wrist.
Flexor digitorum superficialis	Describes action Digit = finger	Humerus, ulna and radius	Middle phalanges	Flexes middle phalanges of each finger.

The posterior compartment muscles originate on the humerus and insert on the metacarpals and phalanges and are extensors. These muscles include the extensor carpi radialis longus, extensor carpi radialis brevis, extensor digitorum, extensor digiti minimi, extensor carpi ulnaris, abductor pollicis longus, extensor pollicus brevis, extensor polllicis longus and extensor indicis. Not all of these muscles are discussed here.

Extensor carpi radialis longus	Describes action	Humerus	Metacarpal 2	Extends and abducts wrist.
Extensor carpi radialis brevis	Describes action	Humerus	Metacarpal 3	Extends and abducts wrist.
Extensor digitorum	Extensor = extends Digit = finger	Humerus	Phalanges of the four fingers	Extends fingers.
Extensor carpi ulnaris	Describes action	Humerus and ulna	Metacarpal 5	Extends and adducts wrist.

Thenar and hypothenar eminences
The thenar eminence is a raised area of firm tissue found on the radial side of the palm of the hand, beneath the thumb. It is composed of the muscles that move the thumb. Namely, the abductor pollicis brevis, flexor pollicus brevis, opponens pollicis and adductor pollicis. Not all of these muscles are discussed here. The hypothenar eminence is an area of soft tissue found on the ulnar side of the palm, beneath the little finger. It is composed of the muscles that move the little finger. Namely, the palmaris brevis, abductor digiti minimi, flexor digiti minimi brevis and opponens digiti minimi. These are not discussed here.

Abductor pollicis brevis	Pollex = thumb	Flexor retinaculum and scaphoid	Thumb	Abducts the thumb.
Flexor pollicis brevis	Pollex = thumb	Flexor retinaculum and capitate, trapezium, trapezoid and metacarpal 1 bones	Thumb	Flexes the thumb.

Muscles of the leg

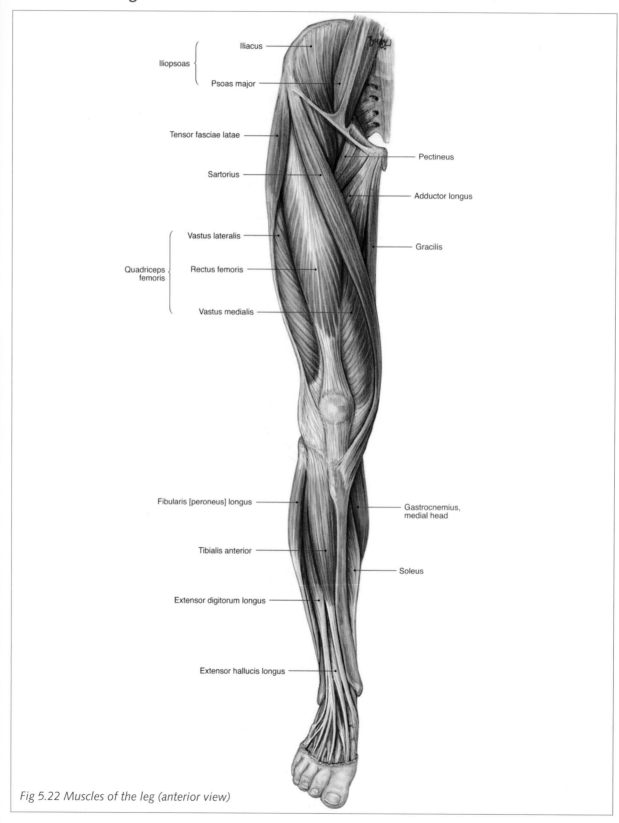

Iliacus

Iliopsoas

Psoas major

Tensor fasciae latae

Pectineus

Sartorius

Adductor longus

Vastus lateralis

Gracilis

Quadriceps femoris

Rectus femoris

Vastus medialis

Fibularis [peroneus] longus

Gastrocnemius, medial head

Tibialis anterior

Soleus

Extensor digitorum longus

Extensor hallucis longus

Fig 5.22 Muscles of the leg (anterior view)

Muscles of the leg

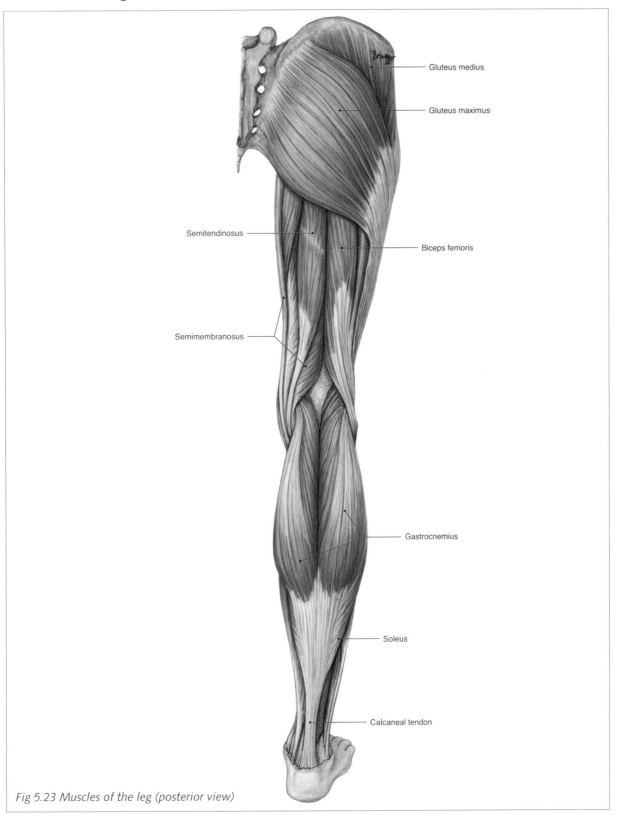

Gluteus medius

Gluteus maximus

Semitendinosus

Biceps femoris

Semimembranosus

Gastrocnemius

Soleus

Calcaneal tendon

Fig 5.23 Muscles of the leg (posterior view)

Muscles of the leg and foot

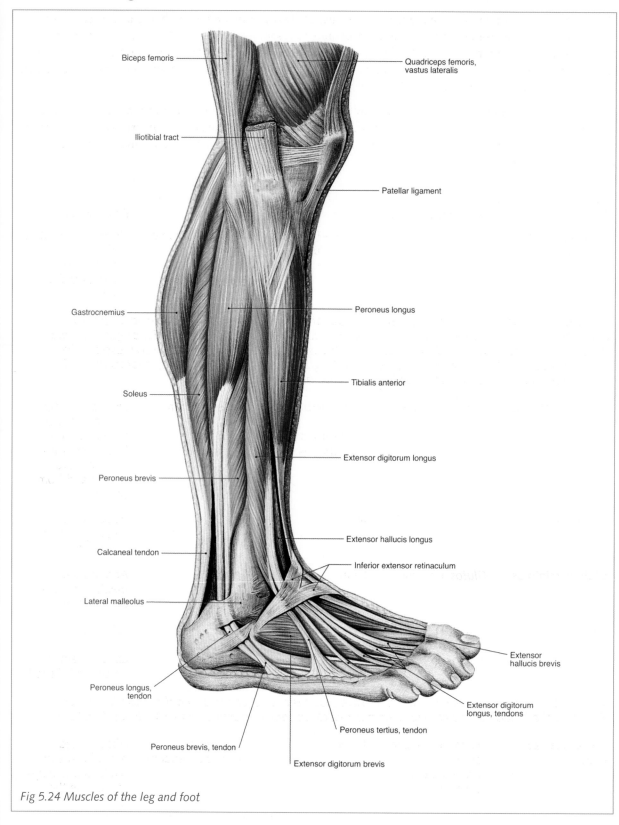

Biceps femoris

Quadriceps femoris, vastus lateralis

Iliotibial tract

Patellar ligament

Gastrocnemius

Peroneus longus

Tibialis anterior

Soleus

Extensor digitorum longus

Peroneus brevis

Extensor hallucis longus

Calcaneal tendon

Inferior extensor retinaculum

Lateral malleolus

Extensor hallucis brevis

Peroneus longus, tendon

Extensor digitorum longus, tendons

Peroneus brevis, tendon

Peroneus tertius, tendon

Extensor digitorum brevis

Fig 5.24 Muscles of the leg and foot

Muscles of the hip joint

The muscles of the hip joint move the femur. They generally originate on the hip bones and insert on the femur. Some of them cross both the hip and knee joints to insert on the tibia and fibula. The hip joint is a ball-and-socket joint that allows for flexion, extension, abduction, adduction, circumduction and rotation. In addition to functioning in movement, the muscles of the lower limb are important in maintaining posture and stability. These muscles are generally large and powerful.

The psoas major and iliacus muscles are the only anterior muscles that move the femur and, because they are located in the hip and lower back region, they are often considered muscles of the trunk. Together they form the iliopsoas which is the main hip flexor and lower back stabiliser. Lower back and hip problems often develop if this muscle is hypertonic.

MUSCLES OF THE HIP JOINT				
Name	Meaning of name	Origin	Insertion	Basic actions
Iliopsoas (Iliacus and Psoas major)	*Psoa* = loin	Iliacus – ilium Psoas – lumbar vertebrae	Iliacus – tendon of psoas major Psoas – femur	Flexes the hip joint and rotates the femur laterally. Also flexes the trunk.
The posterior muscles that move the femur include the gluteus maximus, gluteus medius, gluteus minimus, piriformis, obturator internus, obturator externus, superior gemellus, inferior gemellus and the quadratus femoris. Not all of these muscles are discussed here. The gluteus maximus muscle is the heaviest muscle in the body and, together with the gluteus medius and minimus, it forms the buttock. Imbalances in these muscles can lead to lower back, hip and knee problems.				
Gluteus maximus	*Glutos* = buttock	Ilium, sacrum, coccyx, aponeurosis of sacrospinalis	Femur, iliotibial tract	Extends the hip joint and rotates the femur laterally. Raises the trunk from forward flexion and from sitting.
Gluteus medius	*Glutos* = buttock	Ilium	Femur	Abducts and medially rotates the hip joint
Gluteus minimus	*Glutos* = buttock	Ilium	Femur	Abducts and medially rotates the hip joint.
Piriformis	Describes shape *Pirum* = pear, pyramid	Sacrum	Femur	Laterally rotates hip joint and abducts femur when hip is flexed.
The tensor fasciae latae is a deep lateral muscle that has a long tendon called the fascia lata tendon. This joins the tendon of the gluteus maximus muscle to form a structure known as the iliotibial tract.				
Tensor fasciae latae	*Tensor* = to make tense *Fascia* = band *Latus* = wide	Ilium	Tibia (via the fascia lata tendon)	Abducts and medially rotates hip joint and helps to stabilise the knee.

Muscles of the thigh

The muscles located in the thigh generally originate in the hipbones, cross the knee joint and insert into the leg (tibia and fibula). They permit flexion, extension and slight rotation of the leg and are categorised into medial, anterior and posterior compartments. These compartments are separated by deep fascia.

The medial compartment of the thigh is called the adductor compartment because the muscles within it all adduct the femur. These muscles include the adductor magnus, adductor longus, adductor brevis, pectineus and gracilis muscles and they are the muscles affected when you pull or strain your groin. Not all these muscles are discussed in the table below.

MUSCLES OF THE THIGH				
Name	Meaning of name	Origin	Insertion	Basic actions
Adductors (Adductor longus, brevis, magnus)	*Adduct* = to bring closer to the midline	Longus – pubic bone Brevis – pubic bone Magnus – pubic bone and ischium	Femur	Adduct and laterally rotate hip joint.
Gracilis	*Gracilis* = slender, graceful	Pubic bone	Tibia	Adducts and medially rotates hip joint and aids flexion of knee joint.

The anterior compartment of the thigh is called the extensor compartment because all the muscles within it extend the leg. Some of the muscles also flex the thigh (namely, the rectus femoris and sartorius). The anterior compartment includes the quadriceps femoris and sartorius muscles. The quadriceps femoris (often referred to as the quads) is made up of four distinct parts: the rectus femoris, vastus lateralis, vastus intermedius and vastus medialis. Their tendon, the quadriceps tendon, attaches to the patellar.

Quadriceps femoris (Rectus femoris, vastus lateralis, vastus medialis, vastus intermedius)	*Quadriceps* = 4 heads Group of 4 muscles	Vastus group – femur Rectus femoris – ilium	Patella and tibia	Extend knee joint. Rectus femoris also helps to flex the hip joint.
Sartorius	*Sartor* = tailor (contracts when you sit in the cross-legged position of a tailor)	Ilium	Tibia	Flexes, abducts and laterally rotates the hip joint and flexes the knee joint.

The posterior compartment of the thigh is called the flexor compartment because its muscles flex the leg. Most of them also extend the thigh. The flexor compartment is composed of three muscles with long, string-like tendons that are collectively called the hamstrings. These muscles are the biceps femoris, semitendinosus and semimembranosus.

Hamstrings: Biceps femoris, semitendinosus, semi-membranosus)	*Hamme* = back of leg *Stringere* = to draw together group of three muscles	Biceps femoris – ischium, femur Semitendinosus – ischium. Semimembranosus – ischium	Biceps femoris – fibula and tibia Semitendinosus – tibia. Semimembranosus – tibia	Extend hip joint and flex knee joint.

Muscles of the lower leg and foot

The muscles located in the leg generally originate in the fibula and tibia (sometimes in the femur or knee capsule), cross the ankle joint and insert on the foot. Like the muscles of the thigh, the muscles of the leg are divided into three compartments which are separated by deep fascia. The ankle joint is a hinge joint that allows for dorsiflexion and plantar flexion of the foot.

The anterior compartment of the leg contains the muscles that dorsiflex the foot. Its tendons are held firmly to the ankle by the transverse ligament of the ankle (superior extensor retinaculum) and the cruciate ligament of the ankle (inferior extensor retinaculum). The muscles of the anterior compartment of the leg include the tibialis anterior, extensor hallicus longus, extensor digitorum longus and the peroneus tertius. Not all these muscles are discussed in the table below.

MUSCLES OF THE LOWER LEG AND FOOT				
Name	Meaning of name	Origin	Insertion	Basic actions
Tibialis anterior	Describes location	Tibia and interosseous membrane	Foot	Dorsiflexes foot and inverts and supports the medial arch of the foot.
Extensor hallucis longus	Describes action Hallucis = big toe	Fibula and inter-osseous membrane	Big toe	Dorsiflexes and inverts foot and extends big toe.
Extensor digitorum longus	Describes action Digit = toe	Tibia, fibula and interosseous membrane	4 outer toes	Dorsiflexes and everts foot and extends toes.
The posterior compartment of the leg contains the muscles that plantar flex the foot and comprises the gastrocnemius, soleus, plantaris, popliteal, flexor hallicus longus, flexor digitorum longus and tibialis posterior. The tendons of the gastrocnemius, soleus and plantaris muscles join to form the Achilles Tendon (calcaneal tendon) which is the strongest tendon in the body.				
Gastrocnemius	Gaster = belly Kneme = leg	Femur and knee capsule	Heel via the Achilles tendon	Plantar flexes the foot and aids flexion of the knee. Is a postural muscle.
Soleus	Soleus = sole of foot, fish shaped	Fibula and tibia	Heel via the Achilles tendon	Plantar flexes the foot.
Flexor hallucis longus	Hallucis = big toe	Fibula	Big toe	Plantar flexes and inverts the foot and flexes the toes.
Flexor digitorum longus	Digit = toe	Tibia	4 outer toes	Plantar flexes and inverts the foot and flexes the toes.
The lateral compartment of the leg contains the muscles that plantar flex and evert the foot. These are the peroneus longus and brevis. Only the peroneus longus is discussed here.				
Peroneus longus	Perone = fibula	Fibula	Metatarsal 1, cuneiform 1	Plantar flexes and everts the foot.

Common Pathologies Of The Muscular System

Disorders of muscles, bursae and tendons

The health of the muscular system is essential to the functioning of the body as a whole and it can be negatively affected by injury, overuse and infection. Certain disorders, such as muscular dystrophy, are inherited. In general, muscular disorders can be accompanied by pain, tenderness, inflammation and limited movement.

Anatomy and physiology in perspective – How do skeletal muscles heal?

Skeletal muscles do not have much potential to divide but, if injured, they can be replaced by new cells derived from dormant stem cells called satellite cells. However, if there is more damage to the muscle than the satellite cells can cope with, fibrosis occurs. This is the replacement of muscle fibres by scar tissue, which is a fibrous connective tissue that does not allow much movement and which can restrict movement at joints.

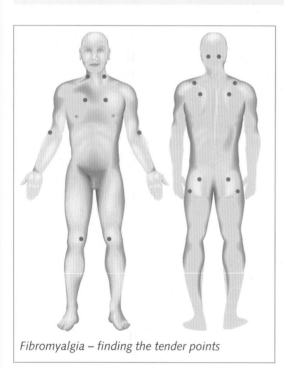

Fibromyalgia – finding the tender points

Fibromyalgia

Fibromyalgia is the term used to describe a group of disorders that are all characterised by aching stiffness and pain in the soft tissue (muscles, tendons and ligaments) coupled with pain resulting from gentle finger pressure applied at specific 'tender spots'. Different types of fibromyalgia exist and their causes are not always known although it tends to affect women more than men. They are all usually aggravated by physical or mental stress, fatigue, strain or overuse. *Lumbago* is fibromyalgia of the lumbar region. It can be caused by general tension, a slipped disc or a strained muscle or ligament.

Fibrositis (muscular rheumatism)

Fibrositis, also called muscular rheumatism, is inflammation of the fibrous connective tissue and is characterised by inflammation, pain and stiffness.

Ganglion cyst

A ganglion cyst is a fluid-filled growth that usually develops near joints or tendon sheaths on the hand or foot. The cause of ganglion cysts is unknown and although they can be removed by surgery, they can also disappear over time.

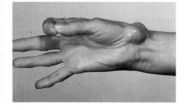

Ganglion cyst

Muscular dystrophies

Muscular dystrophies are a group of inherited muscle-destroying diseases that lead to muscle weakness. They are generally characterised by a progressive atrophy of the skeletal muscle due to the degeneration of individual muscle fibres.

Myasthenia gravis

Myasthenia gravis is an autoimmune disease characterised by a weakness of skeletal muscles. It is a result of impaired communication between nerves and muscles and is more common among women. Symptoms may include drooping eyelids, weak eye muscles, double vision, extreme fatigue, difficulty in speaking and swallowing and weakness of the arms and legs. Myasthenia gravis often occurs in exacerbations (periods in which symptoms worsen) and symptom-free periods.

Poliomyelitis (polio)

Poliomyelitis, commonly referred to as Polio, is a contagious viral infection of the nervous system that results in muscle weakness and sometimes paralysis. Its initial symptoms include a general feeling of malaise, fever, headaches, muscle pain and a stiff neck and back. These symptoms may progress into weakness or paralysis of the muscles.

Rupture

A rupture is the tearing of muscle fascia. It is generally accompanied by pain and swelling.

Spasm

A spasm is an abnormal, involuntary muscular contraction which may occur as part of another disorder or as a localised condition. A cramp is a type of spasm, as is a tic (twitching of the eyelid or facial muscles). *Cramp* is a common muscular disorder that can affect healthy, active people as much as it affects other groups of people. It is characterised by a sudden, painful contraction of a muscle or group of muscles and it can happen during sleep and during or after exercise. There are a number of theories as to what causes cramps including: insufficient stretching before exercise; an inadequate blood flow to the muscles; low blood levels of electrolytes; and excess intake of caffeine or nicotine.

Strain

A strain is the overstretching of a muscle. It is characterised by pain, swelling and sometimes restricted movement.

Tetanus (lockjaw)

Tetanus, or lockjaw, is an infectious bacterial disease characterised by muscle stiffness, spasms and rigidity in the jaw and neck. If not treated, these spasms can move into the back and chest and eventually affect the entire body.

Sports injuries and repetitive strain injury (RSI)

Sports and repetitive strain injuries (RSI) are most often caused by overuse of specific muscles. Muscles may be overused during sport, exercise or when repeating movements while performing a regular activity (for example typing). Muscles can also be overused when structural abnormalities, such as an unequal length of legs, place extra stress on other parts of the body.

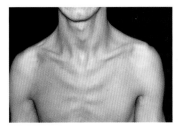
Muscular dystrophy (anterior)

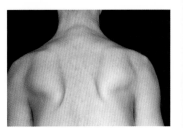
Muscular dystrophy (posterior)

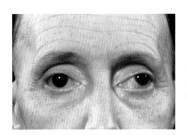
Myasthenia gravis

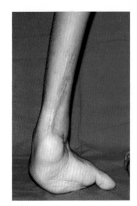
Poliomyelitis (polio)

Anatomy and physiology in perspective
Most sports injuries are best treated as soon as possible with the **RICE** method. This involves:
Rest, Ice, Compression, Elevation.
The injured area is rested immediately and ice is applied to the area to reduce swelling and pain. A compress such as a bandage is applied and the area is then elevated to help limit further swelling. In addition to this, it is important to give the injured area enough time to heal itself properly before returning to the sporting activity.

Carpal tunnel syndrome

Carpal tunnel syndrome is the compression of the median nerve as it passes through the wrist. It is often caused by repetitive, forceful use of the wrist when it is in the wrong position and is also common in pregnancy, diabetes, rheumatoid arthritis and people who have an underactive thyroid gland. Carpal tunnel syndrome is characterised by numbness and tingling of the thumb and first three fingers and the thumb also tends to be weak. The little finger is often symptom-free.

Frozen shoulder

Frozen shoulder is the inflammation of the shoulder joint and it is characterised by chronic, painful stiffness. It has many causes such as injury to the shoulder or a stroke. It can also develop slowly for no apparent reason.

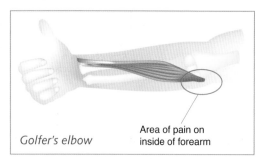

Golfer's elbow

Area of pain on inside of forearm

Golfer's elbow (medial epicondylitis)

Golfer's elbow, also called medial epicondylitis or forehand tennis elbow, is damage to the tendons on the inside of the elbow that bend the wrist toward the palm. Symptoms include pain on the palmar side of the forearm from the elbow toward the wrist and it can be caused by movements that bend the wrist toward the palm with excessive force. These can include certain golf swings, tennis serves, throwing javelins or even carrying a heavy suitcase.

Housemaid's knee (pre-patellar bursitis)

Housemaid's knee, or pre-patellar bursitis, is the inflammation of the bursa in front of the knee cap. It is characterised by inflammation and pain and is usually a result of frequent kneeling.

Shin splints

Shin splints, or shin splint syndrome, is the term used to describe pain along the tibia. It may occur in the anterior and lateral muscles of the shin (anterolateral shin splints) or the posterior and medial muscles (postero-medial shin splints). Where the pain is felt will depend on the muscles affected. Generally, shin splints is initially characterised by pain on movement. If left untreated, the pain will occur when the shins are touched. Shin splints can result from running on hard surfaces with poorly supportive shoes or from having an imbalance in the size of opposing muscles.

Tendonitis

Tendonitis is the inflammation of a tendon and it is usually a result of overuse or repetitive use. However, it can sometimes be caused by a bacterial infection or result from another disorder of the musculo-skeletal system. It is characterised by inflammation, pain and tenderness when moved or touched. *Achilles tendonitis* is inflammation of the Achilles tendon and it is common in runners as it can be caused by excessive uphill or downhill running. It can also be caused by a number of different functional abnormalities including tight hamstrings and high arches. It is essential that the injury is rested and allowed to heal fully before exercise is resumed. *Tenosynovitis* is tendonitis accompanied by inflammation of the tendon sheath.

Tennis elbow (lateral epicondylitis)

Tennis elbow, also called lateral epicondylitis or backhand tennis elbow, is damage to the tendons of the lateral, or outer border, of the elbow. Symptoms include pain in the elbow and on the outer, back side of the forearm and it is often caused by improper backhand tennis techniques (hence the name), having weak shoulder and wrist muscles or repetitive extension of the wrist.

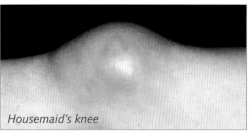
Housemaid's knee

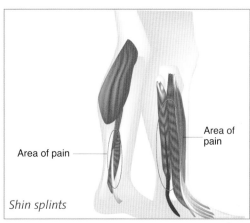
Area of pain
Area of pain
Shin splints

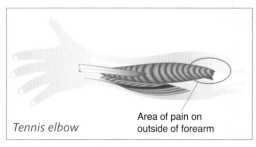
Area of pain on outside of forearm
Tennis elbow

NEW WORDS	
Action potential	an electrical charge that occurs on the membrane of a muscle cell in response to a nervous impulse.
Aerobic	requiring oxygen.
Agonist	the muscle responsible for causing a movement.
Anaerobic	not requiring oxygen.
Antagonist	the muscle that opposes the movement caused by the prime mover. It relaxes and lengthens in a controlled way to ensure the movement is performed smoothly by the prime mover.
Aponeurosis	a flat, sheet-like tendon that attaches muscles to bone, to skin or to another muscle.
Atony	the lack of muscle tone.
Atrophy	the wasting away of muscles.
Autorhythmic cells	muscle or nerve cells that generate an impulse without an external stimulus, i.e. they are self-excitable.

NEW WORDS CONT...

Conductivity	the ability of muscle cells to move action potentials along their plasma membranes.
Contractility	the ability of muscles to contract and shorten.
Depression (of the shoulders)	dropping the shoulders downwards.
Elasticity	the ability of muscles to return to their original shape after contracting or extending.
Elevation (of the shoulders)	lifting the shoulders upwards.
Excitability	the ability of muscle or nerve cells to respond to stimuli.
Extensibility	the ability of muscles to extend and lengthen.
Fascia	connective tissue that surrounds and protects organs, lines walls of the body, holds muscles together and separates muscles.
Fatigue (of muscles)	a muscle's inability to respond to stimulus or maintain contractions.
Fibrosis	the replacement of connective tissue by scar tissue.
Fixators	muscles that stabilise the bone of the prime mover's origin so that it can act efficiently.
Glycolysis	the cellular process through which glucose is split into pyruvic acid and ATP.
Hypertonia	an increase in muscle tone. Muscles are described as hypertonic.
Hypotonia	the loss of muscle tone. Muscles are described as hypotonic.
Insertion	the point where a muscle attaches to the moving bone of a joint.
Irritability	the ability of muscle or nerve cells to respond to stimuli.
Myoglobin	a protein that binds with oxygen and carries it to muscle cells.
Myology	the study of muscles.
Origin	the point where a muscle attaches to the stationary bone of a joint.
Prime mover	the muscle responsible for causing a movement.
Protraction	a forward movement of the shoulder or mandible on a plane parallel to the ground.
Retraction	the opposite of protraction. A backward movement on a plane parallel to the ground.
Stabilisers	muscles that stabilise the bone of the prime mover's origin so that it can act efficiently.
Striated	having the appearance of light and dark bands or striations.
Synergists	muscles that help the prime mover.
Tendon	a strong cord of dense connective tissue that attaches muscles to bones, to the skin or to other muscles.
Thermogenesis	the generation of heat in the body.
Tone (tonus)	the partial contraction of a resting muscle.

Study Outline

Functions of the muscular system

Functions of the muscular system include locomotion, maintenance of posture, movement of substances, regulation of organ volumes and heat production.

> **SHLOP**
> Substances moved, Heat produced, Locomotion, Organ volume controlled, Posture

Muscle tissue

Types of muscle tissue
1. There are three different types of muscle tissue: skeletal, cardiac and smooth (visceral).
2. Skeletal muscle is striated, voluntary and functions in moving the skeleton, maintaining posture and generating heat.
3. Cardiac muscle is striated, involuntary and functions in pumping blood around the body.
4. Smooth (visceral) muscle is non-striated, involuntary and functions in moving substances within the body and regulating organ volume.

> Suzie Violently Strips, Clair Innocently Strips, Sarah Innocently does Nothing.
>
> Skeletal, Voluntary, Striated – Cardiac, Involuntary, Striated – Smooth, Involuntary, Non-striated.

Structure of a skeletal muscle

Connective tissue: the outer protection of a muscle
1. Fascia separates muscles into different functional groups and supports the nerves, blood and lymphatic vessels that serve the muscles.
2. Muscles are surrounded, protected and reinforced by three types of connective tissue: epimysium, perimysium and endomysium.
3. The epimysium is the outermost layer of connective tissue that encircles the whole muscle.
4. The perimysium surrounds bundles of muscle fibres called fascicles.
5. The endomysium surrounds each individual muscle fibre within a fascicle.
6. All three types of connective tissue extend beyond the muscle to become tendons or aponeuroses. These join the muscles to the periosteum of bones or to the skin or other muscles.

Muscle fibres: the inner cells of a skeletal muscle

1. The cell of a muscle is called a fibre. Its membrane is the sarcolemma, its cytoplasm is the sarcoplasm and its endoplasmic reticulum is the sarcoplasmic reticulum.
2. Muscle fibres are multi-nucleated and also have many mitochondria.
3. Muscle fibres contain myofibrils which are the contractile elements of the muscle. They are made up of myofilaments and are arranged into sarcomeres.

Sarcomeres: the basic functional units of a skeletal muscle

1. A sarcomere contains the filaments that move to overlap one another and cause a muscle to shorten.
2. They contain three types of filaments: thick, thin and elastic filaments.
3. Thick filaments contain myosin.
4. Thin filaments contain actin, tropomyosin and troponin. They also contain the myosin-binding sites for myosin.
5. Elastic filaments contain titin (connectin) and help stabilise the position of the thick filaments.
6. Sarcomeres are made up of two bands that give muscles their striated appearance: the A-band and the I-band.
7. The A-band is dark and contains mainly thick filaments. In the centre of the A-band is the H-zone which contains thick filaments only. The H-zone is divided by an M-line of protein molecules that holds the thick filaments together.
8. The I-band is a light area containing thin filaments only.
9. Sarcomeres are separated from one another by Z-discs/lines.

Mechanisms of muscle contraction

How do muscles contract?

1. When a muscle is relaxed the myosin-binding sites on the actin molecules are covered by a tropomyosin-troponin complex and the myosin heads are in an energised state.
2. A nerve impulse triggers the release of acetycholine which triggers a muscle action potential which causes the release of calcium.
3. Calcium binds with the tropomyosin-troponin complex to free up the myosin-binding sites.
4. The myosin heads bind to actin with a power-stroke that draws the thin filaments inwards towards the H-zone. The thick filaments remain in the same place.
5. The muscle shortens (contracts).
6. The myosin heads detach from the actin and attach to another myosin-binding site further along the thin filament.
7. The cycle continues as long as ATP and calcium are present.

How do muscles relax?

1. Acetylcholine is broken down by an enzyme. This stops further muscle action potentials and therefore stops the release of calcium and leads to a decrease in calcium levels.

2. When there is not enough calcium available to bind with the tropomyosin-troponin complex it moves back over the myosin-binding sites and blocks the myosin heads from binding with the actin.
3. Thus, the thin filaments slip back into their relaxed position and no more contraction takes place.

Types of muscular contraction

1. Muscle tone is the constant, partial contraction of a muscle. It gives muscles their firmness and tension and is necessary for maintaining posture.
2. There are two types of muscle contraction: isotonic contractions and isometric contractions.
3. In isotonic contractions, muscles contract, shorten and create movement. Isotonic contractions include concentric contractions which are always towards the centre and involve the shortening of a muscle and eccentric contractions which are always away from the centre and involve the lengthening of a muscle.
4. In isometric contractions, muscles contract but there is no shortening of the muscle and no movement is generated.

Muscle metabolism

1. Muscles need energy in the form of ATP to contract. They obtain this through the phosphagen system or glycolysis.
2. In the phosphagen system, muscles use the small amount of ATP that they store in their own fibres.
3. In glycolysis, muscles break down glucose and convert it into pyruvic acid and ATP. The ATP is used by the muscles while the pyruvic acid (pyruvate) still needs to be broken down.
4. In anaerobic glycolysis, there is not enough oxygen to completely break down the pyruvic acid so it is converted into lactic acid.
5. In aerobic respiration, there is enough oxygen to completely break down the pyruvic acid into carbon dioxide, water, ATP and heat. This process is called cellular respiration or biological oxidation.
6. Muscle fatigue occurs when a muscle can no longer respond to stimulus or maintain its contractions.

Types of skeletal muscle fibres

1. There are three types of skeletal muscle fibres: slow oxidative, fast oxidative and fast glycolytic fibres.
2. Slow oxidative fibres are red, small, have a good oxygen supply, produce ATP aerobically and split it slowly. They are also very resistant to fatigue and are plentiful in muscles of endurance or those that maintain posture.
3. Fast oxidative fibres are red, medium sized, have a good oxygen supply, produce ATP aerobically and split it quickly. They are less resistant to fatigue than slow oxidative fibres and are plentiful in muscles used for walking and running.
4. Fast glycolytic fibres are white, large, have a poor oxygen supply, produce ATP anaerobically and split it very fast. They fatigue easily and are plentiful in muscles used for fast movements such as throwing a ball.

Skeletal muscles and movement

How skeletal muscles produce movement

1. A muscle is usually attached to two articulating bones.
2. The point where the muscle inserts in the stationary bone is called the origin. The point where it inserts in the moving bone is the insertion. When muscles contract they shorten and usually move the moving bone towards the stationary bone.
3. Muscles work in pairs: the prime mover (agonist) causes a movement and the antagonist opposes the movement by relaxing and lengthening in a controlled way to ensure the movement is performed smoothly by the prime mover.
4. Prime movers are supported by synergists and fixators.

> Olives Sell In Markets
>
> Origin stays Still, Insertion Moves

Muscles of the body

The photographs below and on the next two pages will help you to recognise some of the principle muscles of the body.

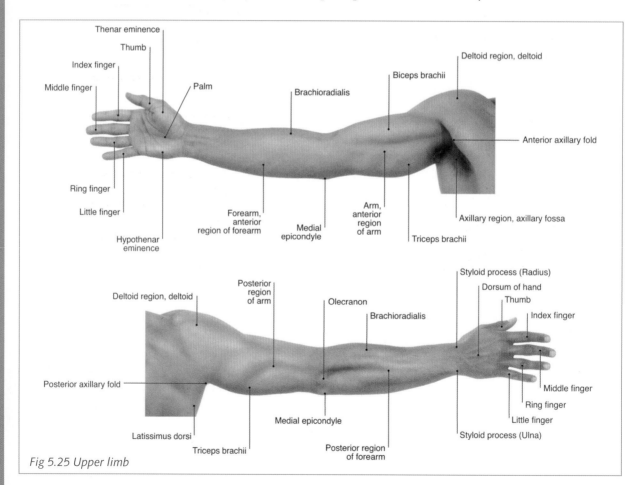

Fig 5.25 Upper limb

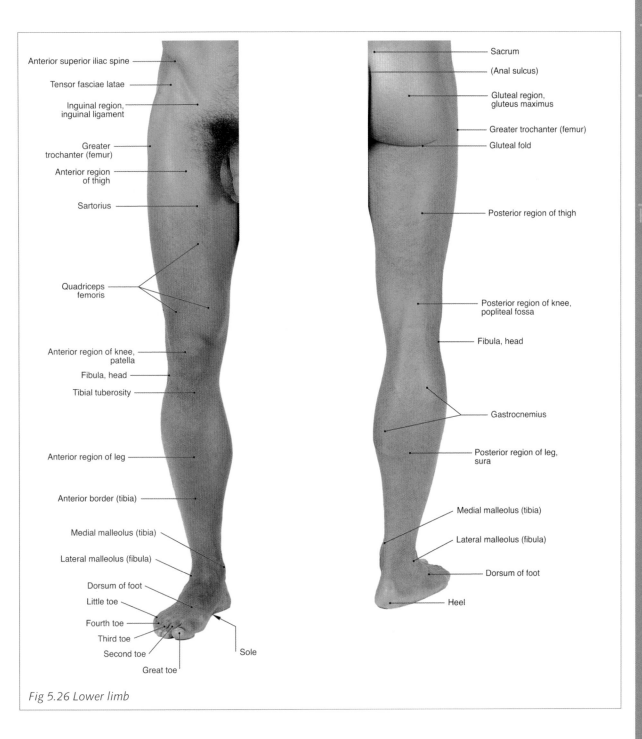

Anterior superior iliac spine

Tensor fasciae latae

Inguinal region,
inguinal ligament

Greater
trochanter (femur)

Anterior region
of thigh

Sartorius

Quadriceps
femoris

Anterior region of knee,
patella

Fibula, head

Tibial tuberosity

Anterior region of leg

Anterior border (tibia)

Medial malleolus (tibia)

Lateral malleolus (fibula)

Dorsum of foot

Little toe

Fourth toe

Third toe

Second toe

Great toe

Sole

Sacrum

(Anal sulcus)

Gluteal region,
gluteus maximus

Greater trochanter (femur)

Gluteal fold

Posterior region of thigh

Posterior region of knee,
popliteal fossa

Fibula, head

Gastrocnemius

Posterior region of leg,
sura

Medial malleolus (tibia)

Lateral malleolus (fibula)

Dorsum of foot

Heel

Fig 5.26 Lower limb

195

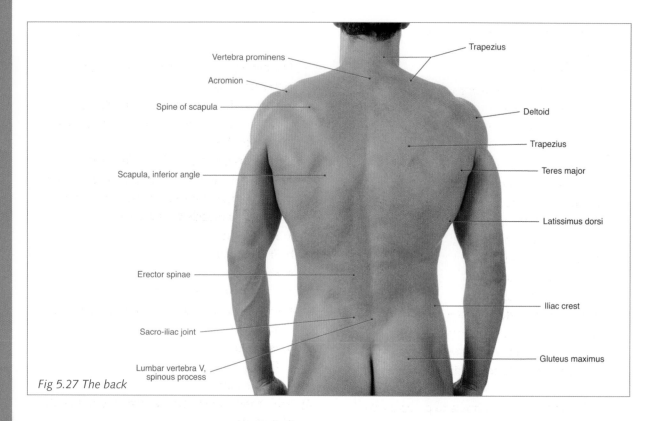

Fig 5.27 The back

Labels on figure:
- Vertebra prominens
- Acromion
- Spine of scapula
- Scapula, inferior angle
- Erector spinae
- Sacro-iliac joint
- Lumbar vertebra V, spinous process
- Trapezius
- Deltoid
- Trapezius
- Teres major
- Latissimus dorsi
- Iliac crest
- Gluteus maximus

Revision

1. Describe the functions of the muscular system.
2. Identify the three different types of muscular tissue.
3. Discuss the differences between the three types of muscular tissue.
4. Describe the different connective tissues associated with muscles.
5. What is a muscle fibre?
6. Describe a sarcomere and identify the different filaments and proteins within it.
7. Describe how a muscle contracts.
8. Describe how a muscle relaxes.
9. Define muscle tone.
10. Identify the two different types of muscular contraction.
11. Identify the differences between concentric and eccentric contractions.
12. Describe how muscles obtain their energy to contract.
13. Identify the difference between aerobic and anaerobic processes.
14. Describe how lactic acid is produced and what its effects on the muscles are.
15. Define muscle fatigue.
16. Describe the three types of muscle fibres.
17. Describe how muscles produce movements at joints.
18. Define the following terms: origin, insertion, prime mover, antagonist, synergist, fixator.
19. Write a list of at least five different muscle characteristics that are often reflected in a muscle's name. For example, a muscles origin and insertion can be reflected in its name (e.g. sternocleidomastoid).
20. List the characteristics of the following common pathologies:
    - Carpal tunnel syndrome
    - Rupture
    - Cramp.

Multiple choice questions

1. **Thermogenesis is:**
   a. The process by which glucose is broken down in the body.
   b. The process by which heat is generated in the body.
   c. The process by which organ volumes are controlled in the body.
   d. None of the above.

2. **Tendons attach:**
   a. Muscles to bones.
   b. Bones to bones.
   c. Aponeuroses to bones.
   d. Fascia to bones.

3. **Which of the following statements is correct?**
   a. Smooth muscle contracts to pump blood around the body.
   b. Skeletal muscle contracts to pump blood around the body.
   c. Cardiac muscle contracts to pump blood around the body.
   d. None of the above.

4. **Which of the following statements is correct?**
   a. Skeletal muscle is striated, voluntary and pumps blood around the body.
   b. All muscles are striated, involuntary and help move substances through the body.
   c. Cardiac muscle is non-striated, involuntary and helps regulate blood pressure.
   d. Smooth muscle is non-striated, involuntary and helps regulate organ volume.

5. **Elasticity is the ability of muscles to:**
   a. Respond to stimuli.
   b. Return to their original shape after contracting or extending.
   c. Extend or stretch.
   d. Shorten or thicken.

6. **The orbicularis oculi muscle functions in:**
   a. Opening the eye.
   b. Opening the mouth.
   c. Closing the eye.
   d. Closing the mouth.

7. **The perimysium is a connective tissue membrane that:**
   a. Surrounds each individual muscle fibre.
   b. Separates muscles into functional groups.
   c. Encircles the entire muscle.
   d. Surrounds bundles of muscle fibres.

8. **Which of the following structures contains the filaments that move to overlap one another and cause a muscle to shorten?**
   a. Sarcoplasm
   b. Sarcolemma
   c. Sarcomere
   d. Sarcoplasmic reticulum.

9. **The pterygoideus lateralis originates on the:**
   a. Sphenoid
   b. Humerus
   c. Tibia
   d. Triquetrum.

10. **Which mineral is necessary for the contraction of muscles?**
    a. Iron
    b. Calcium
    c. Zinc
    d. Boron.

11. **Muscle tone is necessary for:**
    a. Relaxing muscles.
    b. Contracting muscles.
    c. Maintaining posture.
    d. Moving substances around the body.

12. **Which of the following muscles form the rotator cuff:**
    a. Trapezius, rhomboids, levator scapulae.
    b. Biceps femoris, semitendinosus, semimembranosus.
    c. Rectus femoris, vastus lateralis, vastus medialis, vastus intermedius.
    d. Subscapularis, supraspinatus, infraspinatus, teres minor.

13. **Which of the following muscles form the quads:**
    a. Trapezius, rhomboids, levator scapulae.
    b. Biceps femoris, semitendinosus, semimembranosus.
    c. Rectus femoris, vastus lateralis, vastus medialis, vastus intermedius.
    d. Subscapularis, supraspinatus, infraspinatus, teres minor.

14. **Which of the following can cause muscle fatigue?**
    a. A build up of lactic acid.
    b. A sufficient supply of oxygen.
    c. A sufficient supply of glycogen.
    d. A lack of lactic acid.

15. **Which type of muscle fibre is plentiful in the muscles used for throwing a ball?**
    a. Slow oxidative
    b. Fast oxidative
    c. Fast glycolytic
    d. None of the above.

6 The Nervous System

Introduction

Air particles vibrate and we hear words, poetry, even music. Waves of light bend to form colours, shapes and images. Simple molecules are transformed into smells or tastes that remind us of our childhood – that can make us laugh or cry.

All of these sensations, thoughts and emotions are created in one system of our bodies – the nervous system. It is a system that gives us not only a sense of who, what and where we are, but also enables us to survive and to change the environment in which we live.

The study of the nervous system is called neurology and in this chapter we will take a look at how this system is able to control all the other systems of our body and help us control our environment.

Student objectives

By the end of this chapter you will be able to:

- Describe the functions of the nervous system.
- Describe the organisation of the nervous system.
- Identify the different types of neurones.
- Describe the structure of a motor neurone.
- Describe the structure and function of the brain.
- Identify the cranial nerves.
- Describe the structure and function of the spinal cord.
- Identify the spinal nerves and plexuses.
- Describe the structure and function of the sense organs: eyes, ears, nose and mouth.
- Identify the common pathologies of the nervous system.

Functions Of The Nervous System

The nervous system is made up of millions of nerve cells that all communicate with one another to control the body and maintain homeostasis. These cells detect what is happening both inside and outside of the body, interpret these happenings and cause a response. Basically, the nervous system has three functions: sensory, integrative and motor.

Anatomy and physiology in perspective
Your nervous system is not alone in controlling your body and maintaining homeostasis. It works closely with a second control system called the endocrine system which secretes hormones. One of the main differences between the two systems is that the nervous system can respond rapidly to stimuli and works faster than the endocrine system. You will learn more about the endocrine system in Chapter 7.

Sensory function

A stimulus is something that provokes a response, be it a change in the temperature of the air or a pin prick to a finger. *Sensory receptors* pick up stimuli that are taking place both inside and outside the body. All the information gathered by these sensory receptors is called *sensory input*.

Integrative function

Once the nervous system has new sensory input, it *analyses*, *processes* and *interprets* this information in the brain and spinal cord. It is also able to store the information and make decisions regarding it.

Motor function

Having picked up a change in the environment and decided what to do about it, the final function of the nervous system is the ability to *act* or *respond* to a stimulus by glandular secretions or muscular contractions. This is called *motor output*.

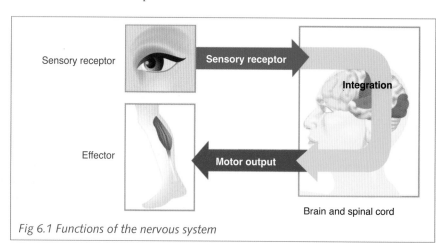

Sensory receptor

Sensory receptor

Integration

Effector

Motor output

Brain and spinal cord

Fig 6.1 Functions of the nervous system

Organisation Of The Nervous System

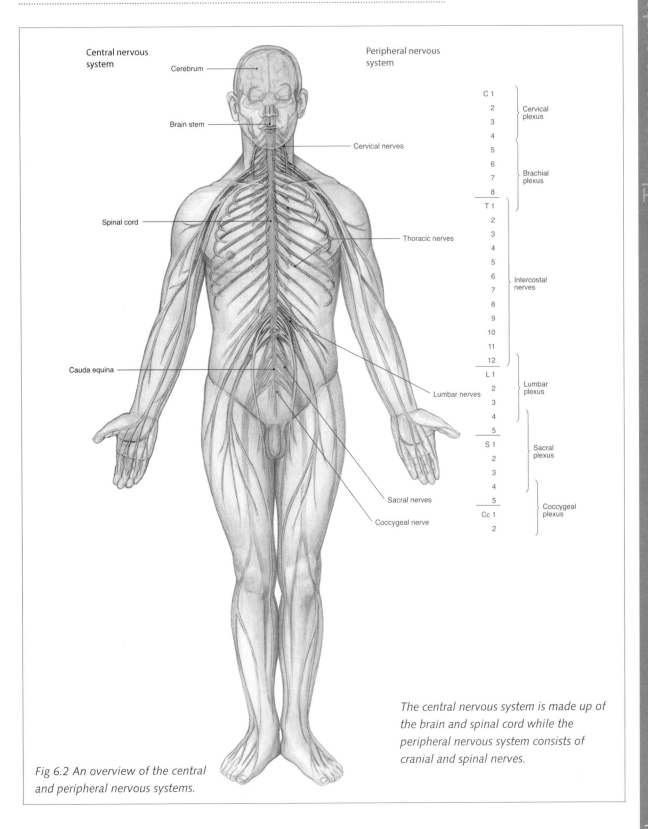

Central nervous system

Cerebrum

Brain stem

Spinal cord

Cauda equina

Peripheral nervous system

Cervical nerves

Thoracic nerves

Lumbar nerves

Sacral nerves

Coccygeal nerve

C 1
2
3
4

Cervical plexus

5
6
7
8

Brachial plexus

T 1
2
3
4
5
6
7
8
9
10
11
12

Intercostal nerves

L 1
2
3

Lumbar plexus

4
5
S 1
2
3

Sacral plexus

4
5
Cc 1
2

Coccygeal plexus

The central nervous system is made up of the brain and spinal cord while the peripheral nervous system consists of cranial and spinal nerves.

Fig 6.2 An overview of the central and peripheral nervous systems.

You may have heard people talking about the 'central nervous system' or the 'autonomic nervous system' or the 'parasympathetic nervous system' and be wondering what exactly is going on. In fact, you may be wondering exactly how many nervous systems you really have. The answer is simple, you only have one nervous system. But, because it is such a complex system, it is divided into a number of different parts. What is important here, before we go any further, is that you understand that all these different parts function together as one coordinated system.

Now let us look at all these divisions. To begin with, the nervous system is divided into the Central Nervous System (CNS) and the Peripheral Nervous System (PNS). The PNS is then subdivided into the somatic and autonomic nervous systems. The autonomic nervous system is then divided into the symphathetic and parasympathetic nervous systems.

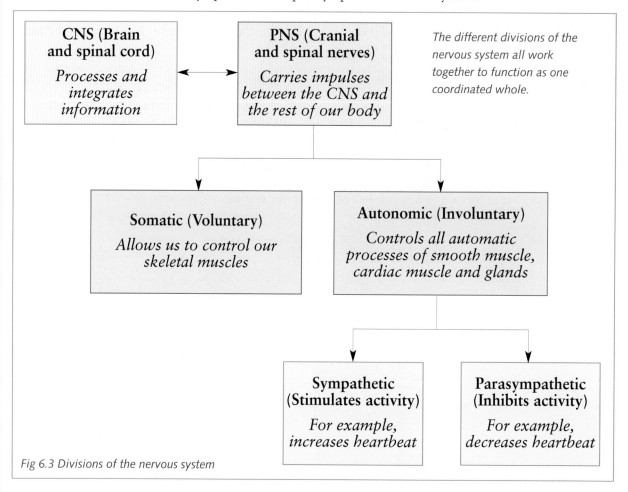

Fig 6.3 Divisions of the nervous system

Central Nervous System

The central nervous system (CNS) is made up of the brain and spinal cord. It functions in analysing and storing information, making decisions and issuing orders. It is where memories are made and stored, emotions generated and thoughts conceived. We will study the brain and spinal cord in detail later in this chapter.

Peripheral Nervous System

The peripheral nervous system (PNS) is made up of nerve cells that reach every part of the body and consists of:

- **Cranial nerves** – Cranial nerves arise from the brain and carry impulses to and from the brain.
- **Spinal nerves** – Spinal nerves emerge from the spinal cord to carry impulses to and from the rest of the body. Spinal nerves contain two types of nerve cells (neurones):
  - *Sensory neurones* (afferent neurones) – Sensory, or afferent, neurones conduct impulses from sensory receptors to the CNS.
  - *Motor neurones* (efferent neurones) – Motor, or efferent, neurones conduct impulses from the CNS to muscles and glands.

The PNS is subdivided into the Somatic Nervous System (SNS) and the Autonomic Nervous System (ANS).

Somatic nervous system

The word *soma* means 'body' and the somatic nervous system (SNS) allows us to control our skeletal muscles. Thus, it is sometimes called the Voluntary Nervous System. It contains:

- Sensory neurones that convey information from the *cutaneous* and *special sense receptors* to the CNS.
- Motor neurones that conduct impulses from the CNS to the *skeletal muscles* only.

We do not always control our skeletal muscles voluntarily. Sometimes they contract involuntarily through what is called a reflex arc.

Did you know?
Have you ever had a doctor (or even a friend) tap you just below your kneecap and your lower leg has jerked up? This is called the 'knee-jerk' and is a reflex-arc response to a stimulus. In other words, it is a response over which you have no control because no message is sent to the brain. Instead, the sensory nerve impulse is transmitted from the knee to the spinal cord where it is then immediately transmitted to motor neurones and back into the knee.

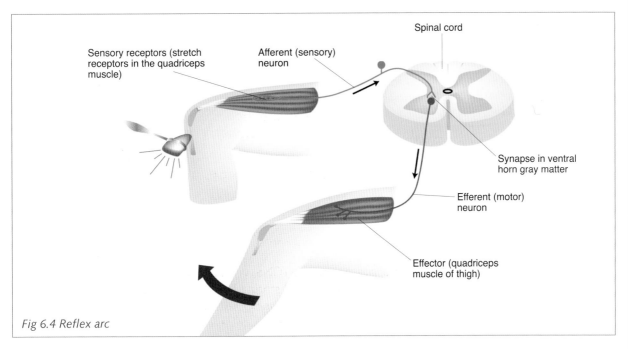

Fig 6.4 Reflex arc

203

Autonomic nervous system

The autonomic nervous system (ANS) controls all processes that are automatic or involuntary. Thus, it is sometimes called the Involuntary Nervous System. It contains:

- Sensory neurones that convey information *from the* viscera to the CNS.
- Motor neurones that convey information from the CNS to *smooth muscle*, *cardiac muscle* and *glands*. The motor portion of the autonomic nervous system has two branches that help the body adapt to changing circumstances. They work in opposition to one another and so are able to counterbalance each other to maintain homeostasis. They are the:
  - *Sympathetic nervous system* – The sympathetic nervous system reacts to changes in the environment by stimulating activity and therefore using energy. For example, the sympathetic nervous system increases the heart beat.
  - *Parasympathetic nervous system* – The parasympathetic nervous system opposes the actions of the sympathetic nervous system by inhibiting activity and therefore conserving energy. For example, the parasympathetic nervous system decreases the heart beat.

The chart opposite highlights some of the opposing effects of the sympathetic and parasympathetic nervous systems.

Nervous Tissue

The nervous system contains only two types of cells: neuroglia and neurones.

Neuroglia

Neuroglia, or glia, are smaller and more numerous than neurones and are the 'glue' or supporting cells of nervous tissue. They insulate, nurture and protect neurones and maintain homeostasis of the fluid surrounding neurones. There are different types of neuroglia, but they all have two things in common: they cannot transmit nerve impulses and they can divide by mitosis.

Anatomy and physiology in perspective
Brain tumours called gliomas are made out of neuroglia. This is because, unlike neurones, neuroglia can divide by mitosis.

Neurones

Neurones are the cells responsible for the sensory, integrative and motor functions of the nervous system. They can differ in size and shape, but they all share one essential characteristic: they transmit impulses or electrical signals to, from or within the brain.

Effects of the sympathetic and parasympathetic nervous systems

SYMPATHETIC STIMULATION	STRUCTURE	PARASYMPATHETIC
Pupil dilated	Iris muscle	Pupil constricted
Vasoconstriction	Blood vessels in head	No effect
Secretion inhibited	Salivary glands	Secretion increased
Rate and force of contraction increased	Heart	Rate and force of contraction decreased
Vasodilation	Coronary arteries	Vasoconstriction
Bronchodilation	Trachea and bronchi	Bronchoconstriction
Peristalsis reduced, sphincters closed	Stomach	Secretion of gastric juice increased
Glycogen to glucose conversion increased	Liver	Blood vessels dilated, secretion of bile increased
Adrenaline and noradrenaline secreted into blood	Adrenal medulla	No effect
Peristalsis reduced, sphincters closed	Large and small intestines	Secretions and peristalsis increased, sphincter relaxed
Smooth muscle wall relaxed, sphincter closed	Bladder	Smooth muscle wall contracted, sphincter relaxed

Types of neurones

Neurones can be functionally classified according to the type of information they carry and the direction in which they carry that information. The chart below shows this classification.

NEURONE	INFORMATION CARRIED	DIRECTION	
		FROM	TO
Sensory/ afferent	Sensory nerve impulse	Skin, sense organs, muscle, joints, viscera.	CNS
Motor/efferent	Motor nerve impulse	CNS	Muscles or glands (called effectors)
Association/ inter-neurones	These are not specifically sensory or motor neurones. Rather, they connect sensory and motor neurones in neural pathways.		

Did you know?

Neurones can vary from 1mm to 1m in length and can conduct impulses at speeds ranging from 1m to more than 100m per second.

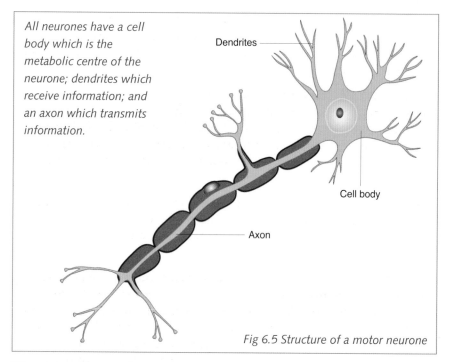

All neurones have a cell body which is the metabolic centre of the neurone; dendrites which receive information; and an axon which transmits information.

Dendrites

Cell body

Axon

Fig 6.5 Structure of a motor neurone

Structure of a motor neurone

All neurones have three parts: a cell body, dendrites and an axon. Here we will look at the structure of a *typical* motor neurone.

Cell body

The cell body is the metabolic centre of the neurone and is quite similar to a generalised animal cell in that it consists of a nucleus, cytoplasm and organelles such as mitochondria, Golgi complex and lysosomes. However, the cell body of a motor neurone does not include centrioles and therefore division via mitosis is not possible. The cell body also includes organelles that are specific to neurones only:

- **Nissl bodies** – these are made of rough endoplasmic reticulum and are the site of protein synthesis.
- **Neurofibrils** – these form the cytoskeleton of the cell to maintain its shape.

Anatomy and physiology in perspective
Most neurones cannot divide by mitosis so once they die they cannot be replaced. Thus, you are born with all your neurones and if they die through ageing, or are killed by head injuries, alcohol or drugs, the damage is permanent.

Dendrites
Dendrites are the receiving or input portion of the cell. The word dendro means 'tree' and dendrites are branching processes that project from the cell body. They are short and unmyelinated (this will be discussed shortly). Neurones can have many dendrites projecting from their cell body.

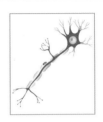

Axon
The axon is the transmitting portion of a cell. It transmits nerve impulses away from the cell body and towards another neurone, a muscle fibre or a gland cell. An axon looks like a long tail and the end of it divides into many fine processes called axon terminals.

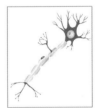

- **Axon terminals** – these are found at the end of the axon and contain membrane-enclosed sacs called *synaptic vesicles*.
- **Synaptic vesicles** – these sacs store chemical substances that influence other neurones, muscle fibres or gland cells. These substances are called neurotransmitters and they are released into an extracellular space called the *synaptic cleft*.

Some axons are myelinated. This means they are covered in a myelin sheath which protects and insulates the axon. This insulation speeds up the conduction of the nerve impulse. Myelin sheaths are created by neuroglia as follows:

- **Schwann cells** (*neurolemmocytes*) are neuroglia found in the PNS. They wrap themselves around small segments of a single axon so that the segment becomes completely enclosed by multiple layers of Schwann cell membrane. The cytoplasm and nucleus of the Schwann cells forms the outermost layer and is called the *neurolemma (neurilemma)*. The membrane of the Schwann cells forms the innermost layer and is known simply as the *myelin sheath*.
Between each Schwann cell enclosed segment is a small gap that is not covered. These gaps are called *nodes of Ranvier (neurofibril nodes)*. They are found at intervals along the sheath.
- **Oligodendrocytes** are neuroglia found in the CNS. They myelinate parts of many axons together and myelinated neurones in the CNS do not have a neurolemma.

Did you know?
The amount of myelin surrounding our neurones is responsible for the speed of the conduction of nerve impulses. It increases from birth to maturity and this is why infants respond more slowly to stimuli than adults do.

Anatomy and physiology in perspective

Multiple sclerosis (MS) is a disorder in which myelin sheaths are destroyed. This slows down the conduction of nerve impulses and can eventually lead to the short-circuiting of impulses. This is why symptoms of MS can include muscular weakness, abnormal sensations and double vision (MS is discussed in more detail in the pathology section at the end of this chapter).

Did you know?

Myelin is a whitish colour and it is what gives the white matter found in the brain and spinal cord its appearance. White matter consists of the myelinated processes of neurones while grey matter contains unmyelinated structures such as neuroglia, cell bodies, axon terminals and unmyelinated axons and dendrites.

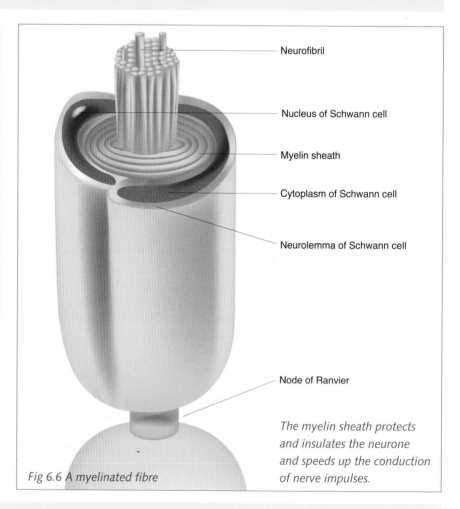

Neurofibril

Nucleus of Schwann cell

Myelin sheath

Cytoplasm of Schwann cell

Neurolemma of Schwann cell

Node of Ranvier

The myelin sheath protects and insulates the neurone and speeds up the conduction of nerve impulses.

Fig 6.6 A myelinated fibre

Anatomy and physiology in perspective

Although most neurones cannot divide and thus can never be replaced, the axons and dendrites of myelinated neurones in the PNS can be repaired as long as there has been no damage to the cell body. This is because the Schwann cells in the neurolemma form a tube that aids and guides regeneration of the axon and dendrites.

However, because there is no neurolemma in myelinated cells of the CNS, the axons and dendrites of these cells cannot be repaired. Thus, injury of the brain or spinal cord is usually permanent.

Transmission of a nerve impulse

A neurone has two main characteristics:

- Irritability is the ability to respond to a stimulus and convert it into an impulse.
- Conductivity is the ability to transmit an impulse from a neurone to another neurone, muscle or gland.

A neurone responds to a stimulus and converts it into an impulse via electrical means. It then conducts the impulse chemically to another neurone. Thus, nerve impulses are transmitted *electrochemically*. We will firstly look at the electrical transmission and then at the chemical transmission.

Electrical transmission across the plasma membrane

A nerve impulse is transmitted across the plasma membrane of an unmyelinated axon or across the nodes of Ranvier on a myelinated axon. For ease of learning, we will consider transmission across the plasma membrane only, but before we look at exactly what happens, here is some basic information that will help you understand the process:

- Ion – This is an electrically charged molecule.
- **Sodium ions (Na^+)** – These are the chief extracellular ions and are therefore usually found outside of the plasma membrane.
- **Potassium ions (K^+)** – These are the chief intracellular ions and are therefore usually found on the inside.

A nerve impulse is generated and propagated as follows:

- An inactive plasma membrane has a resting membrane potential that is *polarised*. This means that there is an electrical voltage difference across the membrane with the external voltage being positive and the internal one being negative. Remember, that the main external ions are sodium while the main internal ones are potassium.
- The dendrites of the neurone are *stimulated*.
- This stimulus causes ion channels in a small segment of the plasma membrane to open and allow the movement of sodium ions into the cell. This causes the inside of the cell to become positive and the outside negative. This is called *depolarisation*.
- Depolarisation causes the membrane potential to be reversed and this initiates an *action potential (impulse)*.
- When one segment of the membrane becomes depolarised, it causes the segment next to it to be depolarised and so a wave of depolarisation is propagated down the length of the plasma membrane.
- This is how the action potential travels to the end of the neurone from where it is chemically transmitted.

The speed of nerve transmission

The speed of propagation of an impulse is not related to the strength of the stimulus. It is proportional to the diameter of the fibre, the presence of myelin and the temperature. The larger the diameter, the more myelin present and the warmer the temperature, the faster the impulse will be propagated.

Did you know?
Have you ever wondered how local anaesthetics work? They block the sodium channels of the plasma membrane of an area. Therefore, nerve impulses cannot pass and the area anaesthetised is sensation-free.

This means that the fibres in our body that let us know about potential danger have the largest diameter and therefore the quickest rate of conduction (between 12 to 130 metres per second). These include fibres that relay impulses associated with touch, pressure, position of joints, heat, cold and skeletal muscles.

The fibres in our body that are related to the autonomic system and therefore the involuntary, automatic functioning of the body have the slowest rate of conduction (between 0.5 to 2 metres per second). They have the smallest diameters and are usually unmyelinated and include fibres that relay impulses to the heart, smooth muscles and glands.

Anatomy and physiology in perspective
Propagation of an impulse can be slowed down or even partially blocked by low temperatures. This is why applying ice to a painful area reduces the sensation of pain.

Chemical transmission across the synapse

Transmission between nerves can be either electrical or chemical. Electrical transmission is very fast and allows for the two-way transmission of an impulse across a gap junction. It takes place between neurones in the CNS, viscera, smooth muscle and cardiac muscles.

Chemical transmission, on the other hand, is slower and only allows for one-way transmission of an impulse. It occurs across a synapse which is a small space filled with extracellular fluid between the end of a neurone and another neurone, muscle or gland.

When the action potential reaches the end of the neurone it causes vesicles containing neurotransmitters to open up and release a neurotransmitter into the synaptic cleft.
- Two to three different neurotransmitters can be present in a single neurone. Neurotransmitters can also be excitatory or inhibitory and their action depends on the receptors to which they bind. Examples of neurotransmitters include:
  - Acetylcholine
  - Dopamine
  - Noradrenaline (norepinephrine) – this is both a neurotransmitter and a hormone.
- The neurotransmitter diffuses across the synaptic cleft and binds to the receptors of the next neurone, muscle or gland where it now acts as a stimulus.
- The neurotransmitter in the synaptic cleft is then quickly removed by diffusion, enzyme degradation or being actively transported back into the cells where it is recycled.

Did you know?
Synapses are the site of action for many drugs that affect the brain – these can be either therapeutic or addictive drugs.

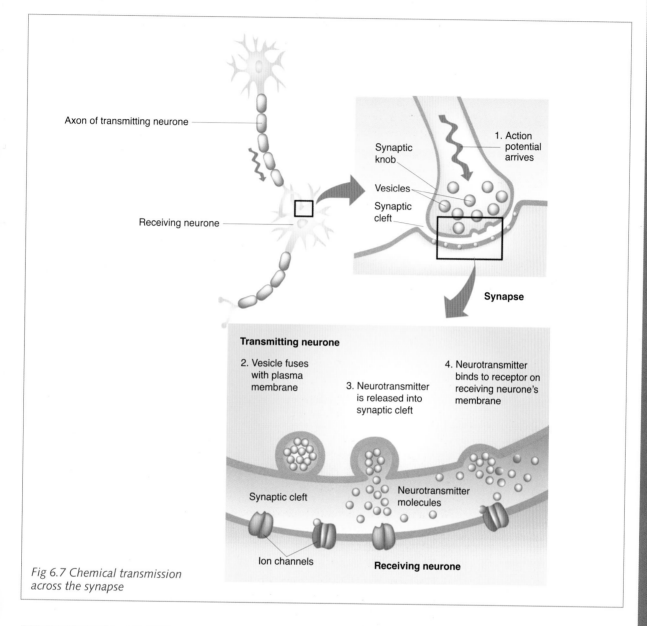

Fig 6.7 Chemical transmission across the synapse

Anatomy and physiology in perspective

The drug cocaine blocks the transporters that cause the removal of the neurotransmitter from the synaptic cleft. Therefore, dopamine remains in the synaptic cleft and causes an excess stimulation of certain regions of the brain. This results in what is described as a sense of euphoria.

Infobox

Nerves

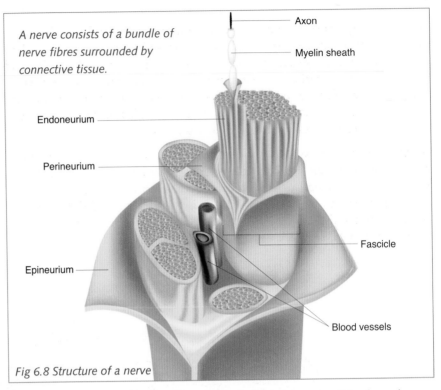

A nerve consists of a bundle of nerve fibres surrounded by connective tissue.

Axon

Myelin sheath

Endoneurium

Perineurium

Epineurium

Fascicle

Blood vessels

Fig 6.8 Structure of a nerve

A neurone is a single nerve cell consisting of dendrites, a cell body and an axon. A nerve is composed of the axons and dendrites of a group or bundle of neurones bound together by connective tissue. The structure of a nerve is very similar to that of a muscle. It consists of a bundle of nerve fibres surrounded by connective tissue. The:

- **endoneurium** surrounds each individual nerve fibre.
- **perineurium** surrounds groups of nerve fibres called fascicles.
- **epineurium** binds the fascicles together and is the outer covering of the nerve.

Nerves are classified into three different types:

- **Sensory or afferent nerves** – these contain sensory fibres that carry sensory impulses towards the CNS.
- **Motor or efferent nerves** – these contain motor fibres that carry motor impulses away from the CNS.
- **Mixed nerves** – these contain both sensory and motor fibres. All spinal nerves are mixed nerves.

Some important vocabulary associated with nerves includes:

- Nerve fibre – the term 'fibre' refers to the processes that project from a nerve body, these being dendrites and axons. However, the term fibre is usually used to refer to the axon and its sheath only.
- Ganglion (plural = ganglia) – a bundle or knot of nerve cell bodies.
- Tract – tracts are bundles of fibres that are not surrounded by connective tissue. They are found in the CNS and some of them interconnect regions of the brain.

The Brain

What is it that makes us so different from other mammals? Physically, we are generally one of the weaker mammals: our skin is thin and tears easily; our jaws are relatively small and weak; and our senses of smell, hearing and sight are not as powerful as those of many other mammals. However, we are able to light fires to warm us when we are cold, build homes to protect us from the environment and even create objects with which to kill animals that are stronger and faster than us. How is this so?

What makes humans so unique is our large brain which is approximately six times heavier than that of a dog of the same weight[i]. This brain gives us the ability to not simply adapt to the changes in our environment, but also to adapt our environment to suit us. This is what makes us different from other mammals: thanks to our incredible brains we can sense changes, think about them, make decisions and act on them.

An overview of the brain

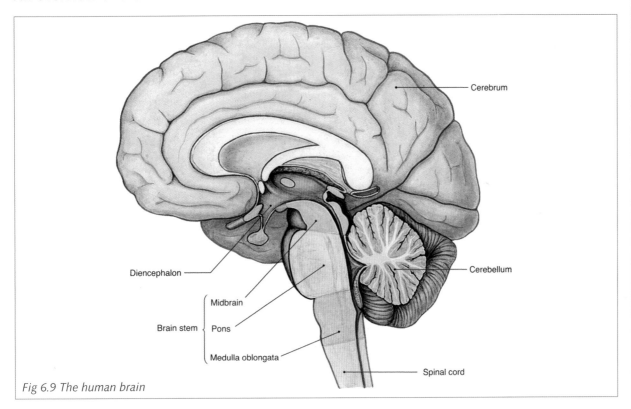

Fig 6.9 The human brain

An adult brain weighs approximately 1.3kg and consists of 100 billion neurones and 1000 billion neuroglia[ii]. It lies in the cranial cavity and is composed of four regions:

- **Brain stem** – this is continuous with the spine and consists of the:
  - Medulla oblongata
  - Pons
  - Midbrain

- **Cerebellum** – this cauliflower-shaped region is found at the back of the head, behind the brain stem.
- **Diencephalon** – this lies above the brain stem and includes the:
  - Epithalamus
  - Thalamus
  - Hypothalamus
- **Cerebrum** – this appears as a cap over the diencephalon and fills most of the cranium. It is divided into two halves, namely the right and left *cerebral hemispheres*. The outer, most superficial layer of the cerebrum is called the *cerebral cortex*. The inner layer consists of *white matter* and *basal ganglia*. The *limbic system* is located between the cerebrum and diencephalon.

Protection of the brain

The brain is protected by the scalp, skull, meninges and cerebrospinal fluid.

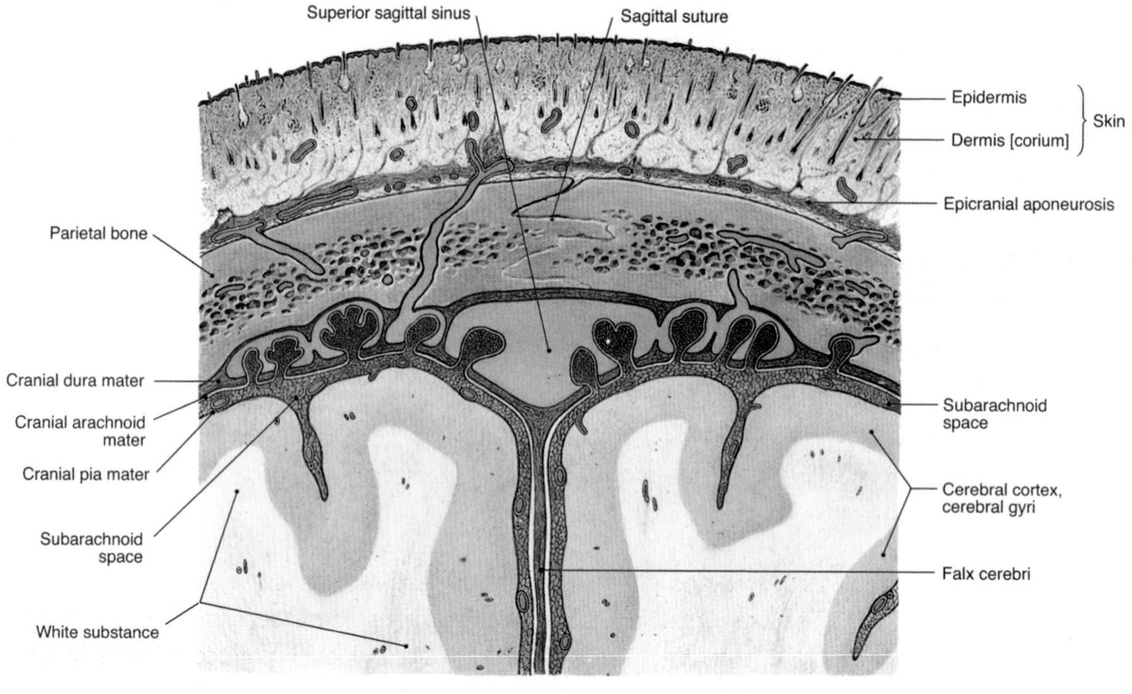

Fig 6.10 Protection of the brain

Did you know?
Meningitis is inflammation of the meninges.

The brain is a vital and sensitive organ and therefore needs maximum protection. The body provides this in a number of ways:

- The hard bones of the *cranium* form a nearly impenetrable wall against the external environment.
- Three layers of connective tissue cover the brain and provide further barriers to the external environment. These are called the meninges.
- The brain itself floats in a fluid called *cerebrospinal fluid*. This acts as a shock-absorber so that the brain does not knock against the hard walls of the cranium. It also provides a barrier against any substances trying to enter the brain from the internal environment.

Cranial meninges

Three layers of connective tissue encircle and protect the brain and spinal cord. They are called the cranial and spinal meninges and are continuous with one another, sharing the same basic structure and names. The meninges are:

- **Dura mater** – the outer covering of the brain, continuous with the periosteum of the cranial bones. It does not leave the skull apart from at the three following folds:
  - *Falx cerebri* – separates the two hemispheres of the cerebrum.
  - *Falx cerebelli* – separates the two hemispheres of the cerebellum.
  - *Tentorium cerebelli* – separates the cerebrum from the cerebellum.
- **Arachnoid** – the middle covering of the brain, separated from the other two meninges by spaces:
  - The *sub-dural space* lies between the dura mater and arachnoid.
  - The *sub-arachnoid space* lies between the arachnoid and pia mater and contains cerebrospinal fluid.
- **Pia mater** – the thin inner covering of the brain. It dips into all the folds and spaces of the brain tissue.

Cerebrospinal fluid (CSF)

This is a clear, colourless liquid that circles the CNS, protecting it and helping to maintain homeostasis. It contains glucose, proteins, lactic acid, urea, cations, anions and a few white blood cells.

Functions of CSF

Cerebrospinal fluid is crucial not only in protecting the CNS, but also in maintaining homeostasis:

- It acts as a *shock-absorber* for the brain and spinal cord.
- It provides the correct *chemical environment* in which neurones can function.
- It acts as a medium for the *exchange of nutrients and waste* between the blood and the nervous tissue and provides a constant supply of oxygen and glucose for nerve cells.

Did you know?
The adult brain comprises only 2% of the weight of the entire body but it uses 20% of the body's resting oxygen consumption. It therefore needs a constant supply of oxygen and even a short interruption of only one to two minutes can impair brain cells. Permanent damage of brain cells will result from no oxygen supply for four minutes.[iii]

Anatomy and physiology in perspective
The brain needs a constant supply of glucose because it is unable to store large amounts of it. If there is a low level of glucose in the blood to the brain then symptoms such as mental confusion, dizziness or even loss of consciousness can result.

Ventricles

CSF flows through and fills cavities in the brain called ventricles. There are four of them:

- **Left and right lateral ventricles** – these two lateral ventricles are found in the hemispheres of the cerebrum, either side of the midline.
- **The third ventricle** – this narrow ventricle is found at the midline below the lateral ventricles.
- **The fourth ventricle** – this ventricle lies below and behind the third ventricle and is found between the brain stem and the cerebellum.

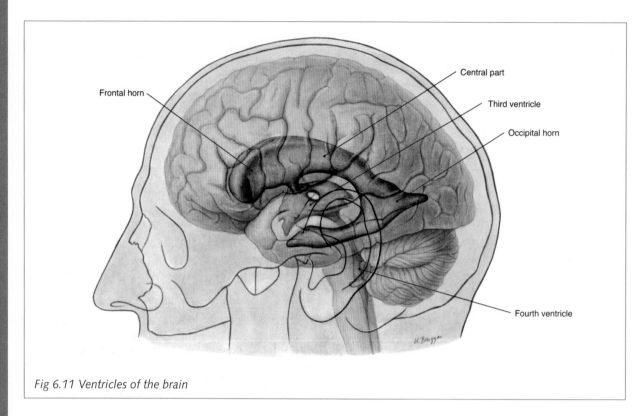

Frontal horn

Central part

Third ventricle

Occipital horn

Fourth ventricle

Fig 6.11 Ventricles of the brain

Did you know?
Although the function of the blood-brain barrier is to protect the brain from harmful substances, it does also keep out potentially beneficial or therapeutic substances such as drugs that could be used to treat brain tumours or disorders of the CNS.

Formation of CSF
CSF is formed from blood plasma at networks of capillaries on the walls of the ventricles. These networks are called the *choroid plexuses* and they contain specialised cells called *ependymal cells* which are found on the capillaries. These cells filter the blood plasma and remove any potentially harmful substances from it. Thus, the CNS is protected from many harmful substances that can be found in the blood and this barrier is referred to as the *blood-brain barrier*.

Absorption of CSF
CSF is formed at the choroid plexuses in the ventricles, circulates through the ventricles, into the subarachnoid space and it then circulates in the central canal of the spinal cord and the subarachnoid space around the brain and spinal cord. It is finally reabsorbed back into the blood through arachnoid villi. CSF is absorbed as quickly as it is formed and this ensures that its pressure remains constant.

Anatomy and physiology in perspective
The blood-brain barrier (BBB) is a selective barrier that protects the brain. However, some substances are helped across it by transporters such as lipids. Thus, lipid-soluble substances such as alcohol, caffeine, nicotine and anesthetics can pass directly into the brain cells.

Regions Of The Brain

The chart below and on the following pages takes a more detailed look at the regions of the brain.

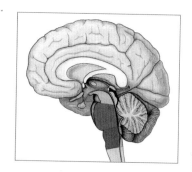

Brain stem

The brain stem is a continuation of the spinal cord and connects the spinal cord to the diencephalon. It consists of the medulla oblongata, pons and midbrain and its main function is to relay motor and sensory impulses between the spinal cord and the other parts of the brain. Running along the length of the brain stem is a mass of grey matter called the reticular formation.

REGION	DESCRIPTION	FUNCTIONS
Medulla oblongata (often referred to as the medulla)	Found at the top of the spinal cord. Approximately 3cm long.	The medulla oblongata contains: <ul><li>All sensory and motor white matter *tracts that connect the brain with the spinal cord*. Most of these tracts cross over from the left to the right and vice versa. Thus, the left side of the brain controls the muscles of the right side of the body and vice versa.</li><li>A cardiovascular centre which regulates the *heartbeat* and *diameter* of *blood vessels*.</li><li>A medullary rhythmicity area which regulates *breathing*.</li><li>Centres for the coordination of *swallowing, vomiting, coughing, sneezing* and *hiccupping*.</li><li>Nuclei of origin for *cranial nerves eight to twelve*. Cranial nerves will be discussed later in this chapter.</li><li>Neurones that function in *precise voluntary movements, posture and balance*.</li></ul>
Pons Varolii (pons)	Lies above the medulla oblongata and in front of the cerebellum. Approximately 2.5cm long.	*Pons* means 'bridge' and the pons acts as a bridge between the spinal cord and brain as well as between different parts of the brain itself. The pons also contains: <ul><li>Nuclei of origin for *cranial nerves five to eight*. (Note: cranial nerve VIII has its origins in both the medulla oblongata and the pons).</li><li>Areas which function with the medullary rhythmicity area to help control *respiration*.</li></ul>
Midbrain (mesencephalon)	Lies between the pons and the diencephalon. Is approximately 2.5cm long.	The midbrain contains white matter tracts and grey matter nuclei that function as: <ul><li>Reflex centres for *movements of the eyes, head and neck in response to visual and other stimuli*.</li><li>Reflex centres for *movements of the head and trunk in response to auditory stimuli*.</li><li>Areas that control *subconscious muscle activities*.</li><li>Areas that function with the basal ganglia and cerebellum to help coordinate *muscular movements*.</li><li>Nuclei of origin for *cranial nerves three and four*.</li></ul>
Reticular formation	This is a mass of grey matter that extends the entire length of the brain stem.	The reticular formation functions in: <ul><li>The *motor control of visceral organs* and in *muscle tone*.</li><li>*Consciousness and awakening from sleep*.</li></ul>

Anatomy and physiology in perspective

The brain stem plays a vital role in many activities and injuries to it can be fatal or, at the least, very serious. Injuries can result from a hard blow to the back of the head or upper neck. If the medulla oblongata is injured then paralysis, loss of sensation, irregular breathing or irregular heart functioning may occur. On the other hand, if the reticular formation is injured a coma may result.

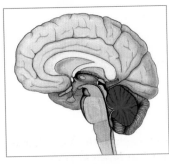

Cerebellum

The cerebellum is the second largest portion of the brain and it looks like a piece of cauliflower. It functions in producing smooth, coordinated movements as well as posture and balance and is what gives us the ability to do complex movements like somersaulting, dancing or throwing a ball.

REGION	DESCRIPTION	FUNCTIONS
Cerebellum	Found at the back of the head, behind the medulla oblongata and pons and beneath the cerebrum.	Its main functions include: • *Coordinating and smoothing complex sequences of skeletal muscular contraction*. It receives input from proprioceptors in muscles, tendons and joints as well as from receptors for equilibrium and visual receptors in the eye. It takes this information and compares the intended movement that has been programmed in the cerebrum with what is actually happening and so is able to smooth and coordinate movements. • *Regulating posture and balance*.

Anatomy and physiology in perspective

Damage to the cerebellum causes clumsy and uncoordinated movements and a person who has damaged it can sometimes appear drunk: they are unable to keep their balance or perform simple coordinated movements such as touching their finger to their nose with their eyes closed. The cerebellum can be damaged by a tumour, stroke or blow to the head.

Diencephalon (Interbrain)

The diencephalon is also called the interbrain and it lies above the brain stem where it is enclosed by the cerebral hemispheres. It contains the *epithalamus*, *thalamus* and *hypothalamus* and has a number of different functions including housing the pituitary and pineal endocrine glands.

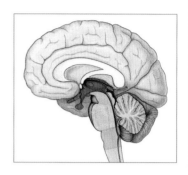

REGION	DESCRIPTION	FUNCTIONS
Epithalamus	Forms the roof of the third ventricle and is found above the thalamus.	The epithalamus contains: • The *pineal gland*, part of the endocrine system (see Chapter 7). • The *choroid plexus* which is the group of capillaries that forms cerebrospinal fluid.
Thalamus	The word *thalamus* means 'inner chamber'. It is approximately 3cm long and makes up about 80% of the diencephalon. It encloses the third ventricle of the brain.	Functions of the thalamus include: • It is the main relay station for sensory impulses to the cerebral cortex. Sensory impulses include those of hearing, vision, taste, touch, pressure, vibration, heat, cold and pain. • It contains an area for the crude appreciation of sensations such as pain, temperature and pressure before they are relayed to the cerebral cortex where the sensations are refined. • It contains nuclei that play a role in voluntary motor actions and arousal. • It contains nuclei for certain emotions and memory as well as for cognition which is the ability to acquire knowledge.
Hypothalamus	Found below the thalamus.	The hypothalamus is one of the main regulators of homeostasis. Sensory input from either the internal or external environment eventually comes to the hypothalamus. It also contains receptors that monitor osmotic pressure, hormone concentrations and blood temperature. In addition, the hypothalamus is connected to the endocrine system. Its functions include: • *Regulating the ANS* – By controlling the contraction of smooth and cardiac muscle and the secretions of many glands, the hypothalamus regulates visceral activities such as heart rate and the movement of food through the gastrointestinal tract. • *Controlling the pituitary gland*: – The hypothalamus releases hormones that control the secretions of the pituitary gland. – It also synthesises two hormones that are transported to, and stored in, the posterior pituitary gland until they are released. These are oxytocin and antidiuretic hormone. • *Regulating emotional behaviour* – Working together with the limbic system (to be discussed shortly), the hypothalamus functions in regulating emotional behaviour such as rage, aggression, pain, pleasure and sexual arousal. • *Regulating eating and drinking* – The hypothalamus controls sensations of hunger, fullness and thirst. • *Controlling body temperature*. • *Regulating sleeping patterns*.

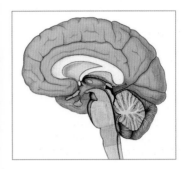

Cerebrum

The cerebrum sits like a large cap covering and basically obscuring the rest of the brain and it is often referred to as the 'seat of intelligence'. It is this area that gives us the ability to read, write, speak, remember, create and imagine.

The cerebrum's outer, most superficial layer is made up of ridges and grooves that look like deep wrinkles. This layer is called the *cerebral cortex* (the word *cortex* means 'rind') and it consists of grey matter. Beneath the grey matter of the cerebral cortex is an inner layer of white matter, the limbic system and the basal ganglia.

The ridges of the cortex are called *gyri* or *convolutions*, its shallow grooves are *sulci* and its deep grooves are fissures. A deep fissure called the *longitudinal fissure* separates the cerebrum into two halves: the right and left hemisphere. Although they appear to be separate, these hemispheres are still connected internally by the *corpus callosum*, which is a large bundle of white matter transverse fibres. Each hemisphere is subdivided into four lobes named after the bones that cover them: *frontal*, *parietal*, *temporal* and *occipital* lobes.

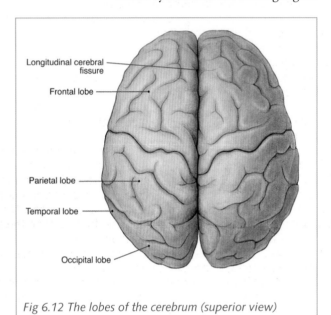

Fig 6.12 The lobes of the cerebrum (superior view)

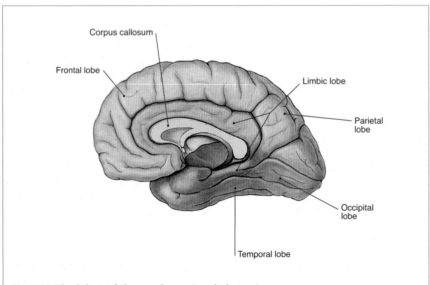

Fig 6.13 The lobes of the cerebrum (medial view)

The nervous system

REGION	DESCRIPTION	FUNCTIONS
Cerebral cortex	Approximately the size and shape of two closed fists held together. It consists of grey matter ridged and grooved and separated into two hemispheres.	The cerebral cortex is large and has many areas related to different functions. A simplistic view of it is that it receives and interprets almost all the sensory impulses of the body and interprets them into meaningful patterns of recognition and awareness. It has: • *Sensory areas* that receive and process information. These are mainly found in the posterior half of the hemispheres: – *General sensory area* for receiving impulses related to touch, proprioception, pain and temperature. The body can be mapped onto this sensory area and the exact area where the sensation is originating can be pinpointed on the area. – *Visual area* for receiving information regarding characteristics of visual stimuli such as shape, colour and movement. – *Auditory area* for receiving information regarding characteristics of sound such as rhythm and pitch. – *Gustatory area* for taste. – *Olfactory area* for smell. • *Motor areas* that output information. These are mainly found in the anterior portion of each hemisphere: – *Area for the voluntary contraction of specific muscles or muscle groups.* – *Speech area* that translates spoken or written words into thoughts and then into speech. • *Association areas* that consist of both sensory and motor areas. These are found mainly on the lateral surfaces of the cerebral cortex: – *Somatosensory association area* receives, integrates and interprets physical sensations. It also stores memories of past experiences for comparison with new information. This enables you to interpret complicated sensations such as being able to determine the shape and texture of an object without actually looking at it. – *Visual association area* enables you to relate present sensations to past experiences and therefore be able to recognise objects. – *Auditory association area* enables you to relate present sensations to past experiences and therefore be able to recognise a sound and determine whether it is speech, music or simply a noise. – *Gnostic area* integrates information and enables you to develop thoughts from a variety of sensory inputs such as smell, taste etc. – *Premotor area* enables you to perform complicated learned motor activities such as writing. – *Frontal eye field area* controls scanning movements of the eye enabling you to perform activities such as scanning a paragraph of writing for a specific word. – *Language areas* coordinate the muscles associated with speech and breathing so that you can speak.
White matter	Lies beneath the cerebral cortex.	The white matter consists of axons that transmit nerve impulses around the cortex and between the brain and spinal cord.
Limbic system	Found on the inner border of the cerebrum, the floor of the diencephalon and encircling the brain stem.	The limbic system is often called the 'emotional brain' because it controls the emotional and involuntary aspects of behaviour. It is the area associated with *pain, pleasure, anger, rage, fear, sorrow, sexual feelings* and *affection*. It also functions in *memory*.
Basal ganglia	These are groups of nuclei found in the cerebral hemispheres. They are interconnected by many nerve fibres.	The basal ganglia receive information from, and provide output to, the cerebral cortex, thalamus and hypothalamus. They control large *automatic movements of the skeletal muscles* and also help regulate muscle tone.

221

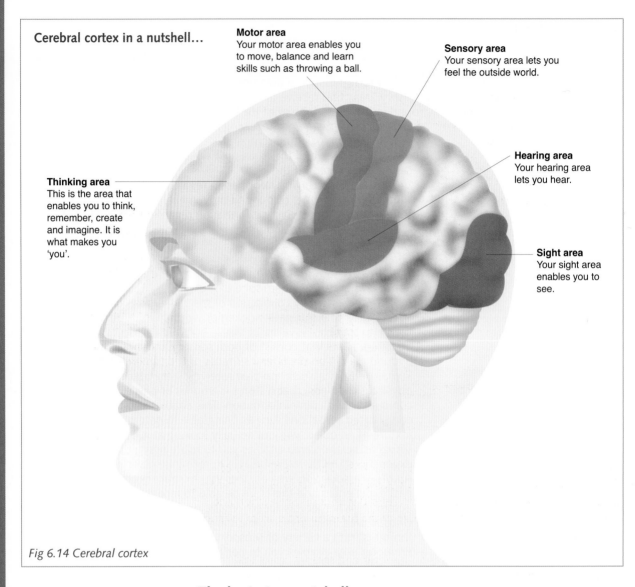

Cerebral cortex in a nutshell...

Motor area
Your motor area enables you to move, balance and learn skills such as throwing a ball.

Sensory area
Your sensory area lets you feel the outside world.

Hearing area
Your hearing area lets you hear.

Thinking area
This is the area that enables you to think, remember, create and imagine. It is what makes you 'you'.

Sight area
Your sight area enables you to see.

Fig 6.14 Cerebral cortex

The brain in a nutshell...

Note – this is a very simplified version of the brain.

Brainstem (medulla oblongata, pons, midbrain):
- Link between the brain and spinal cord
- Autonomic control

Cerebellum
- Movement, posture and balance

Diencephalon (epithalamus, thalamus, hypothalamus):
- Relays sensory impulses to the cerebral cortex
- Regulates homeostasis

Cerebrum (cerebral cortex/grey matter, white matter, limbic system, basal ganglia):
- Emotions and intelligence

Cranial Nerves

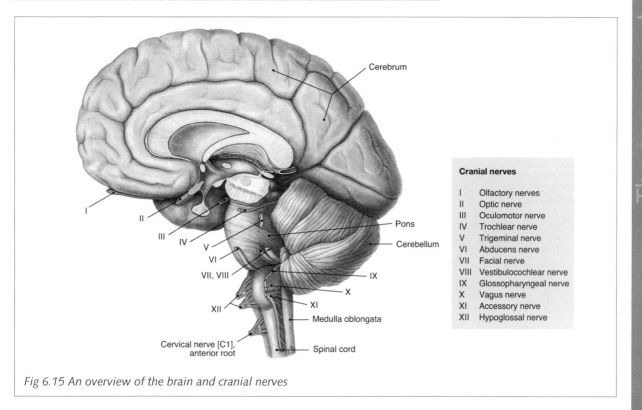

Cerebrum

Pons

Cerebellum

Cranial nerves

I	Olfactory nerves
II	Optic nerve
III	Oculomotor nerve
IV	Trochlear nerve
V	Trigeminal nerve
VI	Abducens nerve
VII	Facial nerve
VIII	Vestibulocochlear nerve
IX	Glossopharyngeal nerve
X	Vagus nerve
XI	Accessory nerve
XII	Hypoglossal nerve

I, II, III, IV, V, VI, VII, VIII, IX, X, XI, XII

Medulla oblongata

Cervical nerve [C1], anterior root

Spinal cord

Fig 6.15 An overview of the brain and cranial nerves

The cranial nerves are considered part of the PNS. There are 12 pairs of them, 10 which originate from the brain stem and 2 which originate inside the brain. They are named according to their distribution or function and numbered according to where they arise in the brain (in order from anterior to posterior).

Most of the cranial nerves contain both motor and sensory fibres and so are mixed nerves. Only three are purely sensory: the olfactory, optic and vestibulocochlear nerves.

The chart on page 225 describes the nerves.

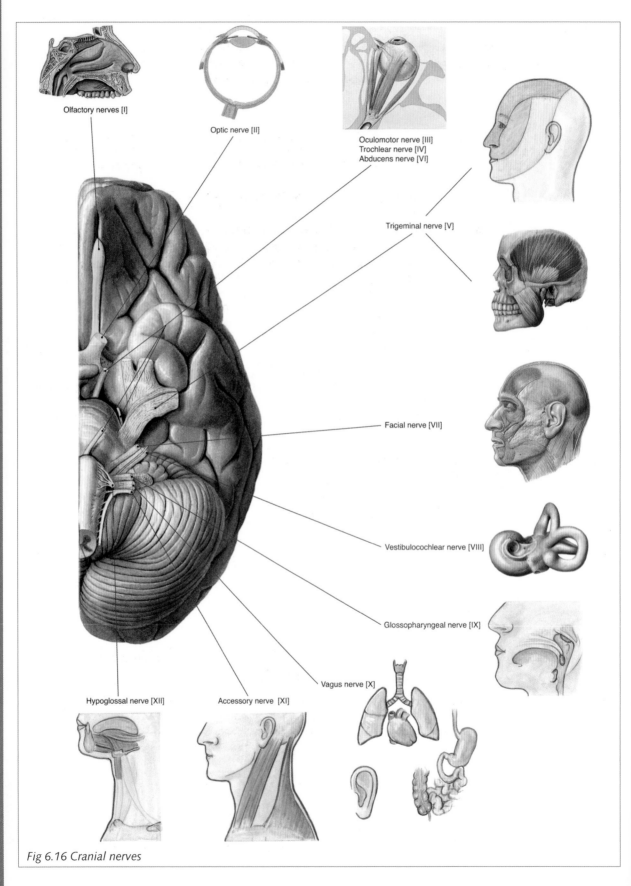

Fig 6.16 Cranial nerves

Olfactory nerves [I]

Optic nerve [II]

Oculomotor nerve [III]
Trochlear nerve [IV]
Abducens nerve [VI]

Trigeminal nerve [V]

Facial nerve [VII]

Vestibulocochlear nerve [VIII]

Glossopharyngeal nerve [IX]

Vagus nerve [X]

Hypoglossal nerve [XII]

Accessory nerve [XI]

Cranial nerves

NUMBER	NAME	FUNCTION
I	Olfactory	Smell
II	Optic	Vision
III	Oculomotor	**Motor function**: Movement of eyelid and eyeball; control of lens shape and pupil size. **Sensory function**: Proprioception in eyeball muscles.
IV	Trochlear	**Motor function**: Movement of eyeball. **Sensory function**: Proprioception in superior oblique muscle of the eyeball.
V	Trigeminal- has 3 branches: the ophthalmic, maxillary and mandibular branches.	**Motor function**: Chewing. **Sensory function**: Sensations of touch, pain and temperature from the skin of the face and the mucosa of the nose and mouth; and sensations supplied by proprioceptors in the muscles of mastication.
VI	Abducens	**Motor function**: Movement of eyeball. **Sensory function**: Proprioception in lateral rectus muscle of the eyeball.
VII	Facial – has 5 branches: the temporal, zygomatic, buccal, mandibular and cervical branches.	**Motor function**: Facial expression; secretion of saliva and tears. **Sensory function**: Proprioception in muscles of face and scalp; taste.
VIII	Vestibulocochlear – has 2 branches: the vestibular and cochlear branches.	**Vestibular branch**: Equilibrium. **Cochlear branch**: Hearing.
IX	Glossopharyngeal	Secretion of saliva.
X	Vagus	**Motor function**: Secretion of digestive fluids and contraction and relaxation of smooth muscle of organs of the thoracic and abdominal cavities. **Sensory function**: Sensory input from the lining of the organs of the thoracic and abdominal cavities.
XI	Accessory – has 2 portions: a cranial and spinal portion.	**Motor function of cranial portion**: Swallowing movements. **Motor function of spinal portion**: Movement of head. **Sensory function**: Proprioception.
XII	Hypoglossal	**Motor function**: Movement of tongue when talking and swallowing. **Sensory function** : Proprioception in tongue.

Spinal Cord

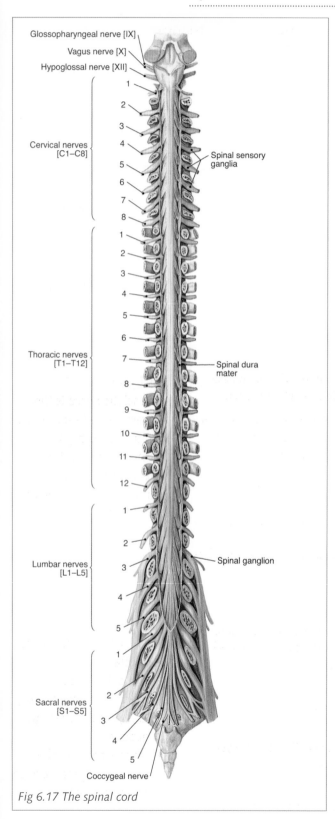

Glossopharyngeal nerve [IX]
Vagus nerve [X]
Hypoglossal nerve [XII]

Cervical nerves
[C1–C8]

Spinal sensory
ganglia

Thoracic nerves
[T1–T12]

Spinal dura
mater

Lumbar nerves
[L1–L5]

Spinal ganglion

Sacral nerves
[S1–S5]

Coccygeal nerve

Fig 6.17 The spinal cord

Approximately 42cm long and on average 2cm in diameter, the spinal cord is continuous with the brain stem and ends just above the second lumbar vertebra (L2).

Functions of the spinal cord

The spinal cord has two main functions that help maintain homeostasis:

1. The spinal cord transports *nerve impulses* from the periphery of the body to the brain and from the brain to the periphery.
2. The spinal cord receives and integrates information and produces *reflex actions* which are predictable, automatic responses to specific changes in the environment.

Anatomy of the spinal cord

Spinal meninges

Like the brain, the spinal cord is protected by the *spinal meninges*. These meninges are continuous with the cranial meninges and are the:

1. **Dura mater** – the outer covering of the spinal cord. Between it and the wall of the vertebral column is a space filled with fat and connective tissue. This is the *epidural space*.
2. **Arachnoid** – the middle covering and is separated from the other two meninges by:
   - The *sub-dural space* which is filled with interstitial fluid.
   - The *sub-arachnoid space* which contains CSF.
3. **Pia mater** – the thin, inner covering that adheres to the tissue of the spinal cord.

Cauda equina

In an adult the spinal cord extends from the medulla oblongata of the brain to the second lumbar vertebra only. However, some of the nerves that arise in the spine do not exit the vertebral column where they arise. Instead, they angle downwards in the canal and only exit the column lower down. These nerves look almost like wisps of hair coming off the end of the spinal cord and so the area is called the *cauda equina* which means 'horse's tail'.

Internal anatomy of the spinal cord

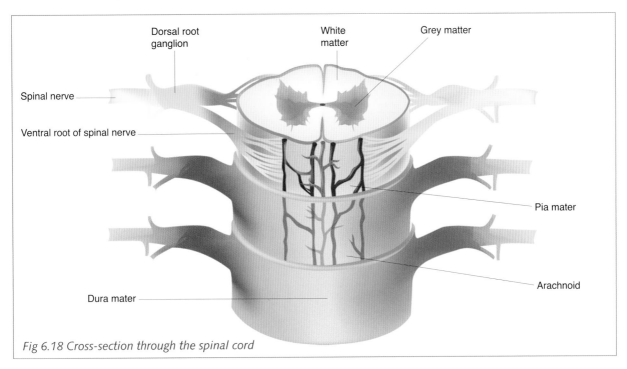

Dorsal root ganglion — White matter — Grey matter — Spinal nerve — Ventral root of spinal nerve — Pia mater — Arachnoid — Dura mater

Fig 6.18 Cross-section through the spinal cord

If you take a cross-section of the spinal cord, you will see that it consists of grey H-shaped (butterfly-shaped) matter surrounded by white matter.

- The *grey matter* receives and integrates information and consists of cell bodies and unmyelinated axons and dendrites of association and motor neurones. Grey matter also contains nuclei where some nerve impulses are processed.
- The *white matter* contains tracts of myelinated fibres that transport impulses between the brain and the periphery.
  - Ascending tracts consist of sensory axons that conduct nerve impulses to the brain.
  - Descending tracts consist of motor axons that conduct nerve impulses to the body.

Spinal nerves

There are 31 pairs of nerves that originate in the spinal cord and emerge through the intervertebral foramina. These nerves form part of the PNS and connect the CNS to receptors in the muscles and glands.

Roots

Each nerve has two points of attachment to the spinal cord. These attachments are called roots and are the:

- Posterior or dorsal root consisting of sensory axons. The dorsal root contains a ganglion of sensory neurone cell bodies.
- Anterior or ventral root consisting of motor axons.

Because spinal nerves have both sensory and motor axons they are classified as mixed nerves.

Did you know?
The sciatic nerve is the largest nerve in the body. It reaches from your lumbar spine right down into your feet and contains nerve fibres up to 1m long.

The nervous system in a nutshell
- The brain is the control centre.
- The spinal cord links the brain to the body and processes reflex actions.
- Nerves connect the brain and spinal cord to the rest of the body.

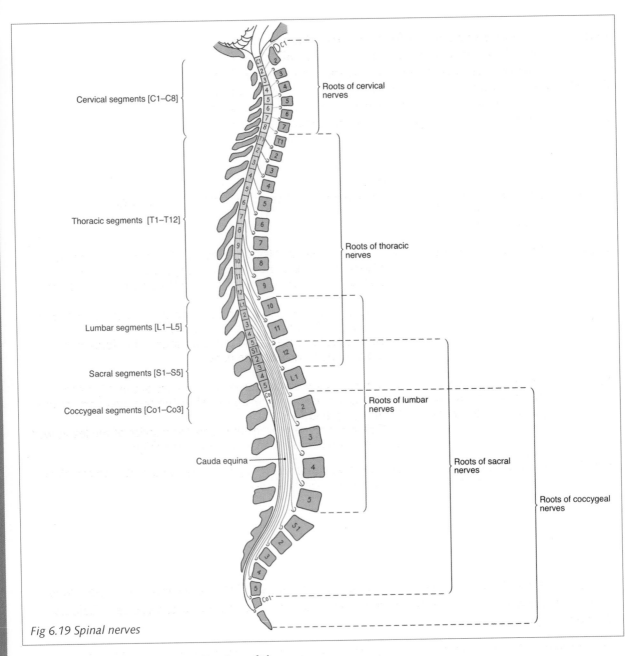

Cervical segments [C1–C8]

Thoracic segments [T1–T12]

Lumbar segments [L1–L5]

Sacral segments [S1–S5]

Coccygeal segments [Co1–Co3]

Cauda equina

Roots of cervical nerves

Roots of thoracic nerves

Roots of lumbar nerves

Roots of sacral nerves

Roots of coccygeal nerves

Fig 6.19 Spinal nerves

Naming of the nerves

The spinal nerves are named and numbered according to where in the vertebral column they emerge. The first spinal nerve starts between the occipital and the first cervical vertebra. There are:

- Eight pairs of cervical nerves
- Twelve pairs of thoracic nerves
- Five pairs of lumbar nerves
- Five pairs of sacral nerves
- One pair of coccygeal nerves.

Note that the lower lumbar, sacral and coccygeal roots are not in line with their origins or corresponding vertebrae because they descend in the form of the cauda equina.

Spinal plexuses

Branches of some of the spinal nerves form networks of nerves on both sides of the body. These networks are called plexuses and all the nerves emerging from a particular plexus will innervate specific structures. The table below discusses these plexuses in more detail. NB nerves T2–T12 do not form a plexus. These are the *intercostal nerves* that serve the muscles between the ribs and the skin and the muscles of the anterior and lateral trunk.

PLEXUS	BODY AREAS SERVED	IMPORTANT NERVES
Cervical	Skin and muscles of the head, neck and top of the shoulders.	**Phrenic nerve** supplies motor fibres to the diaphragm.
Brachial	Shoulder and upper limb.	• **Axillary nerve** supplies the deltoid and teres minor muscles. • **Musculocutaneous nerve** supplies the flexors of the arm. • **Radial nerve** supplies the muscles on the posterior aspect of the arm and forearm. • **Median nerve** supplies the muscles on the anterior aspect of the forearm and some muscles of the hand. • **Ulnar nerve** supplies some of the muscles of the forearm and most of the muscles of the hand.
Lumbar	Abdominal wall, external genitals and part of the lower limb.	• **Femoral nerve** supplies the flexor muscles of the thigh and extensor muscles of the leg as well as the skin over parts of the thigh, leg and foot. • **Obturator nerve** supplies the adductor muscles of the leg as well as the skin over the medial aspect of the thigh.
Sacral	Buttocks, perineum and lower limbs.	**Sciatic nerve** descends through the thigh and splits into the tibial and common peroneal nerves.

Special Sense Organs

In the classroom

Take a moment to think of a place you have been to where you felt safe, relaxed and happy. Now, close your eyes and imagine yourself in that place. Look around you. What can you see? Colours? Shades? Shapes? Movement?

How do you feel there? Warm or cold? Is there a gentle breeze tickling your skin or is the warm sun filtering through you? What can you smell? The sea? Freshly mown lawn? Food cooking? Someone's perfume?

Can you hear anything? Someone talking? Laughter? Singing? Music? The wind in the trees? Water lapping the shore? Sit quietly in this space for a few minutes and enjoy it. Now, how do you feel?

Hopefully this exercise demonstrates how special and powerful our senses are. They bring the world around us to life and connect with our deepest memories. Can you imagine living without them?

The Eyes

Our eyes are one of the brain's vital contacts with the outside world. They enable us to know where we are, what is around us and where we are going. The eyes are thought to contain over a million nerve fibres and more than 70% of the body's total sensory receptors. The information they gather for us is not simply one dimensional – it is far deeper than that – giving us a sense of space, colour, shape, texture and movement.

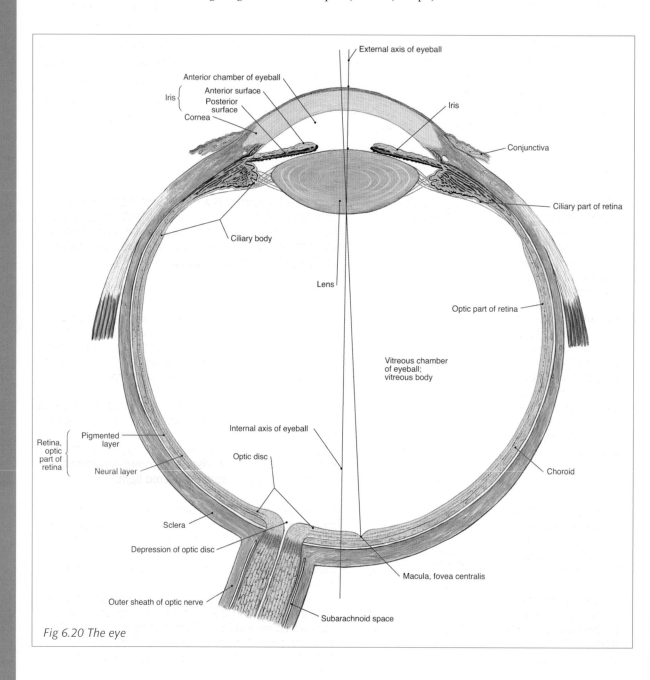

Fig 6.20 The eye

Accessory structures

Surrounding the eyes are the eyelids, eyebrows, eyelashes and eye muscles. All of these work together to protect the eyes from foreign objects, the sun's rays and perspiration. In addition, are *lacrimal glands* that release tears onto the surface of the eye to clean and lubricate the eyeball. The tears are then spread by blinking. Lining the eyelids and covering part of the outer surface of the eyeball is a very thin mucous membrane that helps protect the eyeball. It is called the *conjunctiva*.

Anatomy of the eyeball

The eyeball is approximately 2½ cm in diameter. However, we only see ⅙th of it as the rest is protected by the orbit into which it fits. The eyeball consists of a strong, protective wall and a large inner space that is divided into cavities that are filled with different substances. The wall of the eyeball is divided into three layers: the outer fibrous tunic, the middle vascular tunic and the inner nervous tunic (the retina).

Did you know?

Just as it is the pigments in your skin and hair that determine their colour, so it is the pigments in your iris that determine the colour of your eye. Eyes contain the pigment melanin. People with large amounts of melanin have brown eyes. People with smaller amounts have lighter eyes such as green, grey or blue.

EXTERIOR OF THE EYEBALL		
STRUCTURE	**DIAGRAM**	**DETAILS**
Fibrous tunic – the outermost covering of the eyeball. The front of the fibrous tunic is the cornea and the back of it is the sclera.		**Cornea** – the cornea is an avascular, transparent coat that covers the iris. It is curved and helps focus light. **Sclera** – the sclera is the white of the eye and is made up of dense connective tissue. It protects the eyeball and gives it shape and rigidity.
Vascular tunic – the middle layer of the wall of the eyeball. The front of the vascular tunic is the coloured iris which is surrounded by the ciliary body which then becomes the choroid.		**Iris** – suspended between the cornea and lens and is the coloured portion of the eye. It is shaped like a flattened doughnut and the hole in the centre of it is the *pupil* which regulates the amount of light entering the eyeball. The iris consists of circular and radial smooth muscle fibres. **Ciliary body** – contains processes that secrete aqueous humour and muscles that alter the shape of the lens for near or far vision. **Choroid** – lines most of the internal surface of the sclera, is highly vascularised and provides nutrients to the retina. It also absorbs scattered light.
Nervous tunic (The retina) – the innermost layer of the wall of the eyeball. It lines the posterior ¾ of the eyeball. It consists of a non-visual pigmented portion and a neural portion. The pigmented portion lies between the choroid and the neural portion.		**Pigmented portion** – contains melanin and it absorbs stray light rays. It therefore prevents any scattering or reflection of light within the eyeball and ensures a clear, sharp image. **Neural portion** – contains three layers of neurones that process visual input. In these layers are the *photoreceptors* which are specialised cells that convert light into nerve impulses. These photoreceptors are *rods* which respond to different shades of grey only and *cones* which respond to colour. From the photoreceptors, the information passes through bipolar cells into ganglion cells. The axons of the ganglion cells then extend into the *optic nerve*.

The interior of the eyeball consists of a large space that is divided into two cavities by the lens.

INTERIOR OF THE EYEBALL		
STRUCTURE	**DIAGRAM**	**DETAILS**
Lens – behind the coloured iris and the dark pupil is the transparent, avascular lens.		The lens is responsible for fine tuning of focusing and it is made up of protein.
Anterior cavity – in front of the lens and filled with aqueous humour.		**Aqueous humour** – a watery fluid that nourishes the lens and cornea and helps produce intraocular pressure. The aqueous humour is constantly replaced.
Posterior cavity (vitreous chamber) – the largest cavity in the eyeball and lies behind the lens. It contains the vitreous body.		**Vitreous body** – a jelly-like substance that helps produce intraocular pressure which helps maintain the shape of the eyeball and keeps the retina pressed firmly against the choroid. Unlike the aqueous humour, the vitreous humour is not constantly replaced.

Did you know?

The cones in our eyes only absorb three different wavelengths of incoming light: red, green and blue. These colours are mixed in the brain to produce over 10,000 different colours or shades.

How do we see?

If light passes through substances of different density its speed will change and its rays will bend. This is called refraction and it takes place as light passes through the differing densities of the cornea, aqueous humour, lens and vitreous humour. By the time the light rays reach the retina they have bent to such an extent that the image is reversed from left to right and turned upside down. The photoreceptors on the retina then convert the light into nerve impulses which are transported by the optic nerve to the visual cortex of the brain. Here the impulses are integrated and interpreted into visual images.

Anatomy and physiology in perspective

Excessive intraocular pressure is called *glaucoma*. It can eventually lead to degeneration of the retina and consequent blindness.

The Ears

Our ears do not only enable us to hear and locate a sound, they also help us to balance and thus have a sense of equilibrium.

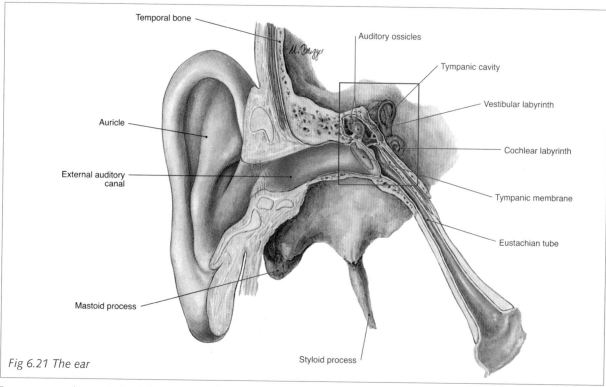

Fig 6.21 The ear

Just as we only see ⅙th of the eye, we also only see a portion of the ear. What we see is only part of the external or outer ear, the rest of the ear extends to just below and behind the eye and includes the middle ear and the internal or inner ear. The chart below and on the next page gives a brief overview of the anatomy of the ear.

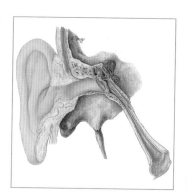

STRUCTURE OF THE OUTER EAR	
The outer, or external, ear collects and channels sound waves inwards. It is composed of the auricle, external auditory canal and eardrum. It also contains *ceruminous glands* which secrete *cerumen* (earwax) which, along with hairs that are also found in the outer ear, cleans and protects the ear and prevents dust and foreign objects from entering it.	
Auricle (pinna)	The part of the ear we see. It is composed of elastic cartilage covered by skin. The upper rim is called the *helix* while the lower lobe is called the *lobule*.
External auditory canal (meatus)	A curved tube approximately 2.5cm long. It carries sound waves from the auricle to the eardrum.
Eardrum (tympanic membrane)	A very thin, semitransparent membrane between the auditory canal and the middle ear. When sound waves hit it, it vibrates, passing them on to the middle ear.

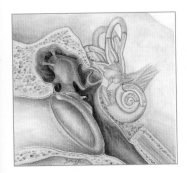

STRUCTURE OF THE MIDDLE EAR

The middle ear is a small air-filled cavity found between the outer ear and the inner ear. It is partitioned from the outer ear by the eardrum and from the inner ear by a bony partition containing two windows: the *oval* window and the *round* window. It contains the three auditory ossicles and is connected to the throat via the Eustachian tube.

Auditory ossicles	Three tiny bones extend across the middle ear. They are the auditory ossicles and each one is named after its shape: • Hammer (malleus) • Anvil (incus) • Stirrup (stapes) The handle of the hammer is attached to the inner surface of the eardrum and when the eardrum vibrates it causes the hammer to move. The hammer hits the anvil which in turn hits the stirrup which is attached to oval window, a membrane covered opening which then transmits the sound wave into the inner ear.
Eustachian tube (auditory tube)	The Eustachian tube connects the middle ear with the upper portion of the throat (nasopharynx) and it functions in equalising the middle ear cavity pressure with the external atmospheric pressure.

STRUCTURE OF THE INNER EAR (LABYRINTH)

The inner, or internal, ear is sometimes also called the labyrinth. It consists of a bony labyrinth filled with a fluid called *perilymph* enclosing a membranous labyrinth filled with a fluid called *endolymph*. This labyrinth is divided into the *vestibule*, *three semicircular canals* and the *cochlea*. Inside the cochlea lies the actual organ of hearing itself: the *organ of Corti*.

Vestibule	The central portion of the bony labyrinth and contains receptors for equilibrium.
Semicircular canals	These project from the vestibule and also contain receptors for equilibrium.
Cochlea	The cochlea is a bony, spiral canal that resembles a snail's shell. It is divided into three channels, one of which is the cochlear duct which is separated from another channel by the *basilar membrane*. Resting on the basilar membrane is the organ of hearing which is called the organ of Corti or the spiral organ.
Organ of Corti	The organ of Corti is a coiled sheet of epithelial cells containing thousands of hair cells. Extensions from these hair cells extend into the endolymph of the cochlear duct and are the receptors for auditory sensations. Over the hair cells is a very thin, flexible membrane called the *tectorial membrane*.

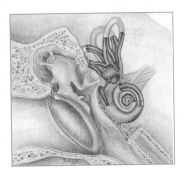

Did you know?

The stirrup (stapes) is the smallest bone in the body and is no longer than a grain of rice. Yet, despite its tiny size, it is essential to our ability to hear. Without this rice-sized bone sound waves could not be transmitted to the inner ear.

Anatomy and physiology in perspective
Have you ever flown in an aeroplane and found your ears begin to ache or you feel as if you have a bubble in your ear and all sounds seem to be coming from far away? This is because the pressure within your middle ear is unequal to the external pressure and so your eardrum is bulging inwards or outwards. To normalise the pressure you can either swallow or cover your mouth and nose and try to blow out at the same time.

How do we hear?

What happens when you drop a stone into a pond of water? It causes ripples to travel over the surface of the water. Sound travels in a similar way. An object vibrates and creates an alternating compression and decompression of air molecules. These are sound waves.

Sound waves travel through the air from their source to the auricle of our ears. The auricle then directs the sound waves into the auditory canal where they are channeled towards the eardrum. Once they hit the eardrum they cause it to vibrate and this sets up a chain reaction of vibrations that eventually causes changes in the pressure of the endolymph in the cochlear duct. As this pressure rises and falls it moves the basilar membrane which vibrates and causes the hair cells of the organ of Corti to move against the tectorial membrane.

The hair cells are mechanoreceptors and they convert this mechanical stimulus into an electrical one. The hair cells synapse with neurones of cranial nerve VIII, the vestibulocochlear nerve which sends the impulses to the medulla oblongata. The fibres then extend into the thalamus from where the auditory signals are projected to the primary auditory area of the temporal lobe of the cerebral cortex.

Did you know?
If you are listening to music through earphones and it is loud enough to be heard by the person standing next to you, then you are damaging the hair cells of your inner ear. This can lead to progressive hearing loss which is often not noticed until extensive damage has been done.

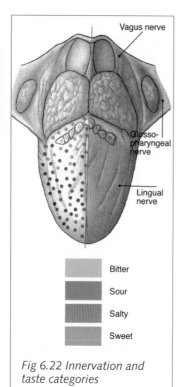

Fig 6.22 Innervation and taste categories

The Mouth

The mouth houses the receptors of taste. These are called gustatory receptors and are found in taste buds which are located mainly on the tongue, the back of the roof of the mouth and in the pharynx and larynx.

A *taste bud* is an oval body that contains *gustatory receptor cells*. Each of these cells has a single hair-like projection that projects from the receptor cell and through a small opening in the taste bud called the *taste pore*.

Once inside the mouth, food is dissolved into saliva and the hairs of gustatory receptors dip into this saliva and are stimulated by the taste chemical within it. Thus, gustatory receptors are chemoreceptors. The taste receptors then send messages to the areas of the brain responsible for taste, appetite and saliva production. On the tongue, taste buds are found in elevations called *papillae*. These elevations give the tongue its rough surface.

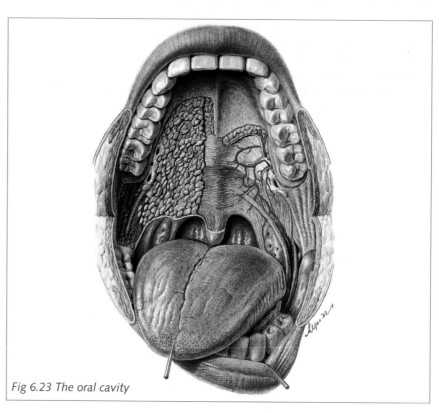

Fig 6.23 The oral cavity

Did you know?

We have four main taste sensations: sour, salty, bitter and sweet. Receptors in some parts of the tongue respond more strongly than others to a specific taste so we can map taste sensations on the tongue. The tip mainly tastes sweet and salty flavours, the back tastes bitter and the sides taste sour.

Anatomy and physiology in perspective

Are we born with a 'sweet tooth'? Most of us do prefer sweet flavours over bitter ones and studies have found that even new born babies prefer sweet to bitter flavours.[iv] This may be because most natural poisons are bitter tasting and the sense of taste may act as a protective mechanism, not allowing something that can be dangerous to the body to enter into the body. On the other hand, a lot of non-poisonous berries and fruits are generally sweet tasting.

The Nose

At the top of the nasal cavity is a small patch about the size of a postage stamp that contains olfactory receptors. Olfactory receptors are neurones that have many hair-like projections extending from their dendrites.

In the connective tissue that supports the olfactory epithelium are *olfactory glands* (Bowman's glands) that produce mucous. When you breathe in, airborne particles dissolve in this mucous and so come into contact with the hair-like cilia of the olfactory receptors. The olfactory cells are chemoreceptors and convert the chemical stimulus found dissolved in the mucous into nerve impulses. These impulses are then sent to the frontal lobe via the limbic system of the brain.

The sense of smell and taste are closely linked because odours from food pass up into the nasopharynx and nasal cavity. This is why you often cannot taste foods if you have a blocked nose from a cold. This is also the reason that wine tasters always smell the bouquet of the wine first before tasting it. The senses of smell and taste are also unique senses in that their impulses travel via the limbic system to the cerebral cortex. This means that they are closely connected to our emotions and memories. For example, the smell of a certain perfume can remind you of someone you have not thought about for years.

Did you know?
Olfactory receptors are unique neurones in that they are continually being replaced. While most mature neurones in the body cannot be replaced, olfactory receptors live for approximately one month before being replaced.

Did you know?
When you sneeze a jet of mucous droplets travels through your nose at approximately 160km/hr. It is not surprising that sneezing clears your nose!

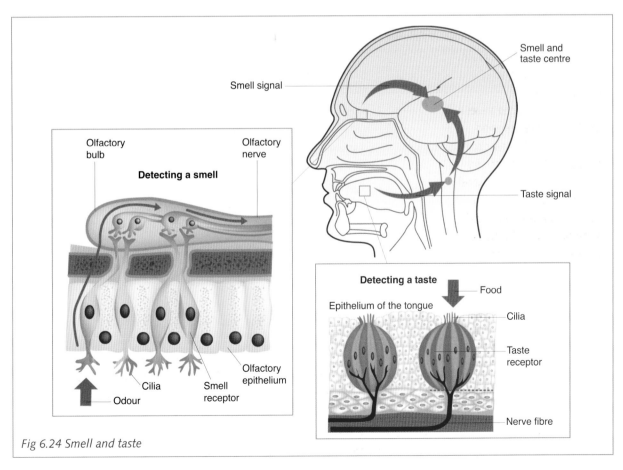

Fig 6.24 Smell and taste

Common Pathologies Of The Nervous System

Because the nervous system controls the entire body, disorders of this system can affect the body in many different ways. Possible indications of nervous system diseases or disorders can include pain, muscle malfunction, changes in sensation, sleep problems and changes in consciousness such as dizziness, fainting, confusion or even dementia.

Alzheimer's Disease

Alzheimer's is the progressive degeneration of brain tissue and is one of the most common causes of dementia. It usually affects elderly people and initial symptoms are memory loss and forgetfulness. Confusion and disorientation then develop and as the disease progresses there can be episodes of paranoia, hallucination and mood change. The cause of Alzheimer's disease is unknown, but it involves a loss of some neurones, the development of plaques on some neurones and the entanglement of the protein filaments within neurones.

Anosmia

Anosmia is the loss of the sense of smell. Generally, one's ability to smell degenerates with age but it can also be affected by blocked nasal passages accompanying colds or it can also be due to problems with the nerves that transmit the sensation of smell to the brain. A loss of the sense of smell also affects one's ability to taste.

Brain tumour

A brain tumour is an abnormal growth in the brain. It can be benign or malignant and symptoms will differ according to the area of the brain in which they are found. Possible early symptoms can include constant headaches, changes in personality, drowsiness, confusion and unusual behaviour. These symptoms can develop to include dizziness, loss of balance, incoordination, nausea and vomiting amongst other symptoms such as a loss of function of the area controlled by the area affected in the brain.

Cerebral palsy

Cerebral Palsy is a disorder characterised by muscular incoordination and loss of muscle control. It results in a lack of balance, abnormal posture, muscle spasticity and speech impairment. It can sometimes also result in mental retardation and is caused by damage to the motor areas of the brain during foetal life, birth or infancy. It is not a progressive disease.

Depression

Depression is characterised by an excessive feeling of sadness that can affect sleeping patterns, appetite and one's ability to concentrate. It can be caused by a number of conditions such as an illness, certain drugs, distressing events such as the loss of a friend or loved one or it can tend to run in families.

Epilepsy

The term epilepsy refers to a group of disorders of the brain characterised by seizures. Seizures are short, recurrent, periodic attacks of motor, sensory or psychological malfunction. The cause of epilepsy can be unknown or can result from a head injury, brain tumour or stroke and seizures are often aggravated by physical or emotional stress or a lack of sleep. Other triggers can be certain drugs, infections, fever and low blood sugar. In rare cases, repetitive sounds, flashing lights or video games can trigger seizures.

Fainting (syncope)
Fainting, a syncope, is the sudden loss of consciousness. It is temporary and brief and usually caused by an inadequate supply of oxygen and nutrients to the brain due to a decrease in blood flow.

Headaches
A headache is a pain felt within the skull and headaches can vary in intensity, frequency and cause. For example, headaches can be related to stress and emotions; can be caused by other disorders of the eyes, nose, throat, sinuses, teeth, jaws, ears or neck; or can be a symptom of a more serious disease.

Cluster headaches are severe, piercing headaches that usually affect the temples or areas around the eye on one side of the head. They don't usually last more than four hours and are not accompanied by nausea or vomiting. However, the pain is so severe that it can cause a person to pace up and down and even bang their head. After the headache the person can sometimes have a droopy eyelid, constricted pupil, runny nose or watery eye on the side that was affected.

Migraines are severe, throbbing headaches that usually affect one side of the head. They can be accompanied by throbbing pain, nausea, vomiting and changes in mood and behaviour. Some people may experience disturbed vision and sensations and most migraines are worsened by movement, light, sounds and smells. Migraines can last from a few hours to a few days.

Tension headaches are generally mild to moderate headaches that affect the whole head. They are not accompanied by other symptoms such as nausea and vomiting. They can be caused by stress or emotional tension.

Insomnia
Insomnia is a disorder in which a person either cannot fall asleep, or cannot stay asleep for an adequate amount of time. Although it is not a disease, it can be a symptom of many different disorders ranging from stress and anxiety to extreme fatigue or even drug use or withdrawal.

Meningitis
Meningitis is inflammation of the meninges that cover the brain and spinal cord. It is usually caused by a bacterial or viral infection although it can also be caused by other conditions such as allergic reactions to certain drugs. Symptoms of meningitis generally include fever, headache, vomiting, weakness and a stiff neck.

Motor neurone disease (MND)
Motor neurone disease is the degeneration of the motor system. It leads to the progressive weakness and wasting away of muscles and eventual paralysis. Both skeletal and smooth muscles, such as those involved in breathing and swallowing, are affected. However, it does not affect the senses. Therefore, people with motor neurone disease can still lead intellectually active lives.

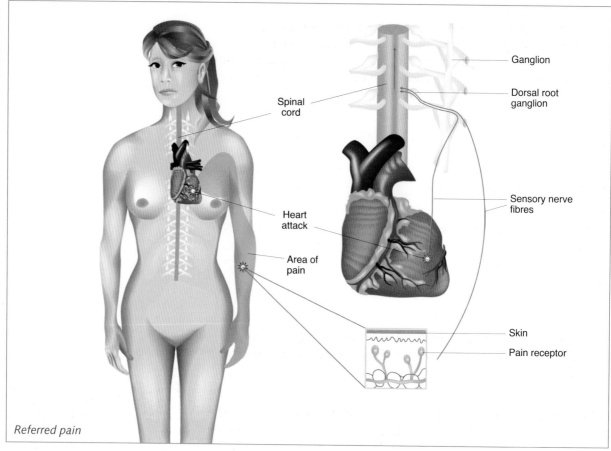

Spinal cord

Heart attack

Area of pain

Ganglion

Dorsal root ganglion

Sensory nerve fibres

Skin

Pain receptor

Referred pain

Myalgic encephalomyelitis (ME, chronic fatigue syndrome)

Myalgic encephalomyelitis, commonly called M.E. or chronic fatigue, is characterised by extreme, disabling fatigue that is not relieved by rest. It is usually accompanied by symptoms such as poor memory, reduced concentration, sore throats, muscle and joint pain, headaches and a persistent feeling of illness after exercise. A cause is not always identifiable.

Pain and referred pain (synalgia)

Pain is usually a reaction to a stimulus such as heat or an injury. Pain receptors are found throughout the body and they transmit pain sensations to the CNS where they are integrated and responded to. Pain also can produce reflex actions in which the impulse does not reach the brain as it is dealt with in the spinal cord. For example, touching a hot oven causes you to immediately withdraw your hand from the oven. This is a reflex reaction. In addition to sensory stimulus, pain can also be caused by a nerve abnormality or a psychological disorder.

Referred pain, synalgia, is pain that is felt in a different part of the body to where it is produced. For example, pain produced from a heart attack is often felt as if it is coming from the arm. Referred pain arises because sensory nerves from different parts of the body often share nerve pathways in the spinal cord.

Parkinson's disease (PD)

Parkinson's disease is a progressive disorder of the CNS. Its cause is unknown, but its symptoms are thought to be due to an imbalance in neurotransmitter activity. Symptoms include involuntary skeletal muscle contractions such as a tremor or rigidity; impaired motor performance; and slow muscular movements.

Spina bifida

Spina bifida is a birth defect in which the spinal vertebrae do not form normally. It can vary in severity from a newborn having one or two vertebrae abnormally formed to a newborn having his/her spinal cord protruding through the skin. In such severe cases the infant will be disabled.

Stress

Stress is any influence that causes an imbalance in one's internal environment. It is part of everyone's life and in itself it is neither good nor bad. However, it becomes a problem when people fail to manage it adequately. It can break down the body's defences, making it more susceptible to illness and disease.

Nerve disorders

Disorders of the peripheral nerves result in muscular weakness or paralysis if motor nerves are affected, or abnormal sensations if sensory nerves are affected.

Bell's palsy

Bell's palsy is a sudden weakness or paralysis of the muscles on one side of the face. It is the result of the malfunction of the facial nerve (cranial nerve VII) and although its cause is often unknown, it may be caused by an infection. Symptoms of Bell's Palsy include pain behind the ear, weakening or paralysis of the facial muscles on one side of the face and a sense of numbness.

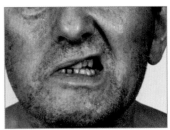

Bell's Palsy

Multiple sclerosis

Multiple sclerosis is a disorder in which patches of myelin and underlying nerve fibres in the eyes, brain and spinal cord are damaged or destroyed. Plaques form on the myelin sheath of the brain and spinal cord and disrupt nerve transmission. This causes weakness, numbness, tremors, loss of vision, pain, fatigue, paralysis, loss of balance and loss of bladder and bowel function. It is most common among women between the ages of 20-40 years and affects them in periods of relapses and remissions. Multiple sclerosis is thought to be an autoimmune disorder.

Neuritis

Neuritis is the inflammation of a nerve or group of nerves. It may result from irritation to the nerve produced by injuries, vitamin deficiency (usually thiamine) or poisons. *Sciatica* is a form of neuritis characterised by severe pain along the path of the sciatic nerve (which runs down the back of the leg). It can be caused by lower back tension, a herniated disc, osteoarthritis or it can result from conditions such as diabetes or pregnancy. It usually only affects one side and symptoms can include pins-and-needles, pain or numbness.

Trigeminal neuralgia

Neuralgia is a pain originating in a nerve and the trigeminal nerve (cranial nerve V) carries sensations from the skin of the face and the mucosa of the nose and mouth to the brain. It is also involved in chewing. Thus, trigeminal neuralgia results in severe, stabbing pain and sensitivity on the face, lips and tongue as well as sensitivity bought on by movements such as chewing. In most cases, its cause is unknown.

Head injuries

Although the brain is protected by the skull and meninges, it can still be injured by car accidents, sports injuries or other falls or accidents. Head injuries can be fatal and differ depending on their severity and where in the brain the injury is:

Concussion – Concussion is the temporary loss of consciousness following a head injury. It is not accompanied by external injuries, but can sometimes produce memory loss, headaches, dizziness, nausea and vomiting.

Contusion – A contusion is the visible bruising of the brain resulting from a direct blow to the head. If the swelling or bleeding is severe, symptoms can include severe headaches, nausea and vomiting as well as impairments of the area affected.

Laceration – A laceration is the tearing of the brain and is usually caused by skull fractures. Symptoms include those of contusions as well as impairments of the area affected. Severe lacerations can lead to further tissue damage due to the swelling of the brain.

Cerebrovascular accident (CVA, stroke)

A cerebrovascular accident, commonly called a stroke, occurs when the arteries to the brain become blocked or rupture and brain tissue consequently dies. Symptoms of strokes depend on the area affected by the stroke, but general early symptoms include a sudden weakness or paralysis of the face and leg on one side of the body, slurred speech, confusion, loss of balance and coordination and sudden severe headaches. Strokes can be caused by a number of things and those at risk are the aged; people with artherosclerosis (the narrowing or blockage of arteries), high blood pressure or diabetes; and smokers. There are three types of strokes:

Ischaemic attack – Ischaemic strokes are caused by a blocked artery.

Transient ischaemic attack (TIA) – This is a mini-stroke caused by a temporary inadequate blood supply to the brain.

Haemorrhagic attack – Haemorrhagic strokes are caused by a ruptured blood vessel.

Spinal cord injuries

Although the spinal cord is well protected by the spinal column, meninges and cerebrospinal fluid, it can still be injured. Injuries can result from accidents such as car accidents or falls or they can also be caused by internal infections, poor blood supply, cord compression or diseases such as multiple sclerosis. Symptoms of spinal cord injury differ according to the area injured and different patterns of numbness, weakness and loss of function will indicate which areas are damaged.

Eye disorders

Disorders of the eyes includes changes in vision, loss of vision, changes in the appearance of the eye and changes in eye sensations.

Cataract
A cataract is a clouding over of the lens of the eye. This is known as opacity and it leads to a loss of vision. Cataracts are common in elderly people but can also be congenital or result from disease or injury to the lens.

Conjunctivitis (pink eye)
Conjunctivitis, commonly called pink eye, is inflammation of the conjunctiva. It can be caused by viral, bacterial or fungal infections and symptoms usually include redness, irritation, light sensitivity and a watery discharge from the eye.

Corneal ulcers
A corneal ulcer is a sore on the cornea. It can be caused by injury to, or irritation of, the cornea and can easily become infected. Symptoms of corneal ulcers include pain, the sense of something being in the eye, sensitivity to bright light and increased tear production.

Glaucoma
Glaucoma is a loss of vision due to abnormally high pressure in the eye. It results from the inability of the eye to drain aqueous humour as quickly as it produces it. The aqueous humour then builds up and this excess pressure damages the optic nerve and causes vision loss. People most at risk of glaucoma are those whose relatives have glaucoma; people with very far or near sightedness; diabetics; or people who have had an eye injury.

Stye (hordeolum)
A stye, or hordeolum, is an infection of the sebaceous glands at the base of the eyelashes. It usually only lasts 2-4 days and is characterised by inflammation, tenderness and pain that then develops into a pus-filled cyst.

Ear disorders

Disorders of the ears such as hearing loss are most common in elderly people as the ears are negatively affected by ageing, wear-and-tear and the cumulative effects of infections and noise. However, young children commonly suffer from ear infections.

Deafness
Deafness is the partial or total loss of hearing. It can affect one ear or both ears and it has many causes such as a mechanical problem in the ear that blocks the sound waves or damage to the hair cells or auditory nerve by infections or injuries. Hearing loss can also be age related.

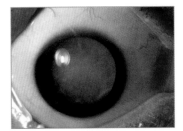
Cataract

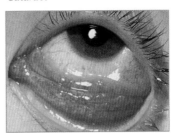
Conjunctivitis

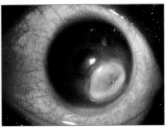
Corneal ulcers

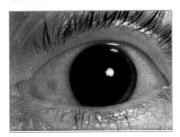
Glaucoma

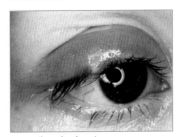
Stye (hordeolum)

Earache

Earache is pain that is thought to originate in the ear. It can be caused by ear infections or result from a blocked Eustachian tube that affects the equalising of ear pressure. Earache can sometimes be referred pain from infections in the nose, throat, sinuses or mouth.

Glue ear (secretory or serous otitis media)

Glue ear, secretory otitis media, is the accumulation of fluid in the middle ear. It is usually caused by otitis media or a blocked Eustachian tube and symptoms include a feeling of fullness in ear or a popping or crackling sound on swallowing. Temporary hearing loss usually accompanies glue ear.

Otitis media

Otitis media is an infection of the middle ear. It can be caused by a bacteria or virus and is often a complication of colds or allergies. Symptoms of otitis media can include fever, pain and a red, bulging eardrum.

Tinnitus

Tinnitus is the sensation of hearing sounds in the ear that have not originated from the external environment. It is most common in elderly people due to age-related hair cell loss but it can also be a symptom of other ear disorders such as injury to the ear, infections or a blocked ear canal or Eustachian tube.

Vertigo

Vertigo is the sensation of one's self or one's surroundings constantly moving or spinning. It is usually accompanied by nausea and a loss of balance. Vertigo can be a symptom of many diseases or caused by a variety of conditions of the inner ear or the vestibular nerve.

NEW WORDS	
Afferent	nerves or neurones that transmit impulses to the CNS.
Conductivity	the ability to transmit an impulse from a neurone to another neurone, muscle or gland.
Efferent	nerves or neurones that transmit impulses from the CNS to the muscles and glands.
Ganglion	a bundle or knot of nerve cell bodies.
Ion	an electrically charged molecule.
Irritability	the ability to respond to a stimulus and convert it into an impulse.
Meninges	the three connective tissue membranes that enclose the brain and spinal cord.
Nerve fibre	the term 'fibre' refers to the processes that project from a nerve body, these being dendrites and axons.
Neurology	the study of the nervous system.
Neurotransmitter	a chemical that transmits impulses across synapses from one nerve to another.
Plexus	a network of nerves (or blood vessels).
Proprioceptor	a specialised nerve receptor located in muscles, joints and tendons that provides sensory information regarding body position and movements.
Tract	tracts are bundles of fibres that are not surrounded by connective tissue.
Viscera	the organs of the abdominal body cavity.

Study Outline

Functions of the nervous system

These include sensory, integrative and motor functions.

Organisation of the nervous system

1. The nervous system is divided into the:
   - **Central Nervous System** (CNS) which is composed of the brain and spinal cord.
   - **Peripheral Nervous System** (PNS) which is composed of the cranial and spinal nerves.
2. The PNS is then divided into the:
   - **Somatic nervous system** which is the voluntary nervous system that controls the skeletal muscles.
   - **Autonomic nervous system** which is the involuntary nervous system that controls the smooth and cardiac muscles and glands. The autonomic nervous system is subdivided into the:
     - **Sympathetic nervous system** which reacts to changes by stimulating activity.
     - **Parasympathetic nervous system** which opposes the sympathetic nervous system and inhibits activity.

Nervous tissue

1. The nervous system contains only two types of cells: neuroglia and neurones.
2. Neuroglia (glia) insulate, nurture and protect neurones.

Neurones

1. Neurones transmit impulses to, from and within the brain. There are three types of neurones:
   - Sensory/afferent – carry sensory nerve impulses from the skin, sense organs, muscles, joints and viscera to the CNS.
   - Motor/efferent – carry motor nerve impulses from the CNS to the muscles or glands.
   - Association/interneurones – connect sensory and motor neurones in neural pathways.

Structure of a motor neurone

1. Neurones are made up of three parts:
   - Dendrites which receive impulses.
   - A cell body which contains the nucleus.
   - An axon which transmits the impulses.
2. Axon terminals are found at the end of axons and contain synaptic vesicles that store neurotransmitters.
3. Some axons are myelinated. This means they are covered in a myelin sheath.

- A myelin sheath protects and insulates the neurone and speeds up the conduction of the impulse.
- Myelin is produced by Schwann cells in the PNS.
- The outermost layer of the myelin sheath is called the neurilemma.
- At intervals along the myelin sheath are gaps called nodes of Ranvier.

Transmission of a nerve impulse

1. Nerve impulses are transmitted electrochemically.
2. Nerve impulses are transmitted across the plasma membrane of an unmyelinated axon or across the nodes of Ranvier of a myelinated axon.
3. An inactive membrane has an electrical voltage across its membrane with mainly sodium ions on the outside of the membrane and potassium ions on the inside. This means it is polarised.
4. The dendrites of the neurone are stimulated and this causes the movement of sodium ions into a segment of the cell.
5. The cell therefore becomes depolarised and this initiates an impulse.
6. The impulse then travels along the membrane in a wave of depolarisation.
7. When the impulse reaches the end of the axon it is chemically transmitted across the synapse by neurotransmitters.

The brain

1. The brain lies in the cranial cavity and is protected by the bones of the cranium, the meninges and cerebrospinal fluid.
2. The three meninges are the outer dura mater, middle arachnoid and inner pia mater.
3. Cerebrospinal fluid functions as a shock-absorber, provides the correct chemical environment for neurones and acts as a medium for the exchange of nutrients and waste.
4. The brain has four major regions:
   - The brain stem consists of the medulla oblongata, pons and midbrain. It is the link between the brain and spinal cord and plays an important role in autonomic functioning.
   - The cerebellum functions in movement, posture and balance.
   - The diencephalon contains the epithalamus, thalamus and hypothalamus and functions in relaying sensory impulses to the cerebral cortex and regulating homeostasis. It also houses the pituitary and pineal glands.
   - The cerebrum consists of the cerebral cortex (grey matter), white matter, the limbic system and basal ganglia. It functions in emotions and intelligence.

S.C.D.C
Stem, Cerebellum, Diencephalon, Cerebrum

Cranial nerves

The 12 pairs of cranial nerves are as follows:
1. **Olfactory** – nose, smell.
2. **Optic** – eyes, vision.
3. **Oculomotor** – eyes, movement.
4. **Trochlear** – eyes, movement.
5. **Trigeminal** (ophthalmic, maxillary and mandibular branches) – eyes, jaw movement, facial skin sensations.
6. **Abducens** – eyes, movement.
7. **Facial** – (temporal, zygomatic, buccal, mandibular and cervical branches) – facial expression, taste, saliva, tears.
8. **Vestibulocochlear** (vestibular and cochlear branches) – ears, hearing, balance.
9. **Glossopharyngeal** – tongue, pharynx, saliva.
10. **Vagus** – thorax, abdomen.
11. **Accessory** (cranial and spinal portions) – head, pharynx and larynx.
12. **Hypoglossal** – tongue, talking, swallowing.

Spinal cord

1. The spinal cord is continuous with the brain stem and ends just above L2.
2. The spinal cord functions in transporting nerve impulses between the brain and the rest of the body and in producing reflex actions.
3. The spinal cord is protected by the vertebral column, cerebrospinal fluid and the spinal meninges which are continuous with the cranial meninges.
4. Internally, the spinal cord contains grey matter which consists of cell bodies that integrate impulses and white matter which consists of myelinated fibres that transport impulses.
5. There are 31 pairs of spinal nerves as follows:
   - 8 pairs of cervical nerves.
   - 12 pairs of thoracic nerves.
   - 5 pairs of lumbar nerves.
   - 5 pairs of sacral nerves.
   - 1 pair of coccygeal nerves.
6. Spinal nerves connect the CNS to the rest of the body.
7. Spinal nerves are mixed nerves. They have a posterior/dorsal root consisting of sensory axons and an anterior/ventral root consisting of motor axons.
8. Branches of spinal nerves form networks called plexuses. All nerves emerging from a specific plexus will innervate specific structures.
9. The spinal plexuses are:
   - The cervical plexus – serves the head, neck and top of shoulders.
   - The brachial plexus – serves the shoulder and upper limb.
   - The lumbar plexus – serves the abdominal wall, external genitals and part of the lower limb.
   - The sacral plexus – serves the buttocks, perineum and lower limbs.
10. There is no thoracic plexus. These nerves form the intercostal nerves that serve the thorax.

Special sense organs

The eyes
1. The eyeball consists of a strong protective wall and a large inner space.
2. The wall of the eyeball is divided into three layers:
   - The outer fibrous tunic contains the cornea which helps focus light and the sclera which is the white of the eye.
   - The middle vascular tunic contains the coloured iris and its pupil; the ciliary body that secretes aqueous humour; and the choroid which provides nutrients to the retina.
   - The inner nervous tunic, also called the retina. It consist of a non-visual pigmented portion that absorbs stray light rays and a neural portion that contains photoreceptors which convert light into nerve impulses.
3. The interior of the eyeball contains the lens which is responsible for the fine tuning of focusing; the anterior cavity which is filled with aqueous humour that nourishes the lens and helps produce intraocular pressure; and the posterior cavity (vitreous chamber) that contains the vitreous body which helps produce intraocular pressure and maintain the shape of the eyeball.
4. Waves of light are bent as they pass through the structures of the eyeball. They then hit the retina where they are converted into nerve impulses and transported by the optic nerve to the brain.

The ears
1. The ear is divided into three regions: the outer, middle and inner ear.
2. The outer ear is composed of the auricle, external auditory canal and eardrum. It channels air waves in towards the middle ear.
3. The middle ear contains the auditory ossicles which are the hammer, anvil and stirrup (malleus, incus and stapes). These bones transmit sound waves to the inner ear.
4. The inner ear is a labyrinth containing the vestibule and semicircular canals which are responsible for the sense of equilibrium and the cochlea which contains the organ of Corti. The organ of Corti (spiral organ) is the organ of hearing.
5. Sound waves are transmitted as vibrations through the ear until they reach hair cells in the organ of Corti where they are converted into nerve impulses and transported by the vestibulocochlear nerve to the brain.

The mouth
1. The mouth contains taste buds which house gustatory receptors (receptors of taste).
2. Taste buds are located mainly in the elevations on the tongue called papillae.
3. Gustatory receptors are cells that have a hair-like projection that makes contact with the stimuli (taste) which has been dissolved in saliva. As tastes are chemicals, gustatory receptors are chemoreceptors.
4. The hairs come into contact with the stimuli and convert it into a nerve impulse which is transported to the brain.
5. The four primary taste sensations are sour, salty, bitter and sweet.

The nose
1. Olfactory epithelium is located at the top of the nasal cavity. It contains olfactory receptors.
2. Olfactory receptors are cells with hair-like projections that make contact with the odour that has dissolved in the mucous that covers the inside of the nose. As odours are chemicals, olfactory receptors are chemoreceptors.
3. The hairs come into contact with the stimuli and convert it into a nerve impulse which is then transported to the brain.

Revision

1. Describe the functions of the nervous system.
2. Identify the divisions/organisation of the nervous system.
3. Describe the central nervous system.
4. Describe the peripheral nervous system.
5. Identify the functions of the somatic nervous system.
6. Identify the functions of the autonomic nervous system.
7. Describe the divisions of the autonomic nervous system.
8. Identify the two types of cells found in the nervous system and explain their differences.
9. Describe the three different types of neurone.
10. Describe the structure of a motor neurone.
11. Explain how an impulse is transmitted across a nerve and from nerve to nerve.
12. Identify the four major regions of the brain.
13. Name the three meninges.
14. Describe the functions of cerebrospinal fluid.
15. Describe the main function of each of the following cranial nerves:
    - Optic, II
    - Trigeminal, V
    - Facial, VII
    - Vagus, X
    - Hypoglossal, XII.
16. Explain what the cauda equina is.
17. How many pairs of spinal nerves are there?
18. Name and describe the four spinal plexuses.
19. Outline how the eye functions.
20. Outline how the ear functions.
21. Outline how the mouth functions.
22. Outline how the nose functions.

Multiple choice questions

1. **Functions of the spinal cord include:**
   a. Controlling the somatic nervous system.
   b. Processing and storing memories.
   c. Producing reflex actions.
   d. Transporting sensory information regarding taste and smell.

2. **How many pairs of cranial nerves are there?**
   a. 8
   b. 10
   c. 12
   d. 14.

3. **Which of the following functions of the cerebrum is correct?**
   a. Gives us the ability to read, write and speak.
   b. Controls homeostasis in the body.
   c. Controls skeletal muscles and helps regulate posture and balance.
   d. Transports all sensory and motor information from the spinal cord to the rest of the brain.

4. **The Eustachian tube is found in the:**
   a. Ears
   b. Eyes
   c. Mouth
   d. Nose.

5. **Which of the following statements is correct?**
   a. The sympathetic nervous system generally inhibits all activity.
   b. The sympathetic nervous system generally conserves energy in the body.
   c. The sympathetic nervous system always conserves energy in the body.
   d. The sympathetic nervous system generally stimulates activity.

6. **Which part of a motor neurone is the receiving or input portion of the cell?**
   a. Cell body
   b. Axon
   c. Dendrite
   d. None of the above.

7. **What are myelin sheaths produced by?**
   a. Neurolemma
   b. Nodes of Ranvier
   c. Neurilemma
   d. Schwann cells.

8. **Which cranial nerves function in swallowing movements?**
   a. Accessory
   b. Abducens
   c. Oculomotor
   d. Vestibulocochlear.

9. **Cranial nerves are part of the:**
   a. Central nervous system
   b. Peripheral nervous system
   c. Sense of smell
   d. Sense of taste.

10. **What type of cells are olfactory receptors?**
    a. Chemoreceptors
    b. Mechanoreceptors
    c. Vascular receptors.
    d. None of the above.

7 The Endocrine System

Introduction

Why is it that some people are exceptionally tall while others are small? Or some people can have children while others can't? What is it that controls such processes in our bodies? It is the same system that controls our blood pressure, immune system, metabolism and even our response to danger. It is the endocrine system.

The study of the glands and hormones of the endocrine system is endocrinology. In this chapter you will learn about the endocrine system and discover how it works together with your nervous system to control all the other systems of your body.

Student objectives

By the end of this chapter you will be able to:

- Describe the functions of the endocrine system.
- Describe the organisation of the endocrine system.
- Explain what a hormone is and how it works.
- Identify the differences between the nervous and endocrine systems.
- Describe the endocrine glands and the hormones they produce.
- Identify the common pathologies of the endocrine system.

Functions Of The Endocrine System

Before looking at the functions of the endocrine system, it helps to understand exactly what endocrine glands and hormones are. There are two types of glands in the body: exocrine and endocrine.

- Exocrine glands – Secrete substances into ducts that carry these substances into body cavities or to the outer surface of the body. Examples of exocrine glands include sudoriferous (sweat), sebaceous (oil), mucous and digestive glands.
- Endocrine glands – Secrete substances into the extracellular space around their cells. These secretions then diffuse into blood capillaries and are transported by the blood to target cells located throughout the body. Substances secreted by endocrine glands are called hormones.

> Hormones are chemical messengers that regulate cellular activity. They are:
> - Secreted by endocrine glands.
> - Transported in the blood.

Functions

The endocrine system has many functions as it affects a variety of cells and tissues in the body. A simplistic view of its functions is that it coordinates body functions such as:

- Growth
- Development
- Reproduction
- Metabolism
- Homeostasis.

It also helps regulate activities of the immune system and the process of apoptosis. This is the normal, ordered death and removal of cells as part of their development, maintenance and renewal.

Types of hormones

Hormones are synthesised from either steroids or amino acids:

- The sex hormones and those of the adrenal cortex are *steroid* hormones.
- Hormones such as insulin and oxytocin are *amino acid* based hormones.

Some molecules, such as noradrenaline, are both hormones and neurotransmitters depending on their action.

Action of hormones

Hormones are transported by the blood to target cells that contain receptors to which the hormones bind. These receptors are very specific and will only bind to a specific hormone, much like two pieces of a puzzle slotting together. Once the hormones have bound to the receptors they act like switches that turn on chemical and metabolic processes within the cell.

General processes stimulated by hormones include:

- Synthesis of new molecules.
- Changes in the permeability of the plasma membrane.
- Transportation of substances into or out of cells.
- Changes in the rates of metabolic reactions.
- Contraction of smooth or cardiac muscle.

Control of hormone release

The secretions of hormones into the bloodstream need to be controlled otherwise there will be an underproduction or overproduction of a hormone that can result in disease. Hormone secretion is controlled by signals from the nervous system, chemical changes in the blood and other hormones.

- **Neural stimulation** – Signals from the nervous system can stimulate the release of hormones into the bloodstream. For example, sympathetic nervous stimulation causes the release of adrenaline.
- **Chemical changes in the blood** – Changes in the levels of certain ions, for example calcium, or nutrients can stimulate the release of hormones. For example, blood calcium levels regulate the secretions of the parathyroid glands.
- **Hormonal stimulation** – The presence of a hormone can stimulate the release of another hormone. For example, hormones secreted by the anterior pituitary gland stimulate the release of other hormones into the bloodstream.

Once hormones have been released into the body, their levels are controlled by a negative feedback mechanism. This is a system in which rising levels of a hormone will inhibit the further release of that hormone.

Neuro-endocrine system

Together the nervous and endocrine systems control all the processes that take place in the body and to a certain extent they control one another. For example, the nervous system can stimulate or inhibit the release of hormones while the endocrine system can promote or inhibit nerve impulses. However, there are some key differences between these two systems as demonstrated in the chart on the following page.

Study tip
When comparing the nervous system to the endocrine system, think of phoning a friend for a quick chat (nervous system) compared with going around for a leisurely dinner (endocrine system).

COMPARISON OF THE NERVOUS AND ENDOCRINE SYSTEMS		
CHARACTERISTIC	**NERVOUS SYSTEM**	**ENDOCRINE SYSTEM**
Messenger	Nerve impulse	Hormone
Transportation	Nerve axons	Blood
Cells affected	Mainly muscles, glands and other neurones	All types of cells
Action	Muscular contractions and glandular secretions	All types of changes in metabolic activities, growth, development and reproduction
Time to act	Very quick (milliseconds)	Slower (from seconds to days)
Duration of effects	Brief	Longer

Did you know?

The pituitary gland is the size and shape of a large pea, yet it is the 'master gland' of the body. Despite its small size, its hormones control most of the other glands of the body. The pituitary gland does, however, have its own master – the hypothalamus.

Endocrine glands and their hormones

The endocrine system coordinates growth, development, reproduction, metabolism and homeostasis.

Hypothalamus

The hypothalamus is not always considered an endocrine gland as it is part of the nervous system. However, it is included in this chapter because of its vital connection to the pituitary gland. The hypothalamus:

- releases a number of hormones that control the secretions of the pituitary gland.
- synthesises two hormones that are then transported to, and stored in, the posterior pituitary gland. These hormones are oxytocin and antidiuretic hormone.

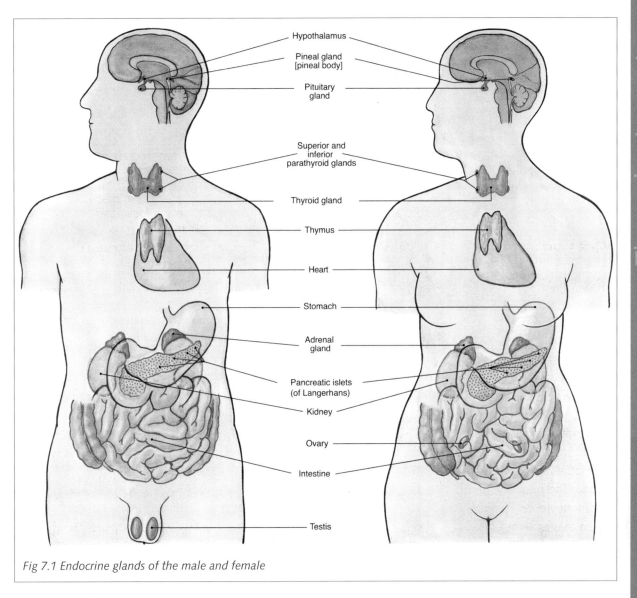

Fig 7.1 Endocrine glands of the male and female

Pituitary gland

The pituitary gland is located in the hypophyseal fossa of the sphenoid bone and is found behind the nose and between the eyes. It is attached to the hypothalamus by a stalk and comprises an anterior and a posterior portion.

Study tip

The charts on the following pages discuss the endocrine glands, their hormones and functions. They also list examples of what can go wrong if the gland is not functioning properly. Most of these disorders are due to either overproduction (**hypersecretion**) or underproduction (**hyposecretion**) of hormones. Some of these disorders are discussed in more detail at the end of this chapter. However, they have been included here to help you understand the functions of the glands.

HORMONES OF THE PITUITARY GLAND

HORMONE	TARGET TISSUE	ACTIONS	DISORDERS AND DISEASES
Anterior pituitary gland All the hormones released by the anterior pituitary gland, except for human growth hormone, regulate other endocrine glands.			
Human growth hormone (hGH) or Somatotropin	Bone, muscle, cartilage and other tissue.	Stimulates growth and regulates metabolism.	**Hyposecretion**: Pituitary dwarfism **Hypersecretion**: Gigantism, Acromegaly
Thyroid-stimulating hormone (TSH) or Thyrotropin	Thyroid gland	Controls the thyroid gland.	**Hyposecretion**: Myxoedema **Hypersecretion**: Graves' disease (NB: These diseases are more commonly caused by a problem with the thyroid gland itself).
Follicle-stimulating hormone (FSH)	Ovaries and testes	**In females**: stimulates the development of oocytes (egg cells or immature ovums). **In males**: stimulates the production of sperm.	**Hyposecretion**: Sterility
Luteinizing hormone (LH)	Ovaries and testes	**In females**: stimulates ovulation, the formation of the corpus luteum and secretion of oestrogens and progesterone **In males**: stimulates the production of testosterone.	**Hyposecretion**: Sterility **Hypersecretion**: Stein-Leventhal syndrome (Polycystic ovary syndrome)
Prolactin (PRL) or Lactogenic hormone	Mammary glands	**In females**: stimulates the secretion of milk from the breasts. **In males**: action is unknown.	**Hypersecretion** in females: Galactorrhoea (abnormal lactation) Amenorrhoea (absence of menstrual cycles)
Adrenocorticotropic hormone (ACTH) or Corticotropin	Adrenal cortex	Stimulates and controls the adrenal cortex.	**Hyposecretion**: Addison's disease **Hypersecretion**: Cushing's syndrome
Melanocyte-stimulating hormone (MSH)	Skin	Exact actions are unknown, but can cause darkening of the skin.	

The endocrine system

HORMONES OF THE PITUITARY GLAND			
HORMONE	TARGET TISSUE	ACTIONS	DISORDERS AND DISEASES
Posterior pituitary gland The posterior pituitary gland does not synthesise hormones. Instead, it stores and releases hormones synthesised by the hypothalamus.			
Oxytocin (OT)	Uterus, mammary glands	Stimulates contraction of uterus during labour and stimulates the 'milk let-down' reflex during lactation.	
Antidiuretic hormone (ADH) or Vasopressin	Kidneys, sudoriferous glands, blood vessels	Antidiuretic effect (i.e. conserves water by decreasing urine volume and perspiration), raises blood pressure.	**Hyposecretion**: Diabetes insipidus

Anatomy and physiology in perspective

A deficiency of human growth hormone results in dwarfism. One of the world's most famous dwarfs was Tom Thumb who was less than 0.9m tall. Gigantism, on the other hand, results from an over-secretion of human growth hormone and one victim of this condition was Robert Wadlow who reached 2.7m tall and weighed 220kg.

Pineal gland

The pineal gland is located in the centre of the brain where it is attached to the roof of the third ventricle. It is made of neuroglia and secretory cells and its physiological role in the body is still unclear.

HORMONES OF THE PINEAL GLAND			
HORMONE	**TARGET TISSUE**	**ACTIONS**	**DISORDERS AND DISEASES**
Pineal gland The pineal gland produces the hormone melatonin which is thought to contribute to setting the timing of the body's biological clock. Its release is stimulated by darkness and inhibited by sunlight.			
Melatonin	Body's biological clock	Causes sleepiness	**Hyposecretion**: Insominia **Hypersecretion**: Seasonal affective disorder (S.A.D.)

Thyroid gland

The thyroid gland is a butterfly-shaped gland found wrapped around the trachea just below the larynx.

HORMONES OF THE THYROID GLAND			
HORMONE	**TARGET TISSUE**	**ACTIONS**	**DISORDERS AND DISEASES**
Thyroid gland The thyroid secretes two hormones that play a vital role in the body's metabolism. These are *tri-iodothyronine* (T_3) and *thyroxine* (T_4). Together they are often referred to as thyroid hormone. The thyroid also secretes *calcitonin*.			
Thyroid hormone (tri-iodothyronine and thyroxine)	Cells and tissues throughout the body	**Controls**: Oxygen use and the basal metabolic rate (the minimum amount of energy used by the body to maintain vital processes). Cellular metabolism. Growth and development.	**Hyposecretion**: Cretinism Myxoedema **Hypersecretion**: Graves' disease Thyrotoxicosis **Thyroid enlargement**: Goitre
Calcitonin (CT) or Thyrocalcitonin	Bone	Lowers blood calcium levels.	No known disorders

Parathyroid glands
The parathyroids are small, round masses of tissue found on the posterior surfaces of the thyroid gland.

HORMONES OF THE PARATHYROID GLAND			
HORMONE	TARGET TISSUE	ACTIONS	DISORDERS AND DISEASES
Parathyroid glands The parathyroid glands release only one hormone, parathormone. It works together with calcitonin from the thyroid gland and calcitriol from the kidneys to control blood calcium levels.			
Parathormone (PTH) or Parathyroid hormone	Bone	Increases blood calcium and magnesium levels. Decreases blood phosphate levels. Promotes formation of calcitriol by the kidneys.	**Hyposecretion**: Tetany Hypocalcemia **Hypersecretion**: Demineralisation of bones

Thymus gland
The thymus gland is located in the thorax, behind the sternum and between the lungs. It is large in infants and reaches its maximum size around puberty. It then begins to decrease with age.

HORMONES OF THE THYMUS GLAND		
HORMONE	ACTIONS	DISORDERS AND DISEASES
Thymus gland The thymus gland plays an important role in the immune system and secretes a number of hormones involved in immunity, including *thymosin*.		
Thymosin	Promotes growth of T-Cells which are a type of white blood cell.	Decreased immunity

Study tip
Insulin works like a key, allowing glucose to enter into cells.

Pancreatic islets

The pancreas is a long organ of approximately 12½-15cm in length. It is found behind and slightly below the stomach and is both an endocrine and an exocrine gland as it also functions in digestion (its digestive functions will be discussed in Chapter 11). Scattered in the pancreas are small groups of endocrine tissue called *pancreatic islets* or *islets of Langerhans*.

HORMONES OF THE PANCREATIC ISLETS			
HORMONE	**TARGET TISSUE**	**ACTIONS**	**DISORDERS AND DISEASES**
Pancreatic islets Pancreatic islets are composed of four types of hormone-secreting cells: alpha cells, beta cells, delta cells and F cells. These cells secrete four different hormones which generally help regulate blood glucose levels.			
Glucagon (from alpha cells)	Liver	Accelerates the breakdown of glycogen into glucose. Stimulates the release of glucose into the blood. Therefore, raises blood-glucose levels.	**Hypersecretion**: Hyperglycaemia
Insulin (from beta cells)	All body cells	Accelerates the transport of glucose into cells. Converts glucose into glycogen. Therefore, lowers blood-glucose levels.	**Hyposecretion**: Diabetes mellitus **Hypersecretion**: Hyperinsulinism
Somatostatin (from delta cells) – This is identical to growth hormone inhibiting hormone secreted by the hypothalamus.	Pancreas	Inhibits insulin and glucagon release. Slows absorption of nutrients from the gastrointestinal tract.	**Hypersecretion**: Diabetes mellitus
NB, the fourth hormone, *pancreatic polypeptide*, secreted by F cells is not discussed here.			

Did you know?
Noradrenaline is also a neurotransmitter.

Adrenal glands

The adrenal glands (also called the *suprarenal glands*) are found above the kidneys. Although an adrenal gland looks like a single organ, it contains two regions that are structurally and functionally different. The outer adrenal cortex surrounds the inner adrenal medulla.

HORMONES OF THE ADRENAL GLANDS

HORMONE	TARGET TISSUE	ACTIONS	DISORDERS AND DISEASES
Adrenal cortex The adrenal cortex produces steroid hormones that are essential to life and loss of them can lead to potentially fatal dehydration or electrolyte imbalances. These hormones are grouped into mineralcorticoids, glucocorticoids and sex hormones.			
Mineralcorticoids (mainly aldosterone)	Kidneys	Regulate mineral content of the blood by: Increasing blood levels of sodium and water. Decreasing blood levels of potassium.	**Hyposecretion** of glucocorticoids and aldosterone: Addison's disease **Hypersecretion** of aldosterone: Aldosteronism
Glucocorticoids (mainly cortisol)	All body cells	Regulate metabolism. Help body resist long-term stressors. Control effects of inflammation. Depress immune responses.	**Hyposecretion** of glucocorticoids and aldosterone: Addison's disease **Hypersecretion**: Cushing's syndrome
Sex hormones (androgens and oestrogens)		Very small contribution to sex drive and libido.	Presence of feminising hormones in males: Gynaecomastia (enlargement of breasts) **Hypersecretion**: Hirsutism
Adrenal medulla The adrenal medulla is innervated by neurones of the sympathetic division of the autonomic nervous system and can very quickly release hormones that are collectively referred to as *catecholamines*. These hormones are, to a large extent, responsible for the fight-or-flight response of the body and they help the body cope with stress.			
Adrenaline (epinephrine) and **Noradrenaline** (norepinephrine)	All body cells	Fight-or-flight response: • Increase blood pressure • Dilate airways to the lungs • Decrease rate of digestion • Increase blood glucose level • Stimulate cellular metabolism	**Hypersecretion**: Prolonged fight-or-flight response

Ovaries and testes

The female sex glands, the ovaries, and the male sex glands, the testes, are called the gonads. The ovaries are almond-sized organs found in the pelvic cavity. The testes are suspended in a sac, the scrotum, outside the pelvic cavity. The ovaries and testes are discussed in more detail in Chapter 13.

HORMONES OF THE OVARIES AND TESTES			
HORMONE	**TARGET TISSUE**	**ACTIONS**	**DISORDERS AND DISEASES**

Ovaries
The ovaries only begin to function properly at puberty when stimulated by FSH and LH from the anterior pituitary gland. The ovaries produce *oestrogens* and *progesterone*.

Oestrogens (includes oestriol, oestrone and oestradiol)	Reproductive system	Stimulate the development of feminine secondary sex characteristics. Together with progesterone, they regulate the female reproductive cycle.	Imbalances in secretions can lead to a range of reproductive disorders.
Progesterone	Reproductive cycle	Together with oestrogens, it regulates the female reproductive cycle. Helps maintain pregnancy.	Imbalances in secretions can lead to a range of reproductive disorders.

Testes
The testes produce both sperm and male sex hormones called *androgens*. The most important androgen is *testosterone*.

Testosterone	Reproductive system	Stimulates development of masculine secondary sex characteristics. Promotes growth and maturation of male reproductive system and sperm production. Stimulates sex drive.	Sterility

Anatomy and physiology in perspective
The fight-or-flight response
When put under short-term stress such as being physically threatened, our sympathetic nervous system responds with a 'fight-or-flight response'. Part of this response is the stimulation of the adrenal medulla which secretes adrenaline and noradrenaline into the blood system. These two hormones prepare the body to either fight or flee a situation by causing changes in the body that result in more oxygen and glucose in the blood and a faster circulation of blood to the brain, muscles and heart.

Long term stress
On the other hand, glucocorticoids such as cortisol are secreted by the adrenal cortex to help you deal with long term stress such as the death of a loved one. Glucocorticoids help protect the body from the long term effects of stress, but they also depress the immune system.

Anatomy and physiology in perspective
There are three groups of hormones that exist in both sexes and their differing proportions determine one's feminine or masculine characteristics. Androgens predominate in men or people with masculine characteristics such as a muscular physique, facial hair and a deep voice. On the other hand, oestrogens predominate in women or people with feminine characteristics such as breasts. Progestogens generally prepare for and maintain pregnancy.

Ageing and our hormones
Surprisingly, our hormones are generally not negatively affected by the process of ageing. However, certain changes in hormonal function do mark our journey through life.

Puberty is a significant time in which a child becomes sexually mature. It generally occurs around the age of 12 in girls and 14 in boys and is stimulated by the gonadotropic hormones of the pituitary gland. It is characterised by the appearance of secondary sexual characteristics in both sexes and the start of menstruation in girls.

Pregnancy is a time in which a woman creates a child and becomes a mother. Many hormones are at work in a woman's body during pregnancy, especially oestrogen and progesterone. Once properly developed in the uterus, the placenta also secretes hormones to help maintain the pregnancy.

Menopause is a very significant time for women as it marks the end of their ability to bear children. It generally occurs between the ages of 45–55 and is characterised by the cessation of ovulation and menstruation.

Common Pathologies Of The Endocrine System

Pathologies of the adrenal glands

Addison's Disease
Addison's disease is caused by the hyposecretion of glucocorticoids and aldosterone by the adrenal cortex. It is characterised by patches of excessive pigmentation, low blood pressure, weakness, tiredness and dizziness on standing. The causes of Addison's disease are not always known and it is sometimes thought to be an autoimmune disorder or due to the destruction of the adrenal cortex by cancer or infection.

Cushing's Syndrome
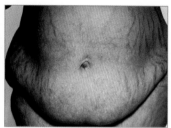
Cushing's Syndrome

Cushing's syndrome is caused by the hypersecretion of corticosteroid hormones by the adrenal cortex. Cushing's syndrome is characterised by fatigue and excessive fat deposits on the face, torso and back. A person with Cushing's syndrome usually has a large, round face coupled with thin skin that bruises or tears easily. The hypersecretion of corticosteroid hormones can be caused by a problem in either the adrenal glands or the pituitary gland.

Pathologies of the pancreatic islets

Diabetes mellitus
Diabetes mellitus is a disorder in which there is an elevation of glucose in the blood (hyperglycaemia). Symptoms of diabetes mellitus include increased thirst and urination, weight loss in spite of increased appetite, fatigue, nausea, vomiting, frequent infections and blurred vision. There are two main types of diabetes mellitus:

Type I diabetes (Insulin dependent diabetes mellitus) – Type I diabetes is most common in people under the age of 20 years and is a hereditary, auto-immune disorder in which the body's immune system destroys its own insulin producing cells.

Type II diabetes (Non insulin dependent diabetes mellitus) – Type II diabetes is more common in people who are overweight and do not eat a healthy diet. In Type II diabetes, insulin levels may be normal but body cells are resistant or less sensitive to it. Therefore, the metabolism of carbohydrates, fats and protein is altered.

Pathologies of the parathyroid glands

Calcium deficiency (Hypocalcemia)
Calcium is an essential element in the body and is necessary for the formation of bones and teeth, muscular contraction, the normal functioning of many enzymes, blood clotting and normal heart rhythm. Calcium deficiency, hypocalcemia, can be caused by low levels of parathormone from the parathyroid glands. This can be due to damaged parathyroid glands, low levels of magnesium, the body not responding properly to parathormone, vitamin D deficiency, kidney damage, insufficient dietary calcium or disorders that affect calcium absorption. Symptoms of calcium

deficiency include confusion, memory loss, depression, muscle aches and spasms, tingling sensations and abnormal heart rhythms. Calcium deficiency can also cause tetany which is characterised by muscle twitches and spasms.

Pathologies of the pineal gland

Seasonal affective disorder (SAD)
Seasonal affective disorder (SAD) is a disorder that usually occurs at the onset of winter and is thought to be caused by a lack of sunlight which leads to an increase in the secretion of melatonin by the pineal gland. SAD is characterised by depression, a lack of interest in one's usual activities, oversleeping and overeating.

Pathologies of the pituitary gland

Acromegaly
Acromegaly occurs in adults and is excessive growth caused by hypersecretion of human growth hormone (hGH). Because the bones of an adult have already stopped lengthening, hypersecretion of hGH does not cause a further growth in height. Instead, it causes the thickening of the bones of the hands, feet, cheeks and jaw as well as the tissues on the eyelids, lips, tongue and nose. The skin also thickens, sweat glands enlarge leading to excessive perspiration and other tissues such as heart tissue and nervous tissue can also be affected.

Diabetes insipidus
Diabetes insipidus is the excessive production of large amounts of very dilute urine. It is characterised by excessive thirst and excessive urination which can quickly lead to dehydration. Diabetes insipidus is caused by a lack of antidiuretic hormone which is produced by the pituitary gland.

Gigantism (Giantism)
Gigantism is excessive growth caused by hypersecretion of human growth hormone (hGH) in children. It causes the abnormal lengthening of the long bones of the arms and legs so that the individual becomes unusually tall. Usually their body proportions remain normal. Gigantism is the opposite of *dwarfism* which can be caused by hyposecretion of hGH in children and which results in a lack of growth.

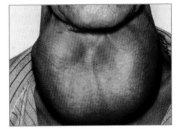

Goitre

Pathologies of the thyroid gland

Goitre
A goitre is an enlarged thyroid gland and it is a symptom of many different thyroid disorders. It is often caused by a lack of dietary iodine which is necessary for the correct functioning of the thyroid gland.

Graves' Disease
Graves' disease is a form of hyperthyroidism. It is a hereditary autoimmune disorder in which the thyroid gland is continually stimulated to produce hormones. Symptoms include an enlarged thyroid gland and oedema behind the eyes which can cause exophthalmos (protrusion of the eyeballs).

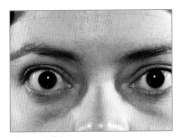

Graves' Disease

Hashimoto's Disease

Hashimoto's disease is an autoimmune disease in which the thyroid gland is inflamed (thyroiditis) and the secretion of thyroid hormones is sometimes affected. Hashimoto's disease occurs when the thyroid gland is attacked by its own antibodies and it is often accompanied by other disorders such as diabetes mellitus.

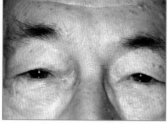

Myxoedema (Hypothyroidism)

Myxoedema (Hypothyroidism)

Myxoedema, or hypothyroidism, is the hyposecretion of thyroid hormones in adults. It is characterised by a swollen, puffy face due to oedema of the facial tissues, a slow heart rate, low body temperature, sensitivity to cold and dry hair and skin. People with myxoedema also feel tired and tend to gain weight easily.

Thyrotoxicosis (Hyperthyroidism)

Thyrotoxicosis is commonly called hyperthyroidism and is the hypersecretion of thyroid hormones. It can be caused by a tumour, overgrowth of the gland, Graves' disease or overstimulation due to an overactive pituitary gland. Thyrotoxicosis causes a speeding up of vital body functions and symptoms include a rapid heartbeat, sweating, loss of weight, anxiety and an intolerance of heat.

NEW WORDS

Apoptosis	the normal, ordered death and removal of cells as part of their development, maintenance and renewal.
Endocrine glands	ductless glands that secrete substances into the extracellular space around their cells. These secretions then diffuse into blood capillaries and are transported by the blood to target cells located throughout the body.
Endocrinology	the study of the endocrine glands and the hormones they secrete.
Exocrine glands	glands that secrete substances into ducts that carry these substances into body cavities or to the outer surface of the body.
Gonads	sex organs that produce mature sex cells.
Hormone	a chemical messenger that regulates cellular activity and is produced by an endocrine gland and transported in the blood.
Hypersecretion	over or excessive secretion.
Hyposecretion	under secretion.

Study Outline

Hormones controlled by the pituitary gland

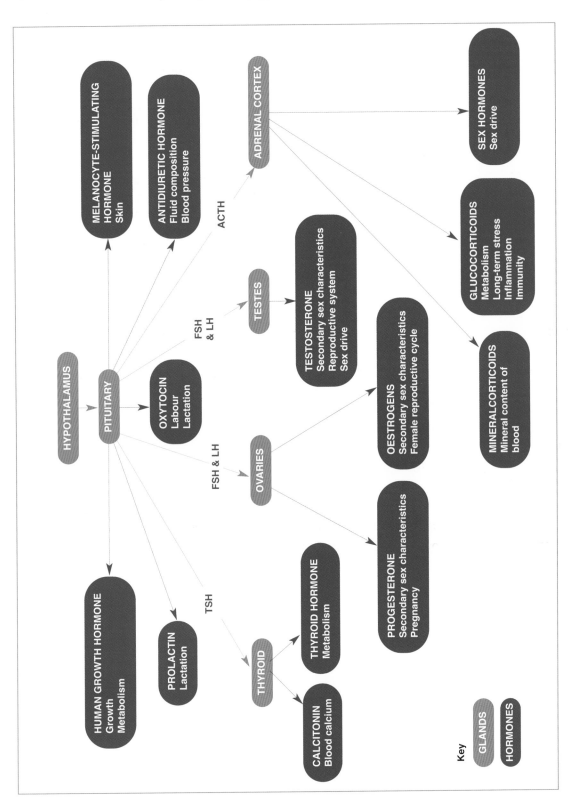

Hormones not controlled by the pituitary gland

Revision

1. Identify the difference between endocrine and exocrine glands.
2. Give two examples of exocrine glands.
3. Describe the functions of the endocrine system.
4. List the endocrine glands of the body.
5. Explain what a hormone is.
6. Explain how a hormone works.
7. Identify the main differences between the endocrine and nervous systems.
8. Explain the role of the hypothalamus in the endocrine system.
9. Draw up and complete a table with the following information for all the endocrine glands:
   * Endocrine gland
   * Location of gland
   * Hormones secreted
   * Actions of hormones
10. Describe the following diseases:
    * Myxoedema
    * Diabetes mellitus
    * Cushing's syndrome
    * Diabetes insipidus

Multiple choice questions

1. **Which of the following statements is correct?**
   a. The adrenal cortex secretes adrenaline.
   b. The adrenal cortex secretes glucagon.
   c. The adrenal cortex secretes mineralcorticoids.
   d. The adrenal cortex secretes insulin.

2. **Goitre is a pathology of the:**
   a. Ovaries
   b. Thyroid
   c. Parathyroids
   d. Adrenals.

3. **Where is the pituitary gland located?**
   a. In the head
   b. In the neck
   c. In the pelvis
   d. None of the above.

4. **Which of the following statements is correct?**
   a. Endocrine glands secrete substances directly into ducts.
   b. Endocrine glands secrete substances directly into target cells.
   c. Endocrine glands secrete substances into the lymphatic system.
   d. None of the above.

5. **Functions of antidiuretic hormone include:**
   a. Decreasing urine production.
   b. Increasing urine production.
   c. Decreasing stress levels.
   d. Increasing stress levels.

6. **Addison's Disease is caused by:**
   a. Hypersecretion of pancreatic hormones.
   b. Hyposecretion of pancreatic hormones.
   c. Hypersecretion of adrenal hormones.
   d. Hyposecretion of adrenal hormones.

7. **The hormone melatonin is secreted by which endocrine gland?**
   a. Pituitary
   b. Pineal
   c. Thyroid
   d. Thymus.

8. **Luteinising hormone is secreted by which endocrine gland?**
   a. Ovaries
   b. Adrenals
   c. Pineal
   d. Pituitary.

9. **The release of which hormone is stimulated by the sympathetic nervous system?**
   a. Somatostatin
   b. Melatonin
   c. Insulin
   d. Adrenaline.

10. **Which two hormones function in controlling blood calcium levels?**
    a. Oxytocin and human growth hormone
    b. Calcitonin and oxytocin
    c. Calcitonin and parathormone
    d. Human growth hormone and parathormone.

8 The Respiratory System

Introduction

If the surface area of your skin is approximately the size of a dining room table, then the surface area of your lungs is the size of a tennis court, however they are only ⅛th of the skin's weight. These unique organs play a vital role in the life of every cell in your body – without them there would be no oxygen and thus no energy.

In this chapter you will explore the respiratory system and follow the movement of air from the atmosphere into the cells of the body.

Student objectives

By the end of this chapter you will be able to:

- Describe the functions of the respiratory system.
- Describe the organisation of the respiratory system.
- Describe the organs and structures of the respiratory system.
- Explain the key stages of respiration.
- Identify the common pathologies of the respiratory system.

Functions of The Respiratory System

Every cell in the body needs oxygen to produce energy in the form of adenosine triphosphate (ATP). Without oxygen cells are unable to produce energy and will therefore die. Oxygen is vital to the survival of a cell yet, ironically, no cells have the ability to store it. They need a constant, new supply from the external environment. In addition, a by-product of cell metabolism is carbon dioxide. This is a waste product which also needs to be constantly removed from the body as an accumulation of it can poison cells.

Two systems work very closely to ensure there is a continuous supply of oxygen to all the cells of the body and a continuous removal of carbon dioxide. These are the respiratory and cardiovascular systems. The respiratory system takes in oxygen from the air we breathe and eliminates carbon dioxide while the cardiovascular system transports these two gases between the respiratory system and the cells of the body.

Gaseous exchange

The primary function of the respiratory system is the intake of oxygen and the elimination of carbon dioxide. This exchange of gases is called respiration and it takes place between the atmosphere, the blood and the cells in different phases:

- **Pulmonary ventilation** – The word *pulmo* refers to the lungs and pulmonary ventilation is another term for breathing. Air is inspired or breathed into the lungs and expired or breathed out of the lungs.
- **External respiration (pulmonary respiration)** – gaseous exchange between the lungs and the blood. In external respiration, the blood gains oxygen and loses carbon dioxide.
- **Internal respiration (tissue respiration)** – gaseous exchange between the blood and tissue cells. In internal respiration the blood loses oxygen and gains carbon dioxide.

NB: Cellular respiration (oxidation) is a metabolic reaction that takes place within a cell. It uses oxygen and glucose and produces energy in the form of ATP. A by-product of cellular respiration is carbon dioxide.

Olfaction

The respiratory system also functions in olfaction, which is the sense of smell. One of its structures, the nose, houses the olfactory receptors and olfaction is discussed in more detail in Chapter 6.

Sound production

Vibrating air particles produce sound. As we breathe air out, air passes through the larynx (voice box) where there are specialised membranes called vocal cords. The air causes these to vibrate and produce sounds which are converted into words by the muscles of the pharynx, face, tongue and lips. The pharynx, mouth, nasal cavity and paranasal sinuses also act as resonating chambers for sound.

Structure Of The Respiratory System

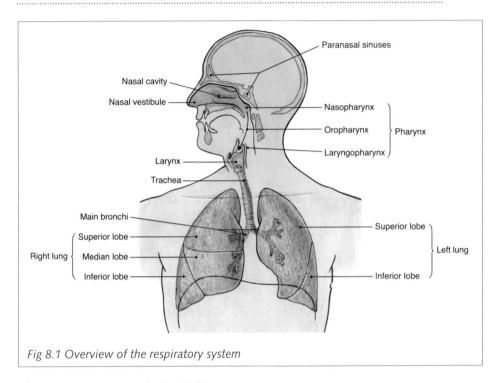

Fig 8.1 Overview of the respiratory system

The respiratory system is divided into two zones:

- **The conducting zone** – a series of interconnecting passageways that allows air to reach the lungs. No gaseous exchange occurs here. The function of the structures of this zone is to transport air to the alveoli and to filter, moisten and warm it. Structures of the conducting zone include the nose, pharynx, larynx, trachea and the bronchi and their smaller branches.
- **The respiratory zone** – this is where the exchange of gases occurs. The alveoli form the respiratory zone.

The Nose

Functions of the nose in a nutshell

- Inhaling air.
- Filtering, warming and moistening air.
- Receiving olfactory stimuli.
- Acting as a resonating chamber for sound.

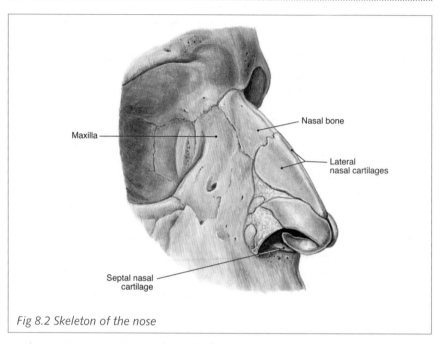

Maxilla

Nasal bone

Lateral nasal cartilages

Septal nasal cartilage

Fig 8.2 Skeleton of the nose

Air enters the respiratory system through the nose which is a framework of bone and hyaline cartilage that is covered by skin and lined internally with a mucous membrane.

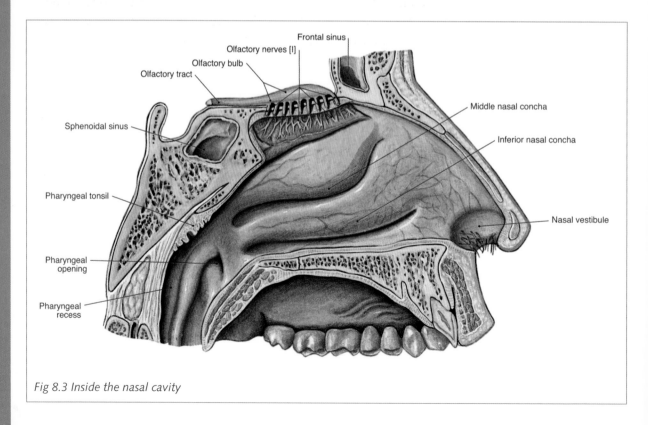

Frontal sinus

Olfactory nerves [I]

Olfactory bulb

Olfactory tract

Sphenoidal sinus

Middle nasal concha

Inferior nasal concha

Pharyngeal tonsil

Nasal vestibule

Pharyngeal opening

Pharyngeal recess

Fig 8.3 Inside the nasal cavity

The nose has the following structures:

- Two external *nares* which are the openings to the nose and are commonly called the nostrils. Inside the nares are coarse hairs that filter out large dust particles.
- A large internal cavity called the *nasal cavity* which is divided into two by the *septum*. The nasal cavity contains:
  - The *olfactory receptors* necessary for the sense of smell.
  - Duct openings from the paranasal sinuses. These allow mucous to drain from the sinuses into the nasal cavity.
  - The nasal conchae (turbinate bones) which look like shelves projecting from the lateral walls of the nasal cavity. The air whirls around them and as it does so, it is warmed by the blood in their capillaries.

  The nasal cavity is lined with:
  - A *mucous membrane* that secretes mucous which moistens the air and traps dust particles.
  - *Cilia* that move the dust-laden mucous down towards the pharynx where it can be swallowed or spat out.
- At the back of the nasal cavity are two internal nares (*choanae*) which are openings that connect the nasal cavity to the pharynx.

Anatomy and physiology in perspective
If the cilia lining the respiratory system are damaged, they cannot move dust-laden mucous out of the system. The only way it can then be removed is through coughing. Substances in cigarette smoke inhibit the movement of cilia and therefore coughing, or smoker's cough, is common amongst people who smoke.

The Pharynx (Throat)

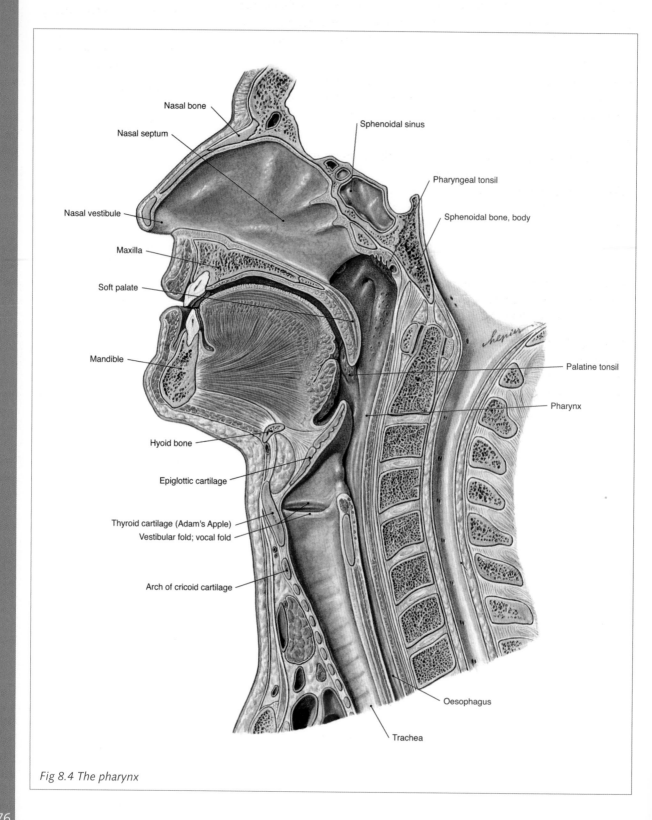

Nasal bone

Nasal septum

Nasal vestibule

Maxilla

Soft palate

Mandible

Hyoid bone

Epiglottic cartilage

Thyroid cartilage (Adam's Apple)

Vestibular fold; vocal fold

Arch of cricoid cartilage

Sphenoidal sinus

Pharyngeal tonsil

Sphenoidal bone, body

Palatine tonsil

Pharynx

Oesophagus

Trachea

Fig 8.4 The pharynx

After being breathed in through the external nares (nostrils) and warmed, moistened and filtered in the nasal cavity, air then passes through the internal nares and into the pharynx (throat). This is a funnel-shaped tube whose walls are made up of skeletal muscles lined by mucous membrane and cilia. The mucous traps dust particles and cilia move the mucous downwards.

The pharynx is divided into three portions which are named after the structure to which they are closest: the nasopharynx (nasal cavity), oropharynx (oral cavity) and laryngopharynx (larynx).

The **nasopharynx**:
- Receives air and dust-laden mucous from the nasal cavity and transports it downwards towards the oropharynx.
- Contains the *pharyngeal tonsil (adenoid)* which functions in immunity.
- Contains openings from the *Eustachian tubes* and exchanges small amounts of air with them to equalise the pressure between the pharynx and middle ear.

The **oropharynx**:
- Receives air from the nasopharynx and air, food and drink from the mouth. It transports these downwards to the laryngopharynx.
- Contains the *palatine* and *lingual tonsils* which function in immunity.

The **laryngopharynx (hypopharynx)**:
- Receives air from the laryngopharynx and transports it into the larynx.
- Receives food and drink from the laryngopharynx and transports it into the oesophagus.

Functions of the pharynx in a nutshell
- Acting as a passageway for air, food and drink.
- Acting as a resonating chamber for sound.
- Housing the tonsils which function in immunity.

The Larynx (Voicebox)

The pharynx transports air to the larynx which is a short passageway between the laryngopharynx and the trachea. The larynx is made up of eight pieces of rigid hyaline cartilage and a leaf-shaped piece of elastic cartilage called the *epiglottis* which protects the opening of the larynx. The first piece of cartilage in the larynx is the *thyroid cartilage*. It gives the larynx its triangular shape. The last piece is the *cricoid cartilage* which connects the larynx to the trachea.

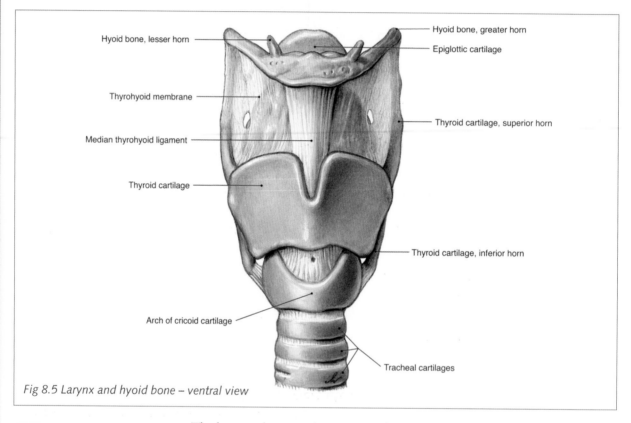

Hyoid bone, lesser horn
Thyrohyoid membrane
Median thyrohyoid ligament
Thyroid cartilage
Arch of cricoid cartilage

Hyoid bone, greater horn
Epiglottic cartilage
Thyroid cartilage, superior horn
Thyroid cartilage, inferior horn
Tracheal cartilages

Fig 8.5 Larynx and hyoid bone – ventral view

The larynx plays two important roles in the respiratory system:

- **It routes air and food into their correct channels** – When we swallow liquids or food, the larynx rises and the epiglottis moves downwards to form a lid over the opening of the larynx. This stops any liquids or food from entering into the larynx. When we are not swallowing, the larynx is in its normal position and the epiglottis does not cover the larynx. Thus, air can move freely into it.
- **It produces sound** – Sound waves are produced by the alternating compression and decompression of air molecules. In the larynx are cords of mucous membrane called the vocal folds (vocal cords) that vibrate when air molecules are forced against them during exhalation. This vibration causes sound.

Infobox

Anatomy and physiology in perspective
If small particles of food, fluid, dust or even smoke do pass into the larynx they are immediately expelled by a cough reflex.

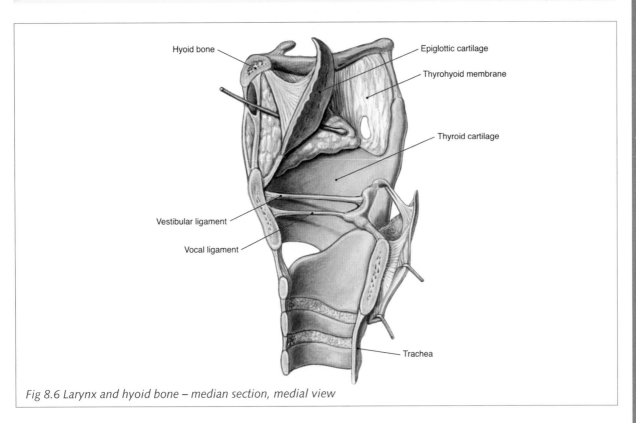

Hyoid bone

Epiglottic cartilage

Thyrohyoid membrane

Thyroid cartilage

Vestibular ligament

Vocal ligament

Trachea

Fig 8.6 Larynx and hyoid bone – median section, medial view

Infobox

Anatomy and physiology in perspective
Laryngitis is inflammation of the larynx. It is often characterised by hoarseness or voice loss because the inflammation prevents the folds from contracting properly or vibrating freely.

In the classroom…

Put your hand on the middle of your throat and swallow. You will be able to feel your larynx rising as you swallow.

Functions of the larynx in a nutshell
- Routing air and food into their correct channels.
- Producing sound.

The Trachea (Windpipe)

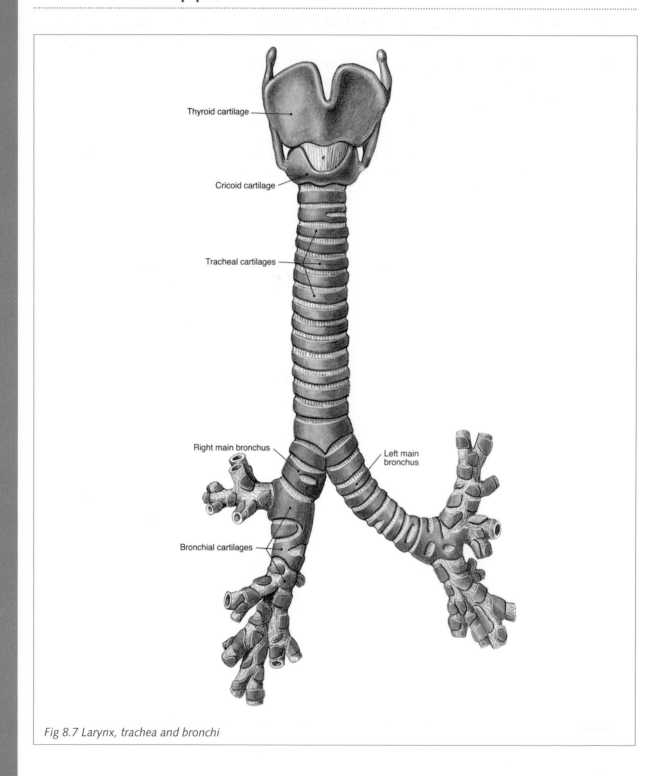

Thyroid cartilage

Cricoid cartilage

Tracheal cartilages

Right main bronchus

Left main bronchus

Bronchial cartilages

Fig 8.7 Larynx, trachea and bronchi

From the larynx, air moves into the trachea or windpipe. This is a long tubular passageway which transports air from the larynx into the bronchi. The trachea lies in front of the oesophagus and is composed of 16–20 incomplete C-shaped rings of hyaline cartilage.

The *open parts* of the C-shape are held together with transverse smooth muscle fibres and elastic connective tissue. This open area lies against the oesophagus and allows for expansion of the oesophagus during swallowing.

The *cartilage parts* of the C-shape are solid so that they can support the trachea and keep it open despite changes in breathing. The trachea is lined with mucous membrane and cilia that move any minute dust particles still in the respiratory system upwards away from the lungs to the throat where they can be swallowed or spat out.

Functions of the trachea in a nutshell
Transports air from the larynx into the bronchi.

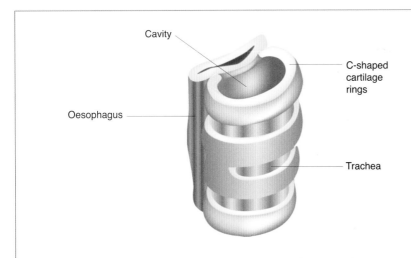

Cavity

C-shaped cartilage rings

Oesophagus

Trachea

Fig 8.8 Relationship of the trachea to the oesophagus

The trachea is composed of incomplete c-shaped rings of cartilage that allow the oesophagus to expand during swallowing.

The respiratory system

The Bronchi

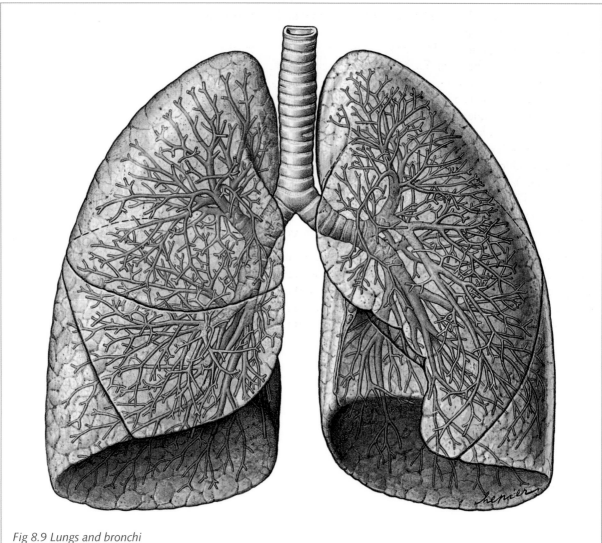

Fig 8.9 Lungs and bronchi

Having travelled down the trachea, air is split into branch-like passageways called *bronchi* (singular = *bronchus*). These bronchi repeatedly divide inside the lungs into smaller and smaller branches that finally carry the air into the alveoli. Their division is similar to the branching of a tree from a large central trunk (trachea) into branches (bronchi), twigs (bronchioles) and finally leaves (alveoli). This continual dividing of the passageways means that air can be transported to literally millions of alveoli where gaseous exchange can finally take place.

Branching of the bronchi is as follows:

- **Primary bronchi** – The trachea divides into the right and left primary bronchi. The right primary bronchus is shorter and wider than the left and carries air into the right lung. The left primary bronchus carries air into the left lung.

- **Secondary bronchi (lobar bronchi)** – The primary bronchi then divide into secondary bronchi. There is one secondary bronchus for each of the lobes of the lungs: the right lung has three lobes and therefore has three secondary bronchi while the left lung has two lobes and therefore has two secondary bronchi.
- **Tertiary bronchi (segmental bronchi)** – The secondary bronchi then divide into tertiary bronchi. Each tertiary bronchi supplies a segment of lung tissue which in itself has many lobules. A *lobule* is a small compartment of tissue containing a lymphatic vessel, blood vessels (an arteriole and a venule) and a branch from a terminal bronchiole. Each lobule is wrapped in elastic connective tissue.
- **Bronchioles** – Tertiary bronchi divide into *bronchioles*.
  - The bronchioles themselves divide repeatedly until they finally become *terminal bronchioles*.
  - Terminal bronchioles subdivide into microscopic branches called *respiratory bronchioles*.
  - The respiratory bronchioles finally divide into *alveolar ducts*.

Similar to the trachea, the bronchi are composed of incomplete rings of cartilage and are lined with mucous membrane. However, as branching takes place, gradual changes occur to the structure of the branches until the cartilage is finally replaced by spiral bands of smooth muscle and the protective mechanism of the cilia is replaced by the action of macrophages.

The Lungs

The word *lunge* means 'light-weight' and although the total surface area of the lungs is approximately 40 times the external surface area of the entire body, they only weigh less than 1kg. This unique surface area is thanks to the branching of the bronchi and bronchioles as described on the previous page.

The two lungs are cone-shaped organs that occupy most of the thoracic cavity. They extend from the diaphragm to slightly above the clavicles and are surrounded and protected by the ribs. The lungs are:

- Different in size and shape. The right lung is shorter, thicker and broader than the left. This is because the diaphragm is higher on the right side to accommodate the large liver beneath it.
- Separated from one another by the heart and the mediastinum, which is a mass of tissue extending from the sternum to the vertebral column.
- Covered with and protected by the pleural membrane. This is a serous membrane made up of two layers:
  - The superficial *parietal pleura* which lines the walls of the thoracic cavity.
  - The deep *visceral pleura* which covers the lungs.

The space between the two layers is called the *pleural cavity*. It contains a lubricating fluid that is secreted by the membranes to reduce friction between them. Thus, they are able to slide freely over one another.

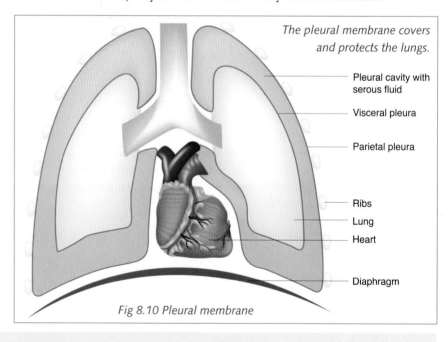

The pleural membrane covers and protects the lungs.

- Pleural cavity with serous fluid
- Visceral pleura
- Parietal pleura
- Ribs
- Lung
- Heart
- Diaphragm

Fig 8.10 Pleural membrane

Anatomy and physiology in perspective
The mediastinum divides the thoracic into two distinct chambers so that if one lung collapses the other one will not be affected and will be able to continue functioning.

The alveoli

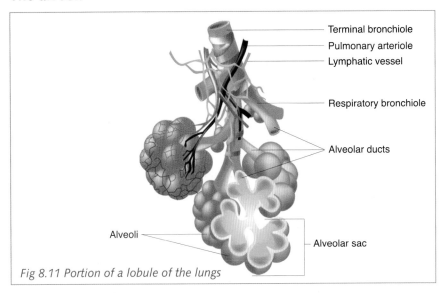

Terminal bronchiole
Pulmonary arteriole
Lymphatic vessel

Respiratory bronchiole

Alveolar ducts

Alveoli

Alveolar sac

Fig 8.11 Portion of a lobule of the lungs

Once air has travelled through the conducting passageways of the respiratory system, it finally arrives at the air sacs, or *alveoli* (singular *alveolus*), inside the lungs:

- Alveoli are cup-shaped pouches where the exchange of gases occurs. It is estimated that there are approximately 300 million alveoli in the lungs and they provide a huge surface area for gaseous exchange.
- The walls of the alveoli are extremely thin and composed mainly of a single layer of squamous epithelial cells. They contain:
  - Cells that secrete *alveolar fluid* which keeps the walls moist.
  - *Macrophages* that remove any fine dust particles and micro-organisms that have not already been removed in the conducting passageways.
  - Alveoli are surrounded by a dense network of blood vessels called *pulmonary capillaries*. The thin walls of the alveoli and the thin walls of the pulmonary capillaries together form the respiratory membrane.
- The *respiratory membrane* is the site of gaseous exchange between the lungs and the blood. It is an extremely thin membrane so that the gases can diffuse rapidly across it.

Blood supply to the lungs

The lungs actually have a double blood supply:

- **Pulmonary arteries** bring deoxygenated blood to the lungs. The blood is oxygenated by the lungs and this oxygenated blood is transported from the lungs to the heart by the pulmonary veins. These are the only veins in the body that carry oxygenated blood.
- **Bronchial arteries** bring oxygenated blood to the lung tissue. Most of this blood is returned to the heart by the pulmonary veins. However, some of it drains into the bronchial veins which transport it to the superior vena cava and then to the heart.

NB: Blood and blood vessels will be discussed in detail in Chapter 9.

Physiology Of Respiration

As mentioned earlier, respiration is the exchange of gases between the atmosphere, blood and cells and it occurs in pulmonary ventilation, external respiration, internal respiration and cellular respiration. We will now look at pulmonary ventilation and external and internal respiration in more detail.

Pulmonary ventilation (breathing)

Pulmonary ventilation is the inspiration and expiration, or inhalation and exhalation, of air. In other words, it is breathing. It is a mechanical process in which air is sucked into the lungs from the atmosphere and then exhaled out and it is dependent on the existence of a pressure gradient between the pressure inside the lungs and outside the lungs.

Boyle's Law

In order to understand the movement of air between the lungs and the atmosphere it helps if you understand Boyle's law, which states that the pressure of a gas in a closed container is inversely proportional to the volume of the container.

In other words, if a gas is put into a closed container the gas molecules will spread out and have space around them so they do not bump into each other or into the walls of the container. This means that the pressure in the container will be low. If the volume or size of the container is then reduced, the gas molecules will have less room to move about and will be closer to one another. They will, therefore, bump into each other and the walls of the container and the pressure will be increased.

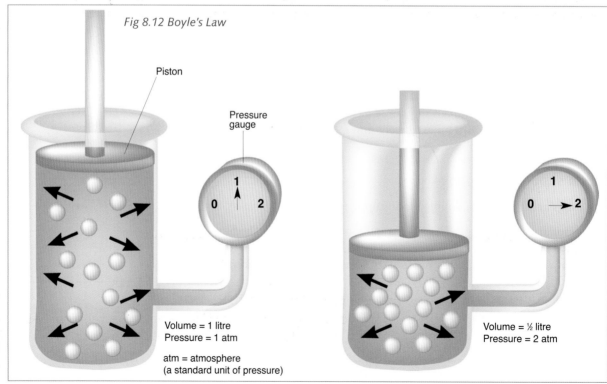

Fig 8.12 Boyle's Law

Piston

Pressure gauge

Volume = 1 litre
Pressure = 1 atm

Volume = ½ litre
Pressure = 2 atm

atm = atmosphere
(a standard unit of pressure)

Inspiration (inhalation)

Inspiration is the movement of air from the atmosphere into the lungs. It occurs when the volume size of the lungs increases, therefore reducing the pressure within the lungs. When the pressure within the lungs is lower than the pressure outside the lungs, air is drawn in to equalise this pressure. Inspiration happens as follows:

- The volume of the lungs (intrapulmonary volume) increases when the size of the thoracic cavity increases. This occurs when:
  - The diaphragm contracts – In its relaxed state, the diaphragm is dome shaped. However, when it contracts it flattens and this increases the vertical dimension of the thoracic cavity.
  - The external intercostal muscles contract – When these muscles contract they pull the ribs upwards and push the sternum forwards. This increases the anterior-posterior dimension of the thoracic cavity.
- An increased intrapulmonary volume causes a decrease in intrapulmonary pressure. When the pressure inside the lungs (intrapulmonary pressure) is less than the atmospheric pressure, a partial vacuum is created.
- A partial vacuum pulls the air into the lungs. Air flows into the lungs to equalise the pressure between the lungs and the atmosphere.

Did you know?
When resting, a healthy adult moves approximately 6 litres of air in and out of their lungs every minute.

Pulmonary ventilation, or breathing, is a mechanical process dependent on the existence of a pressure gradient between the pressure inside and outside the lungs.

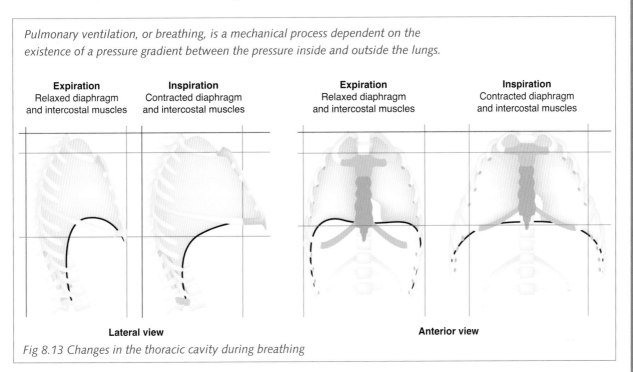

Expiration
Relaxed diaphragm and intercostal muscles

Inspiration
Contracted diaphragm and intercostal muscles

Expiration
Relaxed diaphragm and intercostal muscles

Inspiration
Contracted diaphragm and intercostal muscles

Lateral view

Anterior view

Fig 8.13 Changes in the thoracic cavity during breathing

Anatomy and physiology in perspective
Hiccups are sudden inspirations caused by spasms of the diaphragm. As the air hits the vocal folds a hiccupping sound occurs.

Expiration (exhalation)

Normal expiration – During normal, quiet breathing, expiration is a passive process that does not involve any muscular contraction. After being stretched during inspiration, the muscles of the lungs and chest recoil to their natural state and the volume of the lungs decreases. This increases the pressure within the lungs and once the intrapulmonary pressure is greater than the atmospheric pressure, the air will move out of the lungs to the area of lowest pressure.

Active expiration – Active expiration occurs when the abdominal and internal intercostal muscles contract. These move the ribs downwards and compress the abdominal viscera. This movement forces the diaphragm upwards which reduces the size of the thoracic cavity. Thus, the volume of the lungs decreases, the pressure within them increases and expiration occurs.

External respiration (pulmonary respiration)

Once air has been breathed into the lungs, an exchange of gases takes place between the alveoli and the blood in the pulmonary capillaries. Here oxygen diffuses from the alveolar air into the blood and carbon dioxide moves out of the blood into the alveolar air from where it is expired from the body.

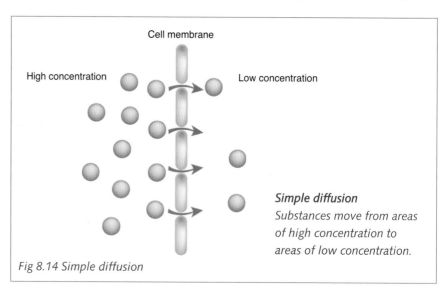

Cell membrane

High concentration

Low concentration

Simple diffusion
Substances move from areas of high concentration to areas of low concentration.

Fig 8.14 Simple diffusion

External respiration occurs as follows:

- The right ventricle of the heart pumps deoxygenated blood through the pulmonary arteries into the pulmonary capillaries surrounding the alveoli.
- Gaseous exchange occurs between the alveoli and the capillaries.
  - Oxygen diffuses into the blood and carbon dioxide diffuses out of the blood.
    - Because the body cells are continually removing oxygen from the blood, there is always more oxygen in the alveoli than in the blood.
    - Therefore, oxygen diffuses from an area of high concentration (alveolar air) to an area of low concentration (blood).

- Likewise, because the cells are continually releasing carbon dioxide into the blood, there is always more carbon dioxide in the blood than in the alveoli. Therefore, carbon dioxide diffuses from an area of high concentration (blood) to an area of low concentration (alveolar air).

- The structure of the capillaries ensures diffusion can take place rapidly between the alveoli and the blood:

  - There are a large number of capillaries surrounding the alveoli and this provides a large surface area for the exchange of gases.
  - The capillaries are also extremely narrow so that only one red blood cell can pass through them at a time. This ensures there is minimal diffusion distance between the capillaries and the alveoli and also ensures the blood flows slowly through them so that there is sufficient time for gaseous exchange to take place.
  - The respiratory membrane between the alveoli and capillaries is very thin to ensure minimal diffusion distance.

- Oxygenated blood is transported to the heart from where it is pumped into systemic circulation. Oxygen attaches to haemoglobin molecules in red blood cells and is transported in the blood as *oxyhaemoglobin*. Minute amounts of oxygen are also transported in blood plasma. Carbon dioxide is transported in blood plasma as *bicarbonate ions*. Small amounts of carbon dioxide are also carried inside red blood cells.

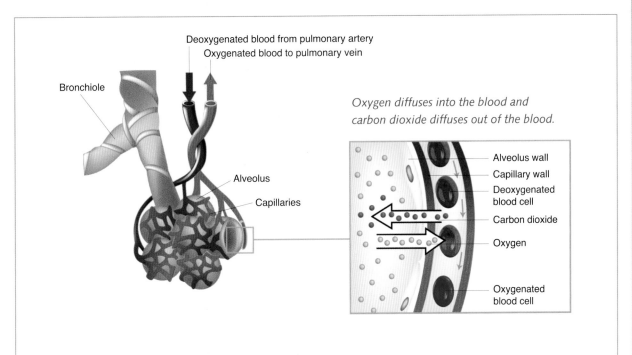

Fig 8.15 Gaseous exchange between the alveoli and capillaries

Anatomy and physiology in perspective
Carbon monoxide combines more easily and strongly to haemoglobin than oxygen does and can cause **hypoxia**, which is the inefficient delivery of oxygen to tissues, or even carbon monoxide poisoning which can be fatal. Carbon monoxide is a colourless, odourless gas found in exhaust fumes from cars as well as fumes from burning tobacco, coal, gas and wood.

Internal respiration (tissue respiration)

Oxygen finally enters cells and carbon dioxide leaves them through the process of internal, or tissue, respiration. This is the exchange of gases between the blood and tissue cells.

In internal respiration, oxygenated blood loses oxygen and thus becomes deoxygenated. It also gains carbon dioxide and returns to the heart from where it is pumped to the lungs to be oxygenated again. Thus begins a new cycle of respiration.

Composition of inspired and expired air

Air is a mixture of oxygen, carbon dioxide, nitrogen, water vapour and a small quantity of inert gases. The percentages of these gases varies in inspired and expired air as shown in the chart below.

GAS	INSPIRED AIR %	EXPIRED AIR %
Oxygen	21	16
Carbon dioxide	0.04	4–4.5
Nitrogen and inert gases	78–79	78–79
Water vapour	Varies	Varies

Common Pathologies Of The Respiratory System

What does smoking really do to your respiratory system?

Smoking decreases your respiratory efficiency in a number of ways.

- The flow of air into and out of the lungs is decreased or hindered by:
  - The constricting effect of nicotine on terminal bronchioles.
  - Irritants present in smoke that increase the secretion of mucous and the swelling of the mucous membrane that lines the respiratory tract.
- The amount of oxygen carried by the blood is reduced by carbon monoxide in smoke which combines with haemoglobin more readily than oxygen does.
- The removal of dust-laden mucous and foreign debris from the system is decreased by irritants present in smoke that decrease the movement of the cilia in the lining of the respiratory tract.
- Smoking also destroys the elastic fibres of the lungs and this can eventually lead to the collapse of small bronchioles.

Finally, smoking causes a number of respiratory disorders including emphysema and lung cancer and is also implicated in cardiovascular diseases such as atherosclerosis and strokes as well as the early onset of osteoporosis.

Chronic obstructive pulmonary disease (COPD)

Chronic obstructive pulmonary disease (COPD) is a group of diseases and disorders that all have some degree of obstruction of the airways. It includes asthma, bronchitis and emphysema and coughing, wheezing and dyspnoea are common symptoms of COPD:

- Coughing – coughs are sudden, explosive movements of air rushing upwards through the respiratory passageways. They clear the passageways.
- Wheezing – a whistling sound produced when the airways are partially obstructed by, for example, a narrowing of the airways, a lodged particle or excess mucous.
- Dyspnoea – laboured or difficult breathing, usually due to partially obstructed airways.
- Sputum – material coughed up from the respiratory tract.

Asthma
Asthma is a chronic, inflammatory disorder in which the airways narrow in response to certain stimuli ranging from pollen and house dust mites to cold air and emotional upsets. Asthma is characterised by periods of coughing, difficult breathing and wheezing and during an asthma attack the person struggles to exhale.

Bronchitis
Bronchitis is inflammation of the bronchi, the airways branching off the trachea. It can be caused by infection or by exposure to irritants such as cigarette smoke. Bronchitis is characterised by excessive mucous secretion which results in a productive cough in which sputum is raised, wheezing and dyspnoea.

Emphysema

The word emphysema means 'blown up' or 'full of air' and it is an irreversible disease in which the alveolar walls disintegrate, leaving abnormally large air spaces in the lungs. These spaces remain filled with air and the person struggles to exhale this air and therefore is constantly exhausted from the effort of trying to breathe out. Emphysema can be caused by cigarette smoke, air pollution or occupational exposure to industrial dust.

Infectious and environmentally related diseases and disorders

Although the respiratory system is equipped to remove many foreign particles before they reach the lungs, it is still vulnerable to infectious and environmentally related diseases because pathogens and irritants can be inhaled directly into the system.

Study tip...

Here is a quick reminder of the different ways in which the respiratory system defends itself from foreign particles:

- Coarse hairs inside the nostrils filter out large dust particles.
- A mucous membrane lines the respiratory passages and traps foreign particles.
- Cilia also line the respiratory passages and move dust-laden mucous towards the pharynx where it can be swallowed or spat out.
- Particles that have not been removed by the mucous and cilia are usually engulfed by macrophages in the alveoli.

Asbestosis

Asbestosis is the scarring of lung tissue due to the inhalation of asbestos dust. It generally only affects people who work with asbestos that has been broken into many pieces (it is, however, illegal in many countries to install asbestos or work with it under unsafe conditions). Asbestosis is characterised by pulmonary fibrosis, which is scarring of the lung tissue, and the thickening of the pleural membrane. Asbestos dust is also known to cause mesothelioma, which is a cancer of the pleural membrane. Symptoms of asbestosis include shortness of breath and difficulty in breathing. It can eventually lead to respiratory failure.

Cor pulmonale

Cor pulmonale is the enlargement and thickening of the right ventricle of the heart. It is due to abnormally high blood pressure in the pulmonary arteries (pulmonary hypertension) which causes the right ventricle to work extra hard and thus thicken and enlarge. Cor pulmonale does not have many symptoms until it is in its advanced stages. Then its symptoms are the same as those of pulmonary hypertension: shortness of breath, lightheadedness, fatigue and chest pain. Cor pulmonale can lead to right-sided heart failure.

Cystic fibrosis

Cystic fibrosis is a genetic disease that causes certain exocrine glands to produce abnormal secretions and it is characterised by the production of thick mucous secretions that do not drain easily. This leads to inflammation and the replacement of healthy tissue with connective tissue. The airways, pancreas, salivary glands, and sweat glands can all be affected and symptoms can include breathing difficulty, pancreatic insufficiency and cirrhosis of the liver.

Hayfever (seasonal allergies)

Hayfever is a seasonal allergy resulting from exposure to pollens and other airborne substances that only appear during certain seasons of the year. It is characterised by itching of the nose, roof of the mouth and back of the throat and eyes as well as sneezing.

Hyperventilation

Hyperventilation is breathing abnormally fast when the body is at rest. It is characterised by dizziness, tingling sensations and tightness across the chest. Prolonged hyperventilation can lead to a loss of consciousness.

Influenza (flu)

Influenza, or flu, is a contagious viral infection of the respiratory system. It is different to a common cold in that it is caused by a different virus and its symptoms are far more severe. Symptoms include a runny nose, sore throat, cough, headache, fever and muscular aches and pains.

Laryngitis

Laryngitis is inflammation of the larynx (voice box) and vocal folds. It is characterised by hoarseness or a loss of voice and symptoms may also include a sore throat, difficulty in breathing and a painful or tickling cough. Laryngitis is usually caused by a viral infection of the respiratory system, for example a common cold or bronchitis, but it can also be caused by excessive use of the voice, an allergy or an irritation of the larynx from substances such as cigarette smoke.

Lung cancer

Lung cancer is cancer of the lungs and it can either originate in the cells of the lungs (primary lung cancer) or metastasise to the lungs from other parts of the body. The symptoms of lung cancer vary according to the type and location of the cancer, but they can include a persistent cough, blood-streaked sputum, wheezing, shortness of breath and chronic pneumonia. Causes of lung cancer can include smoking or exposure to cigarette smoke and occupational exposure to substances such as asbestos.

Methicillin Resistant Staphylococcus Aureus (MRSA)

Methicillin resistant staphylococcus aureus, or MRSA, is a common bacterium that is resistant to many types of antibiotics and that is now responsible for infectious outbreaks in hospitals. The term MRSA is also used more generally to describe a number of different strains of staphylococcus bacteria that are all resistant to one or more conventional antibiotics. Usually patients in hospitals already have lowered immunity and weakened systems and so are more susceptible to infection from bacteria that can easily be passed in a hospital environment. Some patients can show no signs of infection whilst others can have swelling and tenderness at the sites affected (most commonly surgical wounds, burn sites and catheter entry points).

Did you know?
Over 90% of lung cancer patients are smokers[i] and lung cancer is the most common cause of death from cancer in both men and women.

Pharyngitis

Pharyngitis is infection of the pharynx, or throat, and sometimes the tonsils. Symptoms include a sore throat, pain on swallowing and occasionally earache. Pharyngitis can be caused by bacteria or viruses and often accompanies a common cold.

Pleurisy

Pleurisy is inflammation of the pleura and common symptoms include chest pain, rapid and shallow breathing and neck and shoulder referred pain.

Pneumonia

Pneumonia is inflammation of the lungs due to the infection and inflammation of the alveoli and the tissues surrounding them. Pneumonia can be caused by a bacterium, virus or fungus and symptoms can include a sputum-producing cough, chest pain, chills, fever and shortness of breath. Those who are susceptible to pneumonia are the elderly, infants, immuno-compromised individuals, cigarette smokers and people with obstructive lung disease.

Pneumothorax

Pneumothorax is a condition in which there is air between the two layers of pleura. This pocket of air causes the lung to collapse. Symptoms of pneumothorax depend on how much air is between the pleura and how much of the lung has collapsed and can vary from a shortness of breath and chest pain to shock or cardiac arrest. Pneumothorax can occur spontaneously for no apparent reason or it can be caused by lung conditions such as emphysema or cystic fibrosis. It can also be caused by trauma or injury to the lungs.

Pulmonary embolism

Pulmonary embolism is the blocking of the pulmonary artery by an embolus which is material, such as a blood clot, carried in the blood. Pulmonary embolism is generally caused by a blood clot that has formed in a leg or pelvic vein (and occasionally by clots formed in the veins of the arms). These clots can form in people who have been kept still for a long time, for example, those on prolonged bed rest or seated in an aeroplane for many hours. Symptoms of pulmonary embolism will vary according to the extent of the blockage, but it is usually characterised by a sudden shortness of breath, rapid breathing and extreme anxiety.

Pulmonary fibrosis

Pulmonary fibrosis is the thickening and scarring of the lungs and it develops from persistent inflammation of the lung tissue. It can be caused by a number of diseases and it results in the shrinking and stiffening of the lungs and a decrease in their ability to transfer oxygen to the blood. Initial symptoms can include a shortness of breath on exertion, loss of stamina, fatigue and weight loss. As the disease develops it can cause cor pulmonale.

Rhinitis

Rhinitis is the inflammation of the mucous membrane lining the nose. It is characterised by a runny or stuffy nose and often accompanies a common cold or an allergic reaction.

Sarcoidosis

Sarcoidosis is a disease that affects many organs of the body, but primarily the lungs. It is characterised by the presence of granulomas (collections of inflammatory cells) in the lungs, lymph nodes, liver, eyes and skin as well as other organs. Sarcoidosis can have no symptoms or fever, fatigue, chest pain, malaise, weight loss and aching joints may be present. The lymph nodes may also enlarge and night sweats can occur. The cause of sarcoidosis is unknown.

Severe Acute Respiratory Syndrome (SARS)

Severe acute respiratory syndrome, or SARS, is a viral infection with flu-like symptoms such as a fever, headache, sore throat and cough. SARS is believed to be spread through airborne droplets of fluid, for example, through coughing and sneezing. SARS is difficult to treat and because it has an incubation period of approximately ten days, infected people and those who have been in contact with an infected person need to be isolated to ensure it does not spread.

Sinusitis

Sinusitis is inflammation of the sinuses and it is usually caused by an allergy or infection. Symptoms include pain, tenderness and swelling over the affected sinus as well as nasal congestion and post-nasal drip.

Tuberculosis (TB)

Tuberculosis is a contagious infectious disease caused by an airborne bacterium. It can affect almost any organ in the body, but usually affects the lungs because the airborne bacterium is inhaled. The bacteria destroy parts of the lung tissue and it is replaced by fibrous connective tissue or nodular lesions called tubercles. Symptoms can include coughing, night sweats, a sense of malaise and decreased energy and appetite that can lead to weight loss.

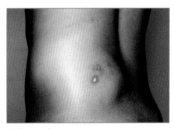

Tuberculosis

Whooping Cough (pertussis)

Whooping cough is a contagious bacterial infection that begins with mild cold-like symptoms and then develops into severe coughing fits. These coughing fits are characterised by a prolonged, high-pitched indrawn breath or whoop at the end of them. They also usually produce large amounts of thick mucous. Although the severity of the fits soon subsides, a persistent cough can linger for many weeks or even months.

Cough	a sudden, explosive movement of air rushing upwards through the respiratory passages.
Dyspnoea	laboured or difficult breathing.
Hypoxia	the inefficient delivery of oxygen to tissues.
Mediastinum	a space in the thorax containing the aorta, heart, trachea, oesophagus and thymus gland. It is found between the two pleural sacs.
Olfaction	the sense of smell.
Oxidation	cellular respiration.
Respiration	the exchange of gases between the atmosphere, blood and cells.
Sputum	material coughed up from the respiratory tract.
Wheezing	a whistling sound produced when the airways are partially obstructed.

Study Outline

Functions of the respiratory system

Functions of the respiratory system include gaseous exchange, olfaction and sound production.

Structure of the respiratory system

The respiratory system is divided into:

1. The conducting zone that allows air to reach the lungs.
2. The respiratory zone where gaseous exchange occurs.

Nose
1. Air enters the system through the nose where it is filtered, warmed and moistened.
2. The nose also receives olfactory stimuli and acts as a resonating chamber for sound.

Pharynx
1. Air then goes into the pharynx (throat) which is divided into the nasopharynx, oropharynx and laryngopharynx.
2. The pharynx is a passageway for air, food and drink and acts as a resonating chamber for sound.

Larynx
1. From the pharynx air goes into the larynx (voicebox) which routes air and food into their correct channels and produces sound.
2. The opening of the larynx is protected by the epiglottis.

Trachea
1. From the larynx, air goes into the trachea (windpipe) which is made of incomplete c-shaped rings of cartilage and which lies in front of the oesophagus.
2. The trachea transports air into the bronchi.

Bronchi

1. The bronchi divide repeatedly to form the bronchial tree.

The Bronchial Tree

Trachea

⇓

Primary bronchi

⇓

Secondary bronchi

⇓

Tertiary bronchi

⇓

Bronchioles

⇓

Terminal bronchioles

⇓

Respiratory bronchioles

⇓

Alveolar ducts

Lungs

1. The lungs are cone-shaped organs that occupy most of the thoracic cavity. They are separated by the mediastinum and protected by the ribs.
2. The right lung is shorter, thicker and broader than the left.
3. The lungs are covered and protected by the pleural membrane which consists of the parietal pleura and visceral pleura.

Alveoli

1. Having travelled through the conducting passageways, air arrives at the alveoli (air sacs) where gaseous exchange takes place.
2. The alveoli are surrounded by a dense network of pulmonary capillaries.
3. The respiratory membrane is made up of the thin walls of the alveoli and pulmonary capillaries and it is the site of gaseous exchange between the lungs and the blood.

Blood supply to the lungs

1. Pulmonary arteries bring deoxygenated blood to the lungs where it is oxygenated and then transported to the heart by the pulmonary veins.
2. Bronchial arteries bring oxygenated blood to the lung tissue.

The physiology of respiration

Pulmonary ventilation (breathing)

1. Pulmonary ventilation is breathing and it includes inspiration and expiration.
2. Pulmonary ventilation is a mechanical process dependent on the existence of a pressure gradient.
3. During inspiration, the diaphragm and external intercostal muscles contract and increase the thoracic cavity.
4. Thus, the volume of the lungs increases and this causes a decrease in the pressure in the lungs.
5. When the pressure in the lungs is less than atmospheric pressure, a vacuum is created and air is sucked into the lungs.
6. Normal expiration is a passive process in which the muscles of the chest and lungs recoil and the volume of the lungs decreases.
7. When the volume of the lungs decreases, the pressure inside the lungs increases and, when it is greater than the atmospheric pressure, air will move out of the lungs to the area of lowest pressure.
8. In active expiration, the abdominal and internal intercostal muscles contract to force the diaphragm upwards. This results in a reduction of the size of the thoracic cavity and an increase in the pressure within them. This leads to expiration.

External respiraton (pulmonary respiration)

1. External respiration is the exchange of gases between the alveoli of the lungs and the blood in the pulmonary capillaries.
2. In external respiration, oxygen diffuses from the alveolar air into the blood and carbon dioxide diffuses from the blood into the alveolar air.
3. Oxygen is transported by haemoglobin in the blood.
4. Carbon dioxide is transported as bicarbonate ions in the blood plasma.

Remember:
External respiration =
blood gains oxygen and loses carbon dioxide.

Internal respiration (tissue respiration)

1. Internal respiration is the exchange of gases between the blood and tissue cells.
2. In internal respiration, oxygen diffuses from the blood into the cells and carbon dioxide diffuses from the cells into the blood.

Remember:
Internal respiration =
blood loses oxygen and gains carbon dioxide.

Composition of inspired and expired air

Inspired air contains:

1. 21% oxygen
2. 0.04% carbon dioxide
3. 78–79% nitrogen and inert gases
4. A variable amount of water vapour.

Expired air contains:

1. 16% oxygen
2. 4–4.5% carbon dioxide
3. 78–79% nitrogen and inert gases
4. A variable amount of water vapour.

Revision

1. Identify the functions of the respiratory system.
2. Identify the two zones into which the respiratory system can be divided.
3. Following the passage of air from the atmosphere into the alveoli, put the following into their correct order:
   - Laryngopharynx
   - Bronchioles
   - Atmosphere
   - Trachea
   - Nasopharynx
   - Bronchi
   - Alveoli
   - Nose
   - Oropharynx
   - Larynx
4. Describe where and how sound is produced.
5. Explain the function of the epiglottis.
6. Compare the differences between the two lungs.
7. Describe the blood supply to the lungs.
8. Identify the differences between pulmonary ventilation, external respiration, internal respiration and cellular respiration.
9. Explain how inspiration and expiration occur.
10. Describe the following disorders:
    - Hyperventilation
    - Cystic fibrosis
    - Rhinitis.

Multiple choice questions

1. **Which of the following statements is correct?**
   a. Tuberculosis is a non-contagious infectious disease caused by an airborne fungus.
   b. Tuberculosis is a contagious infectious disease caused by an airborne bacterium.
   c. Tuberculosis is a non-contagious infectious disease caused by an airborne bacterium.
   d. Tuberculosis is a contagious infectious disease caused by an airborne fungus.

2. **Another name for the nostrils is the:**
   a. Conchaes
   b. Epiglottis
   c. Adenoids
   d. External nares.

3. **Which portion of the pharynx contains openings from the Eustachian tubes?**
   a. Nasopharynx
   b. Oropharynx
   c. Laryngopharynx
   d. Eustachiopharynx.

4. **What is the main function of the respiratory membrane?**
   a. It protects the lungs.
   b. It is the site of gaseous exchange.
   c. It lubricates and moistens the lungs.
   d. It filters out foreign particles.

5. **Which of the following statements is correct:**
   a. During internal respiration blood loses oxygen and gains carbon dioxide.
   b. During internal respiration blood loses carbon dioxide and gains oxygen.
   c. During internal respiration blood loses oxygen and gains nitrogen.
   d. During internal respiration blood loses nitrogen and gains oxygen.

6. **Which of the following is a major by-product of cellular respiration:**
   a. ATP
   b. Nitrogen
   c. Oxygen
   d. Carbon dioxide.

7. **If the volume of a closed container is reduced, what will happen to the pressure of the gas within it?**
   a. It will increase.
   b. It will decrease
   c. It will remain the same.
   d. None of the above.

8. **During inspiration what does the diaphragm muscle do?**
   a. It relaxes.
   b. It contracts.
   c. It stays the same.
   d. None of the above.

9. **Which of the following diseases are classified as Chronic Obstructive Pulmonary Disease (COPD)?**
   a. Cystic fibrosis, hyperventilation, pulmonary fibrosis.
   b. Laryngitis, sinusitis, rhinitis.
   c. Influenza, whooping cough, tuberculosis.
   d. Asthma, bronchitis, emphysema.

10. **How is oxygen transported in the blood?**
    a. It travels freely in the blood.
    b. It is transported as bicarbonate ions in red blood cells.
    c. It attaches to haemoglobin in red blood cells.
    d. It attaches to white blood cells.

9 The Cardiovascular System

Introduction

Every cell in the body needs oxygen to survive and there is only one way they can get it – through the blood. In turn, this life-giving substance depends on a small, fist-sized organ to pump it around the body – the heart. This vital organ is very simple in its structure yet it pumps blood through approximately 100,000 km of vessels to over 60 billion cells. Together, the blood, the heart and its vessels form the cardiovascular system and in this chapter you will discover the important role this system plays in your body.

Student objectives

By the end of this chapter you will be able to:

- Describe the functions of the cardiovascular system.
- Describe blood, its components and functions.
- Describe the heart and explain how it functions.
- Explain what blood pressure is and how it is maintained.
- Describe the different types of blood vessels.
- Explain the different types of circulation.
- Identify the primary blood vessels of the body.
- Describe the common pathologies of the cardiovascular system.

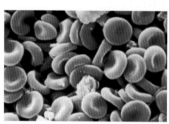

Red blood cells

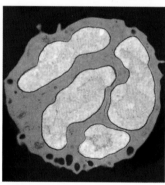

White blood cell

Blood

Functions of blood

Blood is a liquid connective tissue that is slightly sticky, heavier, thicker and more viscous than water. It is a vital substance in the body that functions in:

Transportation
Blood transports:
- Oxygen from the lungs to cells.
- Carbon dioxide from cells to the lungs.
- Nutrients from the gastrointestinal tract to cells.
- Heat and waste products away from cells.
- Hormones from glands to cells.

Regulation
Blood regulates:
- The pH of the body.
- The temperature of the body.
- The water content of cells.

Protection
Blood protects the cells against foreign microbes and toxins and also has the ability to clot and so protect the body against excessive blood loss.

Components of blood

In a laboratory blood can be centrifuged (spun) in a glass tube until it separates into three segments:

- Red blood cells sink to the bottom of the tube.
- White blood cells and platelets form a thin layer on top of the red blood cells. This is called the *buffy coat*.
- A straw-coloured, watery plasma forms a layer on top.

This demonstrates that blood is composed of *blood plasma* and cells called *formed elements*.

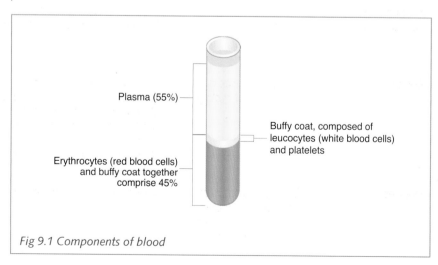

Plasma (55%)

Buffy coat, composed of leucocytes (white blood cells) and platelets

Erythrocytes (red blood cells) and buffy coat together comprise 45%

Fig 9.1 Components of blood

Blood plasma

Fifty-five per cent of blood is composed of a watery, straw-coloured liquid
that contains dissolved substances. This liquid is called blood plasma and
91.5% of it is water while the other 8.5% is dissolved substances (solutes).
The chart below details the components of blood plasma.

COMPONENTS OF BLOOD PLASMA		
COMPONENT	**DESCRIPTION**	**FUNCTION**
SOLVENT		
Water		Functions as a **solvent** and suspending medium for carrying other substances. Also absorbs, transports and releases heat.
SOLUTES		
Proteins	Proteins include albumins, globulins and fibrinogen.	**Albumins** – contribute to the osmotic pressure of the blood and transport fatty acids, some lipid-soluble hormones and certain drugs. **Globulins** – transport lipids and function in defence (antibodies are a type of globulin called gamma globulins or immunoglobulins). **Fibrinogen** – functions in blood clotting.
Electrolytes (ions)	Electrolytes include sodium, potassium, calcium, magnesium, chloride and bicarbonate.	Function in maintaining osmotic pressure, pH buffering and regulation of membrane permeability. Also serve as essential minerals.
Nutrients	Nutrients from the gastrointestinal tract include amino acids, glucose, fatty acids and glycerol.	Serve as nutrients for the cells.
Regulatory substances	Enzymes and hormones.	**Enzymes** – catalyse chemical reactions. **Hormones** – regulate cellular activity.
Gases	Oxygen, carbon dioxide and nitrogen.	**Oxygen** – most oxygen is carried in red blood cells, but a minute amount is also transported in the blood plasma. **Carbon dioxide** – most carbon dioxide is transported in the blood plasma and a small amount is carried by red blood cells. **Nitrogen** – the functions of nitrogen in the blood are not yet known.
Wastes	Waste products of metabolism include urea, uric acid and other substances.	Carried to the organs of excretion.

Formed elements

45% of blood is composed of cells and cell fragments. Together these are referred to as formed elements. 99% of formed elements are red blood cells and only 1% are white blood cells and platelets. The chart on the opposite page details the components of formed elements.

A closer look at red blood cells...

Red blood cells survive in the body for only 120 days. They are produced in the red bone marrow of long bones and are released into circulation at the rate of approximately 2 million red blood cells per second[i].

To ensure there is maximum space within the cells for oxygen transportation, red blood cells do not contain a nucleus and have few organelles. Thus, once in circulation, they cannot reproduce or carry out extensive metabolic reactions. Oxygen is carried by red pigment protein molecules called haemoglobin and each red blood cell contains approximately 280 million haemoglobin molecules.

Red blood cells are also specially shaped to perform their vital job of transporting oxygen. They are shaped like biconcave discs and this shape provides a large surface area for the diffusion of gases into and out of the cell. It also gives red blood cells their flexibility and ability to squeeze through even the smallest blood vessels.

Because red blood cells are constantly squeezing through tiny capillaries their plasma membrane suffers from a great deal of wear and tear and so they do not have a long life span. They are broken down in the spleen and liver where their breakdown products are then recycled.

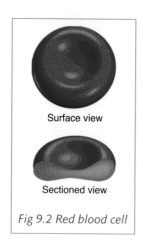

Surface view

Sectioned view

Fig 9.2 Red blood cell

Info box

Anatomy and physiology in perspective

When a blood vessel is injured there needs to be a quick, localised response that can stop the bleeding before there is excessive blood loss. This process is called haemostasis and it occurs in three phases:

Vascular spasm – If damaged, the smooth muscles in the wall of a blood vessel immediately contract to narrow the vessel and thus reduce blood flow through it. This contraction is thought to be caused by reflexes initiated by pain receptors in the vessels as well as by serotonin which is released by platelets once they adhere to the damaged site.

Platelet plug formation – When a blood vessel is damaged, collagen fibres that usually lie under the endothelial cells are exposed and platelets are able to adhere to them. The platelets clump together and release chemicals that attract more platelets to the site. Soon a mass of platelets called a *platelet plug* or white thrombus forms and seals the injury.

Coagulation (blood clotting) – In addition, blood begins to clot or thicken and form a gel at the site of injury. Blood clotting is a complex process that is promoted by the release of chemicals from platelets. This process results in a mesh of fibrin protein fibres in which blood cells are trapped.

COMPONENTS OF FORMED ELEMENTS

COMPONENT	DESCRIPTION	FUNCTION
Erythrocytes (red blood cells) Erythrocytes, or red blood cells, contain a protein called haemoglobin which transports oxygen in the blood.		
Erythrocytes	Biconcave discs with no nucleus and few organelles. Full of haemoglobin.	Transport oxygen.
Leucocytes (white blood cells) Leucocytes, or white blood cells, function primarily in protecting the body against foreign microbes and in immune responses. They do not contain haemoglobin and thus do not have the red colour that haemoglobin gives red blood cells. Thus, white blood cells are a pale 'whitish' colour. Most white blood cells live from a few hours to a few days and they are less numerous than red blood cells (for every one white blood cell in the body there are approximately 700 red blood cells). White blood cells can be categorised into granulocytes and agranulocytes.		
Granulocytes Granulocytes have multilobed nuclei and contain granules in their cytoplasm. Their names represent the colours of the dyes that they take up when stained in a laboratory.		
Neutrophils	Cytoplasm has very fine pale pink granules (*neutro = neutral, takes up both red acid and alkaline methylene blue dyes*).	Engulf and digest foreign particles and remove waste through the process of phagocytosis. Thus, they are referred to as phagocytes. Phagocytes increase rapidly during infection and are attracted in large numbers to areas of infection.
Eosinophils	Cytoplasm has large red-orange granules (*eosino = red acid dye*).	Destroy certain parasitic worms, phagocytise antigen-antibody complexes and combat the effects of some inflammatory chemicals. Eosinophils increase during allergies and infections by parasitic worms.
Basophils	Cytoplasm has large blue-purple granules (*baso = alkaline methylene blue dye*).	The granules in basophils contain histamine which causes the dilation of blood vessels. Basophils release histamine at sites of inflammation and are closely associated with allergic reactions.
Agranulocytes Agranulocytes have a large nucleus and do not contain cytoplasmic granules.		
Lymphocytes	Includes T-cells, B-cells and natural killer cells.	Lymphocytes play an important role in the immune response and are present in lymphatic tissue such as lymph nodes and the spleen. Immunity is discussed in more detail in Chapter 10.
Monocytes	The largest white blood cells.	Some monocytes circulate in the blood and are phagocytes. Other monocytes migrate into the tissues and become macrophages which are large scavenging cells that clean up areas of infection. Monocytes increase in number during chronic infections.
Thrombocytes	Granular, disc shaped cell fragments containing no nucleus.	Function in haemostasis, which is the process by which bleeding is stopped. Thrombocytes form a platelet plug and release chemicals that promote blood clotting.

The Heart

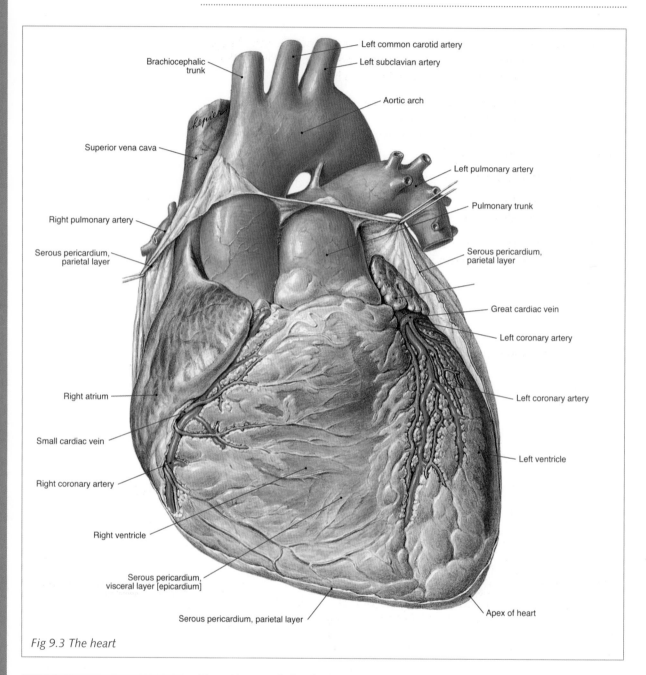

Brachiocephalic trunk

Left common carotid artery

Left subclavian artery

Aortic arch

Superior vena cava

Left pulmonary artery

Right pulmonary artery

Pulmonary trunk

Serous pericardium, parietal layer

Serous pericardium, parietal layer

Great cardiac vein

Left coronary artery

Right atrium

Left coronary artery

Small cardiac vein

Right coronary artery

Left ventricle

Right ventricle

Serous pericardium, visceral layer [epicardium]

Serous pericardium, parietal layer

Apex of heart

Fig 9.3 The heart

Did you know?
Your heart is approximately the same size as your closed fist and weighs only about 250-300g.

Functions of the heart

The heart has one sole function: to pump blood.

Anatomy of the Heart

Overview

The heart has the important job of pumping blood around the entire body. Yet, despite its importance, it is actually an uncomplicated structure. The heart is a hollow, muscular organ divided into two halves:

- The right side of the heart receives *deoxygenated* blood from the body and pumps it to the lungs for oxygenation.
- The left side of the heart receives *oxygenated* blood from the lungs and pumps it to the rest of the body.

Each of these halves both receives and delivers blood and so each has:

- A receiving chamber called an *atrium*.
- A delivering chamber called a *ventricle*.

Separating these chambers are valves. These valves are specially designed to prevent blood from flowing backwards.

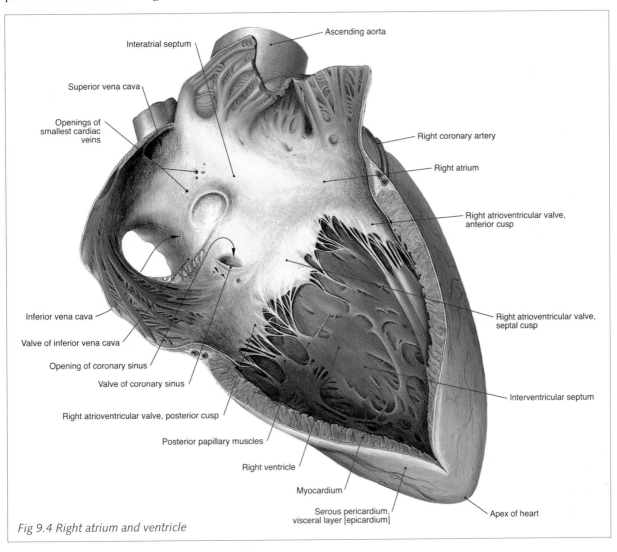

Fig 9.4 Right atrium and ventricle

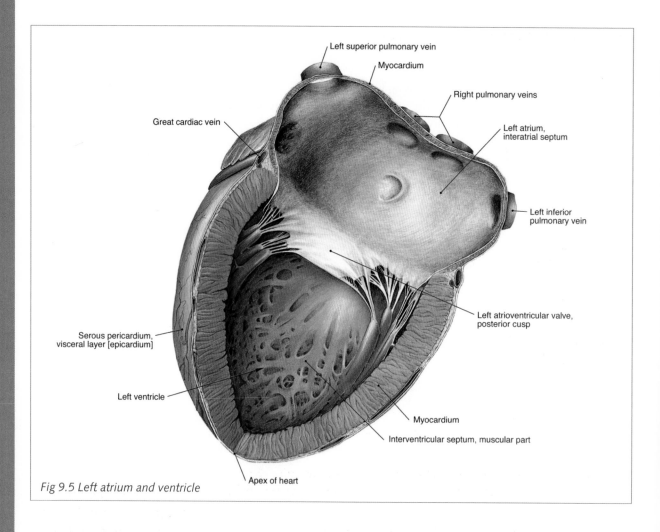

Left superior pulmonary vein

Myocardium

Right pulmonary veins

Great cardiac vein

Left atrium, interatrial septum

Left inferior pulmonary vein

Left atrioventricular valve, posterior cusp

Serous pericardium, visceral layer [epicardium]

Left ventricle

Myocardium

Interventricular septum, muscular part

Apex of heart

Fig 9.5 Left atrium and ventricle

Location

The heart lies in the *mediastinum* which is the partition between the lungs in the thoracic cavity.

- Approximately two-thirds of the heart lies to the left of the median line.
- Its pointed apex rests on the diaphragm at the level of the fifth intercostal space and points towards the left hip.
- Its broader, superior aspect is called the base and it lies beneath the second rib and points towards the right shoulder.

Pericardium: the covering of the heart

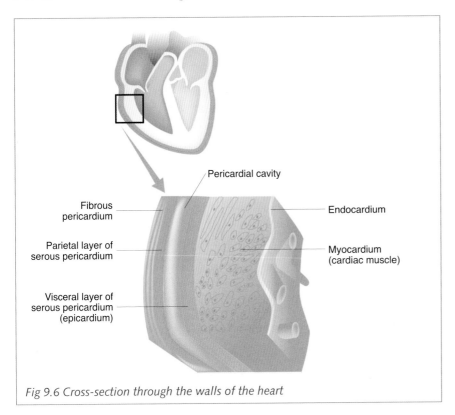

Fibrous pericardium

Parietal layer of serous pericardium

Visceral layer of serous pericardium (epicardium)

Pericardial cavity

Endocardium

Myocardium (cardiac muscle)

Fig 9.6 Cross-section through the walls of the heart

Surrounding and protecting the heart is a triple-layered sac called the pericardium. Similar to the pleural membrane of the lungs, the pericardium is composed of:

- The outer *fibrous pericardium* which is made of tough, inelastic connective tissue that prevents overstretching of the heart and anchors it in the mediastinum.
- The inner *serous pericardium* which is a thin membrane that forms a double layer around the heart.
  - The outer *parietal layer* is fused to the fibrous pericardium.
  - The inner *visceral layer* (epicardium) adheres to the heart.
- The space between the two layers is called the *pericardial cavity*. It contains a lubricating fluid called *pericardial fluid* that is secreted by the membranes and that reduces friction between them as the heart moves.

Study tip

Compare the protective structure and function of the pericardium to that of the pleural membrane of the lungs in Chapter 8.

Did you know?
A heart beats approximately 3 billion times in an average lifetime of 70 years with never a moment's rest.

Epicardium, myocardium and endocardium: walls of the heart

The power of the heart is quite incredible: it contracts about 100,000 times a day and pumps almost 7,500 litres of blood through it every single day for its entire lifetime. Can you imagine how hard it must work? In order to beat tirelessly day-in day-out, the heart is composed of three layers of tissue:

- **Outer epicardium** – also called the visceral layer of the serous pericardium (refer to previous page for more details).
- **Middle myocardium** – the layer that actually contracts to pump blood. It is a thick layer composed of cardiac muscle tissue which is involuntary, striated and arranged in spiral-shaped bundles of branching cells.
- **Inner endocardium** – this is the thin, smooth lining of the inside of the heart. It is composed of flattened epithelial cells and is consistent with the endothelial lining of the blood vessels.

Atria and ventricles: the chambers of the heart

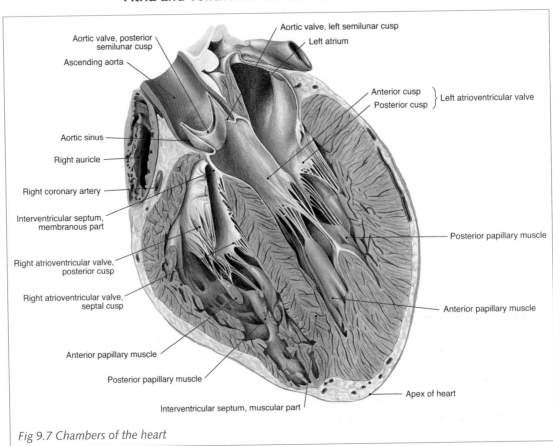

Fig 9.7 Chambers of the heart

The heart is divided into four compartments, or chambers, lined with endocardium:

- **Superior left and right atria** (singular = *atrium*) – The atria are the receiving chambers and they receive blood from the veins. The atria are separated by the *interatrial septum*. The walls of both these chambers are thin because they only need to deliver blood into the ventricles beneath them.

- **Inferior left and right ventricles** – The ventricles are the delivery chambers and they pump blood into the arteries. The ventricles are separated by the *interventricular septum*. The walls of both these chambers are thick because they need to pump blood out of the heart. The walls of the left ventricle are two to four times thicker than those of the right because the left ventricle pumps blood throughout the entire body while the right one only pumps it to the lungs.

Atrioventricular and semilunar valves: the valves of the heart

Blood needs to flow through the heart in one direction only: from the atria into the ventricles and then into the arteries. It must not flow backwards and to prevent this from happening the heart has four valves. These are composed of dense connective tissue covered by endocardium.

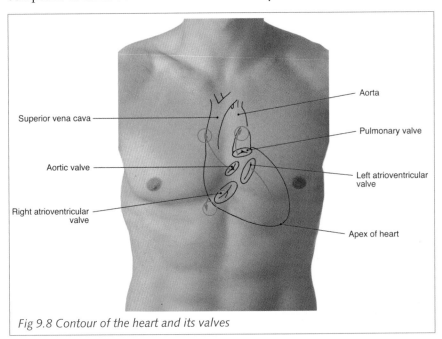

Fig 9.8 Contour of the heart and its valves

- **Atrioventricular valves (AV valves)** – these lie between the atria and the ventricles and prevent blood from flowing back into the atria when the ventricles contract. The valves consist of cusps, or flaps, that are anchored to the inner surface of the ventricles by tendon-like cords called *chordae tendinae*.
  - When the heart is relaxed and the chambers are being filled with blood the valve flaps hang open in the ventricles.
  - When the ventricles contract and the pressure inside the chambers rises, the valves are forced upwards and close. The chordae tendinae prevent them from opening upwards into the atria and thus prevent backflow of blood into the atria.

  The right AV valve is called the *tricuspid valve* as it has three cusps of endocardium. The left AV valve is called the *bicuspid valve (mitral valve)* as it has two cusps.
- **Semilunar valves** – these lie between the ventricles and the arteries and prevent blood from flowing back into the ventricles when they relax.

– The semilunar valves have three cusps that fit together to form a seal when the valves are closed. When the ventricles contract and push blood into the arteries the cusps are forced open and flattened against the walls of the arteries. When the ventricles relax blood starts to flow backwards and fills the cusps so that they close and form a seal that prevents arterial blood from re-entering the heart.

The valve between the right ventricle and the pulmonary trunk is called the *pulmonary semilunar valve*. The valve between the left ventricle and the aorta is called the *aortic semilunar valve*.

Pulmonary and systemic circulation: the flow of blood through the heart

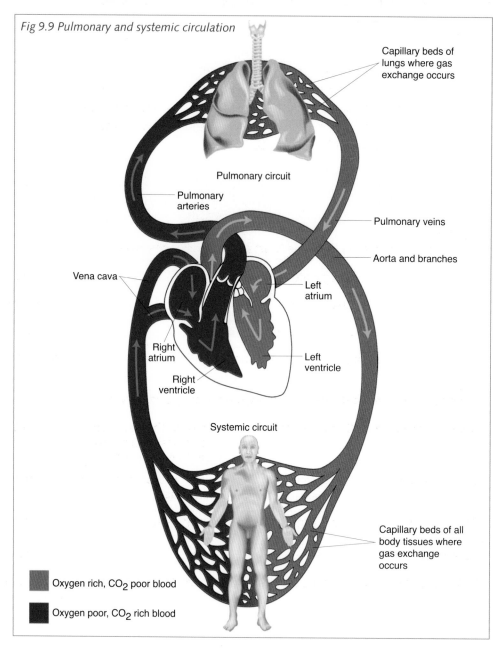

Fig 9.9 Pulmonary and systemic circulation

Capillary beds of lungs where gas exchange occurs

Pulmonary circuit

Pulmonary arteries

Pulmonary veins

Aorta and branches

Vena cava

Left atrium

Right atrium

Left ventricle

Right ventricle

Systemic circuit

Capillary beds of all body tissues where gas exchange occurs

Oxygen rich, CO_2 poor blood

Oxygen poor, CO_2 rich blood

The heart is a double-pump which pumps blood into two different circulations:

- **Pulmonary circulation** – The *right side* of the heart receives deoxygenated blood from the body and pumps it to the lungs where it is oxygenated. This is referred to as pulmonary circulation.
- **Systemic circulation** – The *left side* of the heart receives oxygenated blood from the lungs and pumps it to the rest of the body. This is referred to as systemic circulation.

We will take an in-depth look at blood vessels shortly, but to begin with it helps to have a basic understanding of the principal vessels of the heart. These are outlined in the table below. You will see that the coronary sinus and artery are also mentioned. These will be discussed in the following section on coronary circulation.

> **Study tip**
> Blood flows through a number of different routes in the body and this flow of blood is referred to as 'circulation'. Here we have looked at pulmonary and systemic circulation, but these are not the only routes of blood in the body. Later in this chapter we will look at two other circulations: coronary and hepatic portal.

PRINCIPAL BLOOD VESSELS OF THE HEART			
VESSEL	**BLOOD**	**FROM**	**TO**
Arteries Arteries carry blood *away* from the heart and generally carry *oxygenated* blood.			
Aorta	Oxygenated	Heart	Most of the body
Coronary artery	Oxygenated	Heart	Heart tissue
Pulmonary artery	Deoxygenated	Heart	Lungs
Note: the pulmonary artery is the only artery in the body that transports deoxygenated blood.			
Veins Veins carry blood *towards* the heart and generally carry *deoxygenated* blood.			
Superior vena cava	Deoxygenated	Most of the body superior to the diaphragm, except the alveoli and heart	Heart
Inferior vena cava	Deoxygenated	Body inferior to the diaphragm.	Heart
Coronary sinus	Deoxygenated	Heart tissue	Heart
Pulmonary vein	Oxygenated	Lungs	Heart
Note: the pulmonary vein is the only vein in the body that transports oxygenated blood.			

Blood flow through the heart

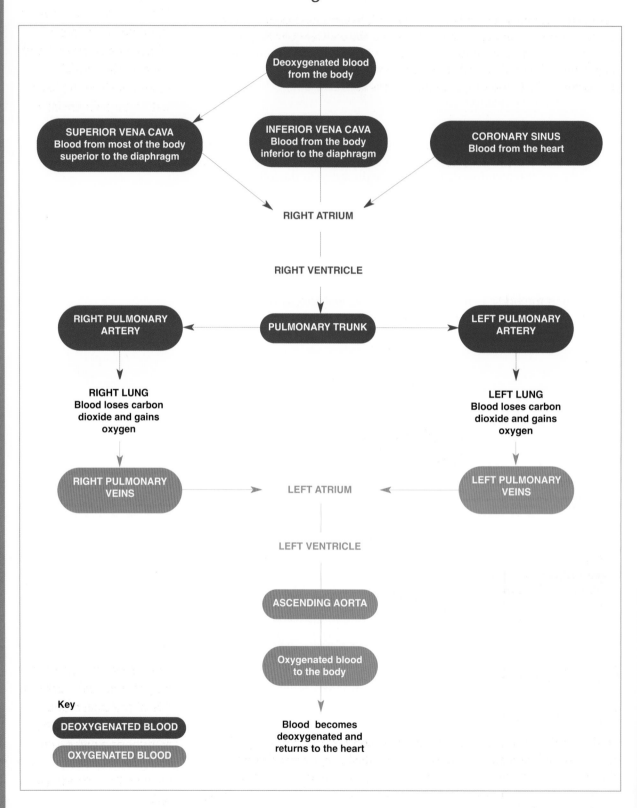

Coronary circulation: the blood supply to the heart

The heart is composed mainly of muscle and, like all muscles, it needs a constant supply of oxygen and nutrients and the constant removal of its waste products in order to function. Thus, it also needs its own blood supply and this supply is called the *coronary circulation (cardiac circulation)*. The blood supply to the heart is as follows:

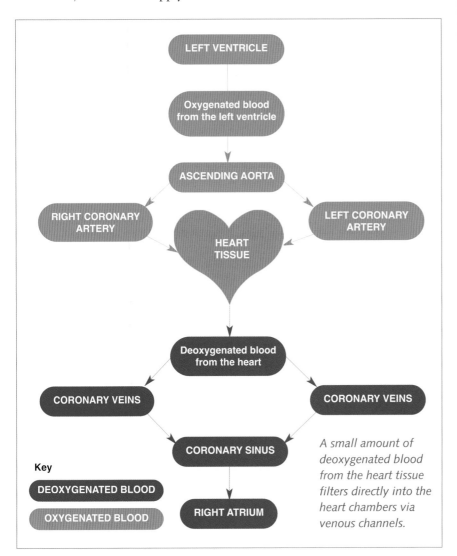

Key
- DEOXYGENATED BLOOD
- OXYGENATED BLOOD

A small amount of deoxygenated blood from the heart tissue filters directly into the heart chambers via venous channels.

Physiology of the Heart

Regulation of the heart rate

Cardiac muscle cells are unique in that they contract independently of nervous stimulation and thus have what is called a *myogenic rhythm*. However, although the cells contract regularly and continuously, the rhythm of their contraction needs to be controlled to ensure the heart functions as a co-ordinated whole.

Did you know?
The word coronary comes from the word *corona* which means 'crown' and the coronary blood vessels encircle the heart just as a crown encircles the head.

Did you know?
Despite its small size and weight, the heart receives 5% of the total blood it pumps. This reflects the amount of oxygen and energy it needs to perform its momentous task.

The cardiovascular system

The rhythm of the heart's contractions is controlled by:

- Specialised cells called autorhythmic cells. These cells form the pacemaker of the heart which is called the *intrinsic conduction system* or *nodal system*.
- The autonomic nervous system and certain hormones also help control the rate at which the heart beats.

Intrinsic conduction system (nodal system)

In the right atrial wall of the heart is a mass of autorhythmic cells that initiates impulses and so starts each heart beat. This is the *Sinoatrial node* (SA node) and it is the heart's pacemaker. The impulses are then conducted through the myocardium (heart muscle) from the right atrial wall down to the apex of the heart and then upwards into the myocardium of the ventricles. This movement of the impulse results in a 'wringing' action which causes the blood to be ejected upwards into the arteries.

Anatomy and physiology in perspective
If the SA node becomes diseased or damaged, other node fibres can become the pacemaker. However, the rhythm they produce may be too slow for efficient functioning of the heart. When this occurs, a device that sends out small electrical currents can be inserted into the heart. This is called an *artificial pacemaker*.

Artificial pacemaker

Impulses are conducted through the heart muscle as follows:

Pacemaker

SINOATRIAL NODE
(SA node)
Function: Initiates impulses and conducts them throughout both atria.
Location: right atrial wall.

ATRIOVENTRICULAR NODE
(AV node)
Function: Receives impulses from SA node and passes them to AV bundle of His. Impulses are slightly delayed here to give the atria time to finish contracting.
Location: atrial septum.

ATRIOVENTRICULAR BUNDLE OF HIS (AV bundle of His)
Function: Is a bundle of fibres that acts as the electrical connection between the atria and ventricles. Receives impulses from AV node and passes them to right and left bundle branches.
Location: interventricular septum.

Blood is ejected upwards

PURKINJE FIBRES
(Conduction myofibres)
Function: Receives impulses from the bundle branches and conducts them to the ventricular myocardium. Conduct impulses to the apex of the heart first and then upwards to the rest of the heart so that the blood is ejected upwards into the arteries.
Location: ventricular myocardium

RIGHT AND LEFT BUNDLE BRANCHES
Function: Conduct impulses towards the apex of the heart. Receive impulses from AV bundle of His and conduct them to Purkinje fibres.
Location: interventricular septum towards the apex of the heart.

Apex of heart

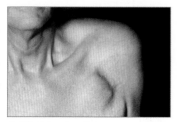

Autonomic and chemical regulation of the heart rate

The number of times the heart beats in one minute is called the *heart rate* and in an average resting male it is approximately 70 beats per minute (bpm) while in an average resting female it is slightly higher, approximately 75 bpm. Although the heart's pacemaker establishes the fundamental rhythm of the heart beat, it can be modified by the nervous and endocrine systems:

- **Autonomic regulation** – In the brain's medulla oblongata is the cardiovascular centre which receives information from sensory receptors such as proprioceptors, monitoring the positions of limbs and muscles; chemoreceptors, monitoring chemical changes in the blood; and baroreceptors, monitoring blood pressure changes in the arteries and veins. Information from higher brain centres such as the cerebral cortex and limbic system also send signals to the cardiovascular system. Once all this information has been interpreted the autonomic nervous system responds either sympathetically or parasympathetically to adjust the heart beat:
  - *Sympathetic stimulation* – Cardiac accelerator nerves extend into the SA node, AV node and most of the myocardium. These nerves stimulate the release of noradrenaline which, acting as a neurotransmitter, increases the heart rate.
  - *Parasympathetic stimulation* – Fibres of the right and left vagus (X) nerves innervate the SA node, AV node and atrial myocardium. These nerves stimulate the release of the neurotransmitter acetylcholine which decreases the heart rate.
- **Chemical regulation** – Different chemicals can also affect the heart rate:
  - *Hormones* such as adrenaline and noradrenaline, released by the adrenal medulla, increase the heart rate. Their release is stimulated by exercise, stress and excitement. Thyroid hormones also increase the heart rate.
  - *Ions* are electrically charged molecules that play an integral role in the production of impulses in both nerve and muscle fibres. If there is an imbalance in their concentrations the heart rate will be affected.
  - Certain drugs and dissolved gases can also alter the heart rate.

Study tip
Refer to Chapter 5 page 149 and Chapter 6 page 209 to revise the roles ions play in the production of impulses.

Anatomy and physiology in perspective
Other factors also influence our heart rates:
Age – Newborn babies have a high heart rate of approximately 120 bpm. As we age our heart rate slows to approximately 70–75 bpm in adulthood and then begins to increase again in old age.
Gender – Males have a lower heart rate than females (70 bpm vs 75 bpm).
Exercise – Physical exercise can increase the heart rate to as much as 150–200 bpm.
Fitness – Trained athletes can have a resting heart rate as low as 40–60 bpm.
Temperature – Increases in body temperature increase the heart rate and decreases in body temperature decrease the heart rate.
Emotions – Excitement, fear and anxiety all increase the heart rate.

Cardiac cycle

Now that you know how often the heart beats and what controls its beating, it is time to look at all the events associated with a heartbeat. This is called the *cardiac cycle* and one cycle, or heartbeat, automatically follows another throughout life.

A cardiac cycle lasts approximately 0.8 seconds and involves the contraction and relaxation of both the atria and of both the ventricles. While the two atria contract, the ventricles relax and while the two ventricles contract, the atria relax. As they alternately contract and relax pressure changes occur and blood flows from areas of higher pressure to areas of lower pressure. Two specific terms are associated with a cardiac cycle:

- Systole – means *contraction*.
- Diastole – means *relaxation*.

- **Atrial systole** – The SA node then stimulates the contraction of the atria and as the right and left atria contract simultaneously, they empty all their contents into the ventricles.
- **Ventricular systole** – Ventricular contraction is then triggered by the AV node and as the ventricles contract blood is pushed up against the AV valves, forcing them shut.
  - The AV valves and the semilunar valves are now shut and the pressure within the ventricles rises.
  - When this pressure is greater than the pressure in the pulmonary trunk and aorta, the semilunar valves are forced open and blood is ejected from the heart into the vessels.
- **Cardiac diastole** (relaxation or quiescent period) – At the end of a heartbeat all four chambers are relaxing and thus are in diastole. At this stage the AV valves are closed and the semilunar valves are open.
  - Because the chambers are relaxing, pressure within them drops and this causes blood to flow from the pulmonary trunk and aorta back towards the ventricles. This backflow of blood causes the semilunar valves to close.
  - The AV valves and semilunar valves are now closed and as the ventricles continue to relax the space inside them expands and the pressure within them drops.
  - When ventricular pressure is below atrial pressure the AV valves open and blood pours into the ventricles.

Anatomy and physiology in perspective
Instead of the 'boom boom boom' sound we associate with a heart beat, hearts actually make a 'lubb-dup' sound. The long, loud 'lubb' is the sound of the AV valves closing and the short, quick 'dup' is created by the semilunar valves snapping shut.

Blood Vessels

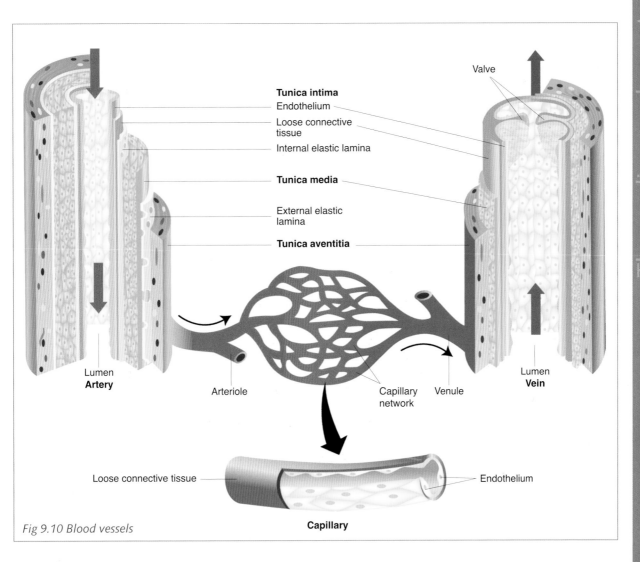

Tunica intima
Endothelium
Loose connective tissue
Internal elastic lamina

Tunica media

External elastic lamina

Tunica aventitia

Valve

Lumen
Artery

Arteriole

Capillary network

Venule

Lumen
Vein

Loose connective tissue

Endothelium

Capillary

Fig 9.10 Blood vessels

Blood is pumped by the heart into vessels that then transport it throughout the body. These vessels form a closed system of tubes that is made up of:

- Vessels that carry blood away from the heart *towards* the tissues: **arteries** and **arterioles.**
- Vessels that branch *throughout* tissues: **capillaries.**
- Vessels that carry blood *away from* the tissues towards the heart: **veins** and **venules.**

Arteries and arterioles

Blood is ejected from the heart into large, thick-walled vessels called arteries.

Structure of arteries

Arteries consist of a *lumen*, which is a hollow centre through which the blood flows, surrounded by a triple-layered wall composed of:

- An outer layer of fibrous tissue called the *tunica adventitia (tunica externa)*.
- A middle layer of smooth muscle and elastic tissue called the *tunica media*.
- An inner layer of squamous epithelium (endothelium) called the *tunica intima (tunica interna)*.

Types of arteries

Elastic (conducting) arteries – large arteries whose walls are relatively thin and whose tunica media consists of more elastic fibres than muscle fibres. This enables the arteries to stretch and recoil as they conduct blood from the heart to the medium-sized arteries. The aorta is an example of an elastic artery.

Muscular (distributing) arteries – medium-sized arteries whose tunica media consists of more smooth muscle fibres and fewer elastic fibres. This large amount of smooth muscle fibres means the walls of muscular arteries are relatively thick and are capable of greater vasoconstriction and vasodilation as they distribute blood to various parts of the body. The brachial artery is an example of a muscular artery.

Arterioles – tiny arteries that deliver blood to capillaries.

The smooth muscle of arteries is arranged circularly around the lumen so that when it contracts it narrows the lumen. This narrowing of the lumen is called vasoconstriction. When the smooth muscle fibres relax the size of the lumen increases and this is called vasodilation. Vasoconstriction and vasodilation are usually controlled by the autonomic nervous system.

Capillaries

Blood from arteries flows into arterioles and then finally into capillaries from where nutrients and wastes can be exchanged with tissue cells through interstitial fluid.

The name capillary derives from the Latin word *capillaris*, which means 'hairlike', and capillaries are microscopic vessels that network through tissues and connect arterioles and venules in what is referred to as *microcirculation*.

The mechanisms of vasoconstriction and vasodilation enable capillaries to help adjust body temperature:

- **Vasoconstriction** – Narrowing of the vessels results in a reduction in the flow of blood through the capillaries. Thus, heat in the blood is conserved and the body is kept warm.
- **Vasodilation** – Widening of the vessels results in an increase of blood to the surface of the skin. From here, heat in the blood is lost through radiation and the body is cooled.

Structure of capillaries

Capillary walls are made up of only a single layer of endothelium and a basement membrane. This extremely thin layer enables substances such as nutrients, oxygen and wastes to pass easily between the interstitial fluid surrounding cells and the blood.

Anatomy and physiology in perspective
The distribution of capillaries varies according to the metabolic needs of different tissues. For example, tissues such as the liver, kidneys and lungs have extensive capillary networks while cartilage, the cornea and the lens of the eye have no capillaries at all.

Veins and venules

Blood from several capillaries drains into venules, which are tiny veins, and then into veins. Veins return blood to the heart.

Structure of veins

The structure of veins is similar to that of arteries in that they consist of a triple-layered wall surrounding a lumen. This wall consists of the tunica adventitia (externa), tunica media and tunica intima (interna). However:

- The walls of veins are much thinner than those of arteries because there is less muscle and elastic tissue in the tunica media.
- The lumen of a vein is usually larger than that of an artery.
- Some veins, such as those in the limbs, have valves to prevent the backflow of blood as it flows upwards towards the heart. These valves are composed of thin folds of tunica intima and form flap-like cusps that project into the lumen of the vein and point towards the heart. If blood moves back towards the feet it causes the cusps to close and so prevent backflow.

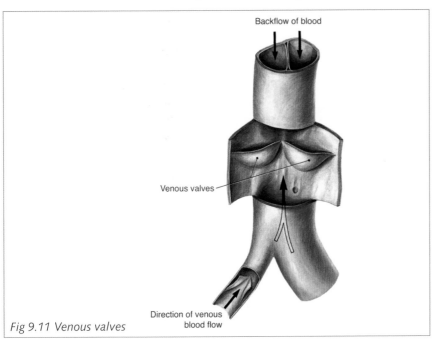

Fig 9.11 Venous valves

Did you know?
Because the walls of arteries are quite thick they remain open when cut, while the thinner walls of veins collapse if cut.

Movement of blood through veins

The flow of blood through vessels is dependent on a pressure gradient: blood flows from an area of high pressure into an area of low pressure. Blood passes from the heart into the arteries at a very high pressure and it flows continually along a pressure gradient, always moving into areas of lower pressure.

However, in the larger veins where there is minimal pressure the pressure gradient is not enough to ensure the flow of blood. Thus blood is moved through the veins by the milking action of skeletal muscles and the movement of the diaphragm during inspiration which causes pressure changes in the thoracic cavity. The blood is also prevented from flowing backwards by valves.

Anatomy and physiology in perspective
Arteries carry blood away from the heart and therefore the blood pressure in them is considerably higher than in veins, which carry blood towards the heart. This can be seen when arteries and veins are cut: blood spurts out of an artery while it flows more slowly out of a vein.

DIFFERENCES BETWEEN ARTERIES AND VEINS		
FEATURES	**ARTERIES**	**VEINS**
Structure	Thick walls No valves Smaller lumen	Thin walls Valves Larger lumen
Function	Carries blood away from the heart	Carries blood towards the heart
Blood pressure	Higher	Lower

Blood pressure

The force exerted by blood on the walls of a blood vessel is referred to as blood pressure. It is the force that keeps blood circulating and it is generated by the contractions of the left ventricle.

Blood pressure is highest in the aorta as this is the artery that receives blood directly from the left ventricle. The pressure then progressively falls as blood moves further and further away from the left ventricle. Thus, pressure is lowest in the veins, specifically the inferior vena cava.

Measuring blood pressure

Blood pressure is measured in millimeters of mercury (mm Hg) and is measured near the large systemic arteries. In a normal, young adult blood pressure is 120/80mm Hg while at rest. The first figure, 120mm Hg, is the systolic pressure and shows the pressure of the blood during systole, or contraction, of the left ventricle. The second figure, 80mm Hg, is the diastolic pressure and shows the pressure of the blood during diastole, or relaxation, of the left ventricle.

When measuring blood pressure it is important to be aware that blood pressure varies slightly according to:

- **Time of day** – blood pressure drops during night-time sleep.
- **Posture** – blood pressure is lower when lying down.
- **Gender** – men usually have a slightly lower blood pressure than women.
- **Age** – blood pressure tends to increase with age.

Factors affecting blood pressure
Blood pressure is increased or decreased by the following factors:

Cardiac output – cardiac output is the amount of blood pumped out of the heart by the left ventricle in one minute. In a normal adult it is approximately 5 litres. This means that approximately 5 litres of blood passes through the entire body in one minute. Cardiac output is affected by the heart rate.

- The heart rate varies with the demands of the body. For example, exercise increases the heart rate. Thus, exercise increases cardiac output and so increases blood pressure. On the other hand, blood loss decreases cardiac output and so decreases blood pressure.
- The heart rate is also affected by the autonomic nervous system, hormones and ions. See page 317 for more details on the Autonomic and Chemical Regulation of the Heart Rate.

Resistance – the opposition to the flow of blood through the vessels. An increased resistance increases the blood pressure. Resistance is affected by:

- **Changes in the tunica media of the arterioles** – The tunica media of arterioles is mainly composed of smooth muscle. If this muscle is replaced by inelastic fibrous tissue then this will cause blood pressure to increase. This occurs with ageing.
- **Blood viscosity** – The viscosity, or thickness, of blood increases with conditions such as dehydration and decreases with conditions such as anaemia or haemorrhaging. An increase in viscosity causes an increase in resistance and thus blood pressure, while a decrease in viscosity causes a decrease in resistance and thus blood pressure.
- **Blood vessel length** – Longer blood vessels increase the resistance and thus increase blood pressure. Longer blood vessels are present in people who are obese as additional vessels are necessary to supply the additional adipose tissue.
- **Blood vessel radius** – A decrease in the radius of a blood vessel increases resistance and thus increases blood pressure. Arterioles change their diameters by vasoconstricting or vasodilating. This enables them to control the amount of blood flowing to a particular organ. For example, an active organ needs more oxygen and nutrients than a resting one. Blood pressure is increased when arterioles vasoconstrict and decreased when they vasodilate. This change in vessel radius is referred to as *systemic vascular resistance (SVR)* or *total peripheral resistance.*

Did you know?
Mean arterial blood pressure (MABP) is a measurement of the average blood pressure of a vessel. At the arteriolar end of a capillary it is approximately 35mm Hg, at the venous end it has dropped to 16mm Hg and by the time the blood enters the right ventricle it is 0mm Hg.

Did you know?
Obese people generally suffer with high blood pressure. This is because for every additional pound of fat the body carries, 300km of extra blood vessels are needed.[ii]

Primary Blood Vessels of the Systemic Circulation

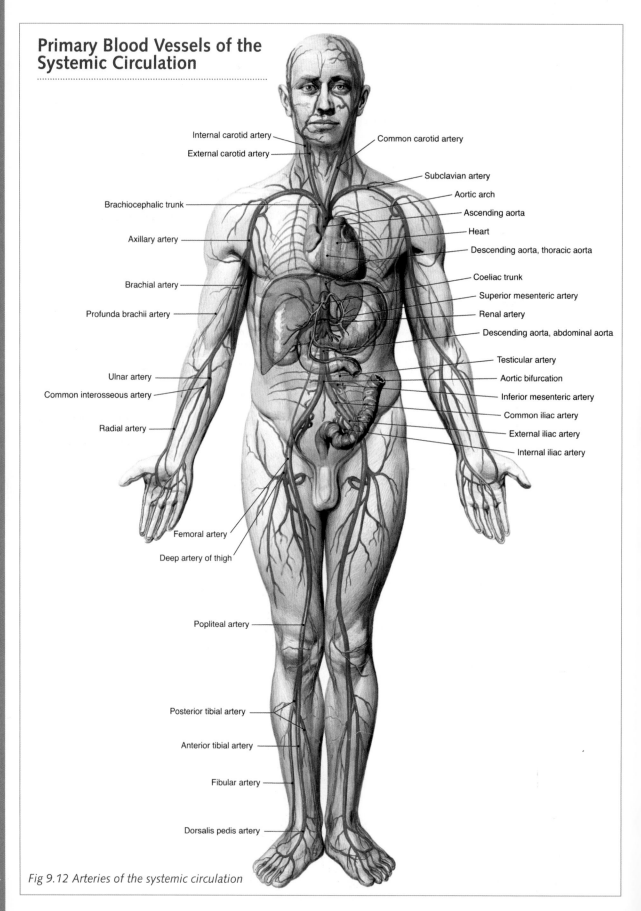

Internal carotid artery

External carotid artery

Common carotid artery

Subclavian artery

Aortic arch

Brachiocephalic trunk

Ascending aorta

Axillary artery

Heart

Descending aorta, thoracic aorta

Brachial artery

Coeliac trunk

Superior mesenteric artery

Profunda brachii artery

Renal artery

Descending aorta, abdominal aorta

Testicular artery

Ulnar artery

Aortic bifurcation

Common interosseous artery

Inferior mesenteric artery

Common iliac artery

Radial artery

External iliac artery

Internal iliac artery

Femoral artery

Deep artery of thigh

Popliteal artery

Posterior tibial artery

Anterior tibial artery

Fibular artery

Dorsalis pedis artery

Fig 9.12 Arteries of the systemic circulation

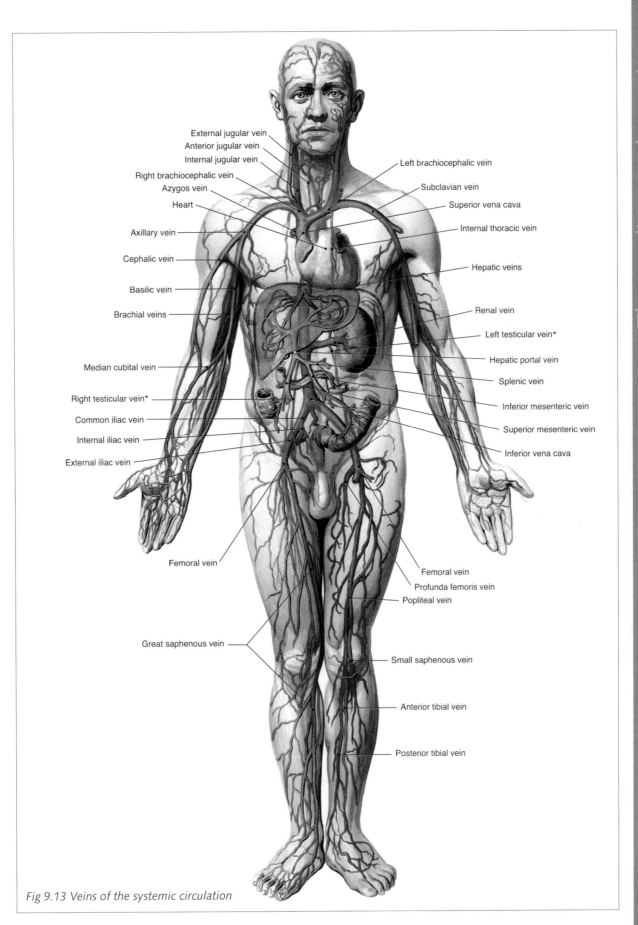

External jugular vein
Anterior jugular vein
Internal jugular vein
Right brachiocephalic vein
Azygos vein
Heart
Axillary vein
Cephalic vein
Basilic vein
Brachial veins
Median cubital vein
Right testicular vein*
Common iliac vein
Internal iliac vein
External iliac vein
Femoral vein
Great saphenous vein

Left brachiocephalic vein
Subclavian vein
Superior vena cava
Internal thoracic vein
Hepatic veins
Renal vein
Left testicular vein*
Hepatic portal vein
Splenic vein
Inferior mesenteric vein
Superior mesenteric vein
Inferior vena cava
Femoral vein
Profunda femoris vein
Popliteal vein
Small saphenous vein
Anterior tibial vein
Posterior tibial vein

Fig 9.13 Veins of the systemic circulation

Systemic circulation in a nutshell

Left ventricle of heart

⇓

aorta

⇓

arteries

⇓

arterioles

⇓

capillaries

⇓

venules

⇓

veins

⇓

superior vena cava/
inferior vena cava/
coronary sinus

⇓

right atrium of heart

Red = oxygenated blood
Black = exchange of
oxygen, carbon dioxide
and nutrients
Blue = deoxygenated
blood

Blood vessels are organised into routes that transport blood throughout the body. The word *systemic* refers to the body as a whole and systemic circulation is the route blood follows from the heart to the tissues and organs of the body and back to the heart. This route is as follows:

- Oxygenated blood from the left ventricle of the heart is ejected into the aorta which then branches into arteries, arterioles and finally capillaries.
  - Blood travelling through the aorta has a high pressure because it has been ejected with force from the heart. As it travels through the systemic circulation it progressively loses this pressure and by the time it returns to the right atrium of the heart it has almost no pressure at all.
  - Blood travelling through the aorta is a bright red colour because it is oxygenated. As it loses its oxygen and gains carbon dioxide it also loses its bright colour and blood travelling through the veins back to the right atrium is a dark red colour.
- At the capillaries nutrients and wastes are exchanged with tissue cells through interstitial fluid. The capillaries connect arterioles to venules.
- Deoxygenated blood then drains into venules, veins and finally into the superior vena cava, inferior vena cava or coronary sinus which return blood to the right atrium.

The aorta

The aorta is the largest blood vessel in the body and it receives oxygenated blood from the left ventricle. Different regions of the aorta are named according to their location or shape as follows:

- **Ascending aorta** – The aorta extends upwards from the left ventricle before arching downwards. This upward extension is the ascending aorta.
- **Aortic arch** – Where the aorta arches downwards is called the aortic arch.
- **Descending aorta** – The aorta then descends downwards behind the heart and through the trunk.
- **Thoracic aorta** – The thoracic aorta follows the spine through the thoracic cavity.
- **Abdominal aorta** – The aorta then descends behind the diaphragm into the abdominopelvic cavity where it is called the abdominal aorta.
- The aorta finally divides into the right and left common iliac arteries which transport blood to the lower limbs.

Superior vena cava, the inferior vena cava and the coronary sinus

The heart ejects blood into only one vessel for systemic circulation: the aorta. However, blood from the systemic circulation returns to the heart via three vessels:

- **Superior vena cava** – begins behind the right first costal cartilage at the union of the right and left brachiocephalic veins and empties its blood into the superior part of the right atrium. It receives blood from veins superior to the diaphragm. This includes the head, neck, upper limbs and thoracic wall.

Did you know?
The diameter of the aorta is approximately the same size as your thumb.

- **Inferior vena cava** – the largest vein in the body and begins at the union of the common iliac veins, above the fifth lumbar vertebra. It ascends through the abdomen and thorax and empties its blood into the inferior part of the right atrium. It receives blood from veins inferior to the diaphragm. This includes the abdominal viscera, most of the abdominal walls and the lower limbs.
- **Coronary sinus** – begins in the coronary sulcus, a groove which separates the atria and ventricles, and empties its blood into the right atrium. It receives blood from the cardiac veins which drain the myocardium of the heart.

Now that you have a basic understanding of the main vessels of the heart, we will look at the primary blood vessels that connect the different regions of the body to the heart.

Study tip
When learning the names of blood vessels, remember that they often tell us which regions or organs the vessels serve or let us know which bones they follow. For example, the brachial artery is found in the arm.

Guide to flowcharts of the blood vessels of the body
Some of the primary blood vessels of the body have been described in flow charts on the following pages. To help you understand these charts please note the following features of the charts:

Colour – the colour of the boxes and arrows represents the type of blood transported. Arteries carrying *oxygenated* blood are coloured red. Veins carrying *deoxygenated* blood are coloured blue.

Coloured boxes – The primary vessels associated with an area are in coloured boxes. These are the vessels you will need to learn for that particular area. All vessels written in black text are there simply to help you picture the flow of the blood.

Direction of arrows – The direction of the arrows represents the direction of blood flow. In general, blood flowing in arteries flows downwards from the heart to the rest of the body and blood flowing in veins moves upwards from the rest of the body to the heart.

Right/left side representation – Most blood vessels of the body are the same on both sides of the body. Thus, for ease of learning, the left and right sides are not noted. However, if there are differences between the left and right sides of the body, then both sides are represented.

The cardiovascular system

Primary Vessels of the Head, Face and Neck

Arteries

The paired arteries supplying the head, face and neck are the *common carotid arteries* and the *vertebral arteries*.

The right common carotid artery arises from the brachiocephalic artery, while the left one arises directly from the aorta. The common carotid arteries divide into the internal and external carotid arteries. The vertebral arteries arise from the subclavian arteries which feed off the aorta.

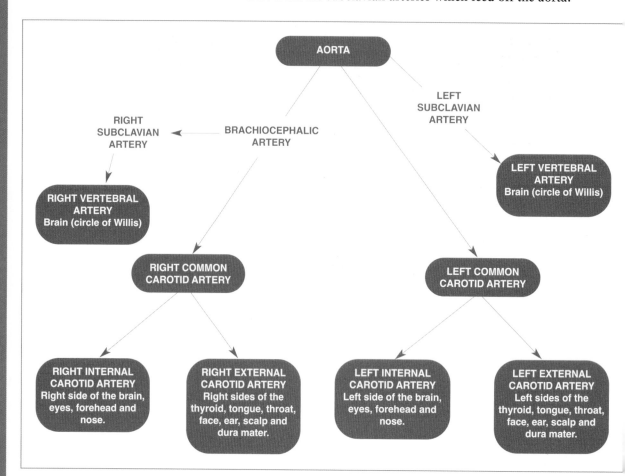

Did you know?

The base of the brain is surrounded by a complete circle of blood vessels called the *circle of Willis (cerebral arterial circle)*. The two internal carotid arteries and the two vertebral arteries contribute to the formation of this circle and this multiple supply of arterial blood ensures that there are alternate routes to the brain should one of the arteries become damaged. In addition, the circle of Willis helps equalise blood pressure to the brain.

Veins

Most blood draining from the head passes into three pairs of veins: the *internal jugular* veins, the *external jugular* veins and the *vertebral* veins.

The internal jugular veins unite with the subclavian veins to form the brachiocephalic veins.

The external jugular veins and the vertebral veins feed into the subclavian veins.

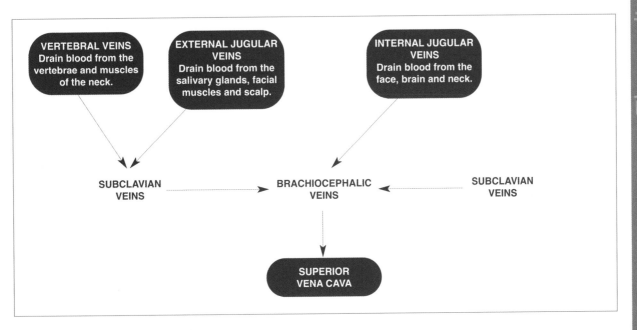

The cardiovascular system

Primary Vessels of the Upper Limbs (Arms and Hands)

Arteries

The right and left subclavian arteries transport blood to the upper limbs. These arteries become the axillary, then brachial, radial, ulnar, palmar and finally digital arteries and are named after the regions through which they pass. Be aware that on the right side of the body blood flows from the aorta into the brachiocephalic artery and then into the right subclavian artery, while on the left side of the body it flows from the aorta directly into the left subclavian artery.

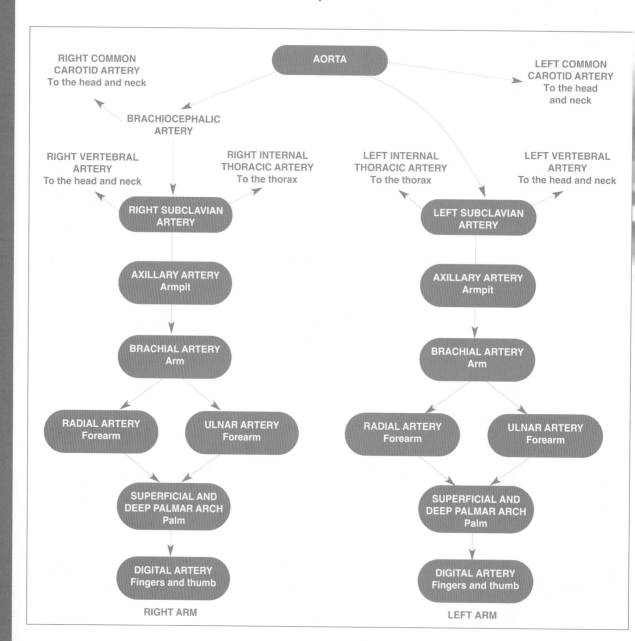

Veins

Veins of the upper limb are divided into two groups: superficial veins and deep veins:

- The superficial veins are the *cephalic, median cubital, basilic* and *median antebrachial* veins.
  - The cephalic vein drains the lateral (radial) aspect of the arm, beginning at the back of the hand and emptying into the axillary vein.
  - At the elbow the cephalic gives off a branch called the median cubital vein which slants upwards to join the basilic vein.
  - The basilic vein begins at the back of the hand and drains the medial (ulnar) aspect of the arm, emptying into the axillary vein.
  - The median antebrachial veins begin at the palm of the hand and ascend on the medial (ulnar) aspect of the forearm and end in the median cubital veins at the elbow.
- The deep veins follow the course of the arteries and have the same names: *dorsal metacarpal, palmar venous arch, ulnar, radial, brachial, axillary* and *subclavian* veins.

Did you know?
Blood is usually drawn from the median cubital vein for blood tests.

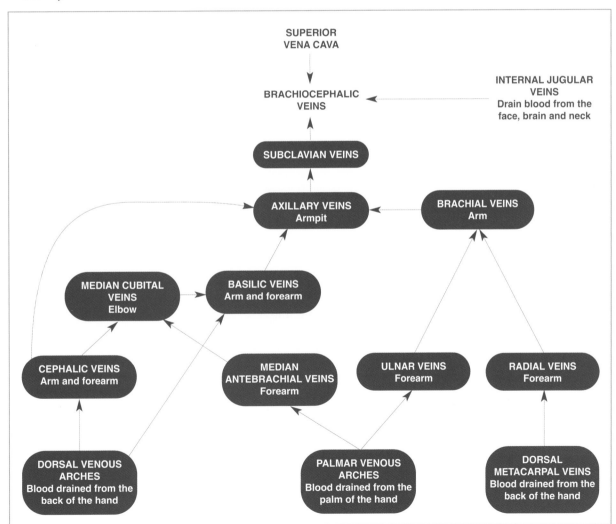

Primary Vessels of the Thorax

Arteries

The *thoracic aorta* gives off paired branches of arteries that supply the thorax. These are divided into two groups and are named after the regions or organs they supply. They include the:

- Arteries that feed the viscera and are collectively called the *visceral branches*:
  - **Pericardial arteries** – supply the pericardium
  - **Bronchial arteries** – supply the lungs
  - **Oesophageal arteries** – supply the oesophagus
  - **Mediastinal arteries** – supply the mediastinum.
- Arteries that feed the body wall structures and are collectively called the *parietal branches*:
  - **Posterior intercostal** and **subcostal arteries** – supply the muscles and skin of the thorax, the mammary glands and the vertebral canal.
  - **Superior phrenic arteries** – supply the diaphragm.

NB: A flow chart of the thoracic arteries has not been included here. This is because most syllabi only require that you learn the thoracic aorta. The other arteries named here have been included for reference only.

Veins

The thorax is drained by a network of veins on either side of the vertebral column called the *azygos system*. This includes the azygos, hemiazygos and accessory hemiazygos veins that receive blood from smaller veins carrying return blood from the parietal and visceral branches of the thoracic aorta. The names of these veins correspond to those of the arteries. The azygos system ultimately feeds into the superior vena cava or the brachiocephalic veins. Both these veins are considered the principal veins of the thoracic region:

- **Superior vena cava** – this is a large vein that receives blood from veins draining the head, neck, upper limbs and thoracic wall.
- **Brachiocephalic veins** – the left and right brachiocephalic veins are formed by the union of the subclavian and internal jugular veins and they unite to form the superior vena cava. The brachiocephalic veins drain blood from the head, neck, upper limbs, mammary glands and superior thorax.

NB: A flow chart of the thoracic veins has not been included here. This is because most syllabi only require that you learn the superior vena cava and brachiocephalic veins. The other veins named here have been included for reference only.

Primary Vessels of the Abdomen

Arteries

The *abdominal aorta* gives off paired and unpaired branches of arteries that supply the abdominopelvic region. Similar to the arteries of the thorax, the abdominal arteries are divided into two groups and named after the regions or organs they supply. They include the:

- Arteries that feed the viscera and are collectively called the *visceral branches*:
  - **Coeliac artery** – divides into three branches:
    Common hepatic artery which branches into smaller arteries supplying the liver, gall bladder and parts of the stomach, duodenum and pancreas.
    Left gastric artery which supplies the stomach and its oesophageal branch supplies the oesophagus.
    Splenic artery which branches into smaller arteries supplying the spleen, pancreas and stomach.
  - **Superior and inferior mesenteric arteries** – supply the small and large intestines.
  - **Suprarenal arteries** – supply the adrenal glands.
  - **Renal arteries** – supply the kidneys.
  - **Gonadal arteries** (testicular or ovarian) – supply the gonads.
- Arteries that feed the body wall structures and are collectively called the parietal branches:
  - **Inferior phrenic arteries** – supply the diaphragm.
  - **Lumbar arteries** – supply the spinal cord and muscles and skin of the lumbar region.
  - **Median sacral arteries** – supply the sacrum, coccyx and rectum.

NB: A flow chart of the abdominal arteries has not been included here. This is because most syllabi only require that you learn the abdominal aorta. The other arteries named here have been included for reference only.

Veins

Blood from the abdomen returns to the heart via the *inferior vena cava*. This large vein receives blood from smaller veins carrying return blood from the parietal and visceral branches of the abdominal aorta. The names of these veins correspond to those of the arteries (*renals*, *gonadals*, *suprarenals*, *inferior phrenics*, *hepatics* and *lumbars*).

However, blood from the digestive organs (the gastrointestinal tract, spleen, pancreas and gall bladder) does not flow directly into the inferior vena cava. It firstly passes through the liver via the *hepatic portal circulation* (see next section). After it has passed through the liver it drains into the hepatic veins and then into the inferior vena cava.

The right and left *common iliac veins* drain most of the blood from the lower limbs and pelvis and unite to form the inferior vena cava.

Hepatic portal circulation

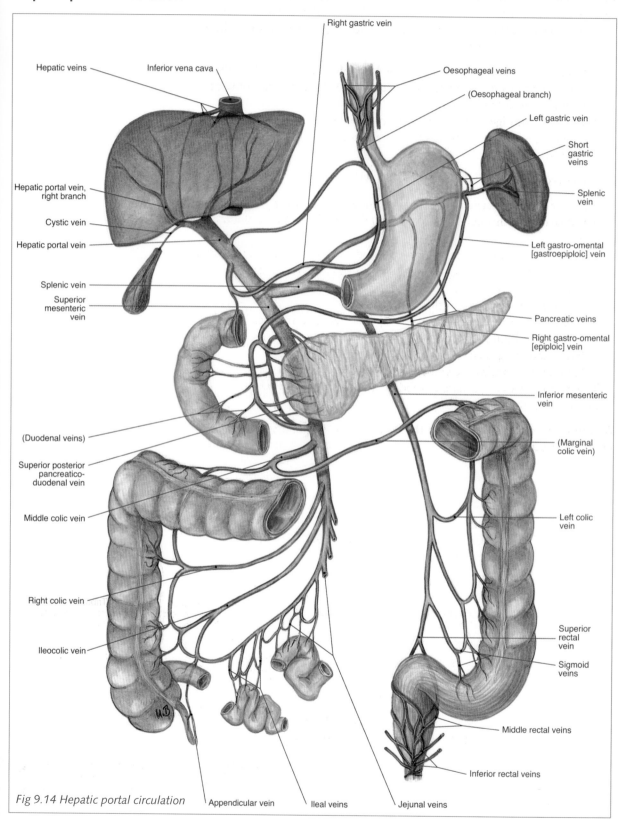

Right gastric vein

Hepatic veins

Inferior vena cava

Oesophageal veins

(Oesophageal branch)

Left gastric vein

Short gastric veins

Splenic vein

Hepatic portal vein, right branch

Cystic vein

Hepatic portal vein

Left gastro-omental [gastroepiploic] vein

Splenic vein

Superior mesenteric vein

Pancreatic veins

Right gastro-omental [epiploic] vein

Inferior mesenteric vein

(Duodenal veins)

(Marginal colic vein)

Superior posterior pancreatico-duodenal vein

Middle colic vein

Left colic vein

Right colic vein

Superior rectal vein

Ileocolic vein

Sigmoid veins

Middle rectal veins

Inferior rectal veins

Fig 9.14 Hepatic portal circulation

Appendicular vein

Ileal veins

Jejunal veins

Blood from tissues usually flows through one capillary bed before it is returned to the heart. However, blood from the digestive organs passes through a second capillary bed at the liver before it is returned to the heart. This is necessary for two reasons:

- Blood from the digestive organs is rich with absorbed nutrients – the liver stores or modifies these nutrients in order to maintain correct nutrient concentrations in the blood.
- Blood from the digestive organs may contain harmful substances – the liver detoxifies the blood to ensure these substances are not transported to the rest of the body.

The liver receives blood from two major vessels:

- The liver receives nutrient-rich, deoxygenated blood directly from the digestive organs through the hepatic portal vein which drains blood from the:
  - **Superior mesenteric vein** – receives blood from veins draining the small intestine, portions of the large intestine, stomach and pancreas.
  - **Splenic vein** – receives blood from veins draining the stomach, pancreas and portions of the large intestine.
  - **Inferior mesenteric vein** – joins the splenic vein and receives blood from veins draining portions of the large intestine.
  - **Right and left gastric veins** – drain the stomach.
  - **Cystic vein** – drains the gall bladder.
- The liver receives oxygenated blood via the hepatic artery which branches off the coeliac artery.

All blood leaves the liver through the hepatic veins which drain into the inferior vena cava.

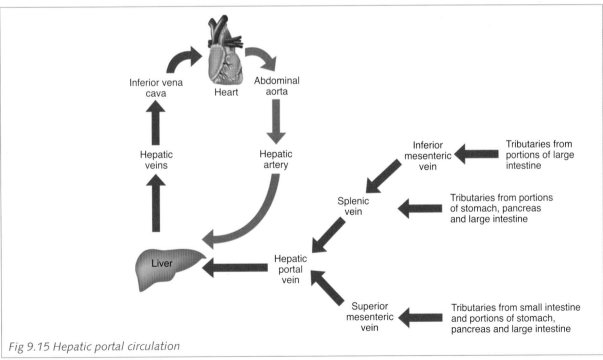

Fig 9.15 Hepatic portal circulation

Primary Vessels of the Pelvis and Lower Limbs (Legs)

Arteries

The abdominal aorta divides into the *right* and *left common iliac arteries* which in turn divide into the *internal* and *external iliac arteries*:

- **Internal iliac arteries** – supply most of pelvic viscera and wall.
- **External iliac arteries** – enter the thigh and become the femoral arteries. As the femoral arteries descend through the leg they become the popliteal arteries which divide into the anterior and posterior tibial arteries.
  - The anterior tibial artery continues over the top of the foot as the dorsalis pedis artery.
  - The posterior tibial which gives off the peroneal branch and then becomes the plantar artery (medial and lateral plantar) supplying the sole of the foot.
  - Together with the dorsalis pedis, the plantar artery and its branches form the plantar arch from which the digital arteries arise.

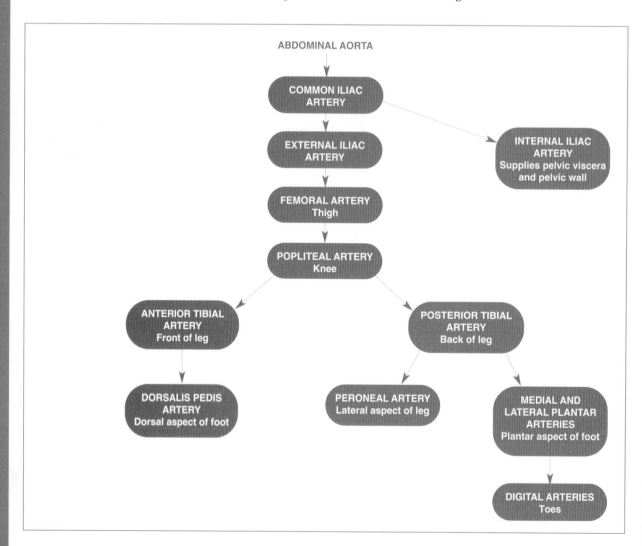

Veins

Veins of the lower limb are divided into two groups: superficial veins and deep veins:

- Superficial veins are the great saphenous and small saphenous veins.
  - **Great saphenous vein** – the longest vein in the body and runs up the inner thigh, beginning at the dorsal venous arch and emptying into the femoral vein in the groin.
  - **Small saphenous vein** – begins behind the ankle joint and ascends up the back of the leg where it joins the popliteal vein behind the knee.
- The deep veins follow the course of the arteries and have the same names: *anterior tibial, posterior tibial, popliteal, femoral, external iliac, internal iliac* and *common iliac* veins.

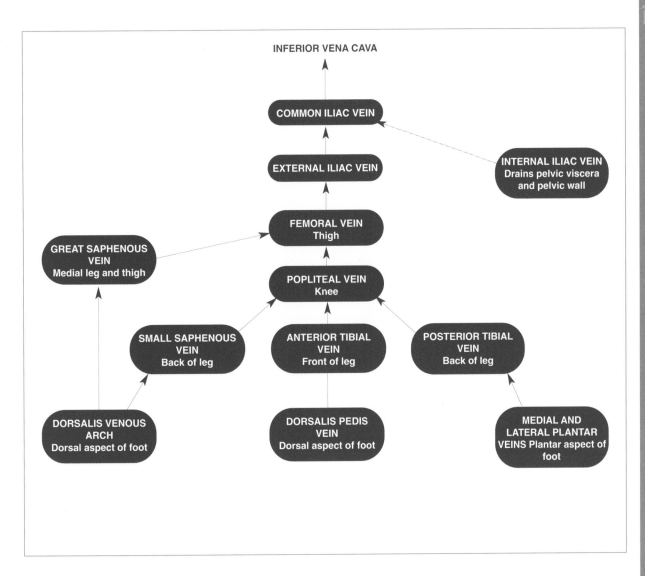

Common Pathologies of the Cardiovascular System

To a certain extent, the health of the cardiovascular system is dependent on one's lifestyle. Leading a healthy lifestyle that ensures you do not become obese or suffer from high blood cholesterol or high blood pressure, avoiding smoking cigarettes and exercising regularly can help to keep your heart and blood vessels healthy.

Blood disorders

Blood disorders are sometimes difficult to diagnose as their symptoms can indicate many other types of disorders. Therefore, it is usually necessary to have blood tests done in order to diagnose blood disorders properly. General symptoms of blood disorders can include fatigue, weakness, shortness of breath, dizziness, excessive bleeding and easy bruising.

Anaemia

Anaemia is a reduction in the oxygen-carrying capacity of the blood and it is characterised by a reduced number of red blood cells, or a reduction in the amount of haemoglobin, in the blood. It can be caused by a loss of blood, a lack of iron or the destruction or impaired production of red blood cells. Symptoms of anaemia include fatigue, paleness, breathlessness on exertion, lowered resistance to infection and an intolerance to the cold. There are many different types of anaemia, including:

- **Iron-deficiency anaemia** – The body needs iron to produce red blood cells and if it is lost through bleeding or if there is a lack of iron due to an inadequate supply or poor absorption of dietary iron, then anaemia can occur. Iron deficiency anaemia is common in menstruating females and also affects infants, pregnant women and the elderly.
- **Pernicious anaemia** (vitamin-deficiency anaemia) – results from low levels of vitamin B12 (cyanocobalamin). This can be caused by either low dietary levels of vitamin B12 or from the inability of the stomach to produce intrinsic factor which is necessary for the absorption of vitamin B12 in the small intestine.
- **Sickle-cell anaemia** – a genetic disorder that is more common in people who live in malarial areas such as sub-Saharan Africa and tropical Asia. In sickle-cell anaemia the haemoglobin molecule is abnormally shaped and bends the red blood cells into sickle, or crescent, shapes. These cells then rupture easily and this destruction of the red blood cells leads to chronic anaemia. Ironically, people with sickle-cell anaemia have a high resistance to malaria.

Blood clotting disorders

Haemostasis is the way in which the body stops bleeding and it involves a series of processes that involves the constriction of blood vessels and the clotting of the blood. This clotting is balanced by the breakdown of the clots once vessels have healed, thus ensuring that the blood remains fluid. However, sometimes the blood does not clot properly or the clots do not dissolve properly and so disorders occur. Some of these disorders include:

- **Haemophilia** – Haemophilia is a hereditary disorder in which blood clots very slowly due to a deficiency of certain clotting factors. The severity of the disorder varies: in mild haemophilia the person may bleed more than expected after injuries or surgery while in severe haemophilia a slight bump can trigger chronic internal bleeding which can be fatal.
- **Thrombophilia** – In thrombophilia the blood clots easily or excessively and patients are at risk of developing blood clots, usually in veins. Thrombophilia can be hereditary or can develop after birth due to a variety of factors.
- **Thrombosis** – a condition in which a blood clot, or thrombus, is produced. If it is large enough, a blood clot can obstruct the flow of blood to an organ:
  - **Coronary thrombosis** – the formation of a blood clot in the coronary artery and it can lead to the obstruction of the flow of blood to the heart.
  - **Deep vein thrombosis** (DVT) – a deep vein clot, called a phlebothrombosis, in the legs and is usually characterised by the leg becoming swollen and tender. The clot may become detached and block the pulmonary artery (pulmonary embolism). Risk factors of DVT include thrombophilia, prolonged bed rest, pregnancy and surgery.
- **Pulmonary embolism** – the blocking of the pulmonary artery by an embolus (which can be a clot). This disorder is discussed in more detail in Chapter 8.

Nosebleeds (Epistaxis)

Bleeding through the nose is commonly caused by nose picking or injury to the nose. However, it can sometimes develop as a side effect of taking aspirin or anticoagulant drugs or it can be associated with fever, high blood pressure or blood disorders. Common nosebleeds are usually easily controlled at home and should subside within ten minutes. If the bleeding is severe and prolonged medical help should be sought.

Heart and blood vessel disorders

Risk factors in heart disease

Statistics show that your lifestyle influences your risk of suffering from heart disease and simple changes to your lifestyle can often reduce your chances of heart disease. Factors that increase your chance of suffering from a disease are called risk factors and the following are major risk factors in heart disease:

- Obesity
- Lack of regular exercise.
- High blood cholesterol level.
- High blood pressure.
- Cigarette smoking.
- Diabetes mellitus.
- Family history of heart disease at an early age.
- Gender – men are more at risk of heart disease than women. However, after the age of 70 years the risk is equal in both genders.

Angina pectoris

Angina pectoris is the temporary sensation of a chest pain that often spreads to the arms or jaw or of suffocation. It is caused by a lack of oxygen to the heart muscle. This pain is usually induced by exercise or emotional distress and relieved by rest. Angina is usually caused by coronary artery disease in which the coronary arteries have become narrowed so that blood flow to the heart is reduced.

Arrhythmia (Abnormal heart rate rhythms)

Abnormal heart rate rhythms are irregular sequences of heartbeats. A normal heart rate is usually 60–100 beats per minute and anything above or below this rate is considered abnormal (except in very fit, young people who can have a heart rate below 60). Symptoms of abnormal rhythms vary from not being felt at all to palpitations (an awareness of the heartbeat) to weakness, light headedness, dizziness and fainting. The most common cause of abnormal heart rates is heart disease although they can also be caused by certain drugs or congenital birth defects.

- **Bradycardia** – an abnormally slow heart rate, usually considered to be below 50 beats per minute. Bradycardia can be triggered by pain, hunger, fatigue, diarrhoea or vomiting.
- **Tachycardia** – an abnormally fast heart rate, it can be triggered by exercise, stress, alcohol, cigarette smoking or stimulant drugs.

Arteriosclerosis

Arteriosclerosis is the hardening of the arteries. It can be related to many different diseases and is a condition in which arterial walls become thicker and less elastic.

Arteriosclerosis

Atherosclerosis

Atherosclerosis is a type of arteriosclerosis in which fatty substances, especially cholesterol and triglycerides, deposit on the inner walls of arteries and develop into atherosclerotic plaques which obstruct the flow of blood through the arteries and cause the arteries to harden. Atherosclerosis is initially symptom free until the artery has narrowed by more than 70%. Then symptoms will vary according to where the narrowing or blockage occurs in the body. For example, blockage of coronary arteries can cause angina pectoralis or a heart attack while blockage of the carotid arteries to the brain can result in a stroke. Risk factors for atherosclerosis include cigarette smoking, high cholesterol levels, high blood pressure, diabetes mellitus, obesity, a lack of exercise and high blood levels of an amino acid called homocysteine.

Coronary artery disease

In coronary artery disease the main arteries to the heart become narrowed and blood flow to the heart is reduced. Causes of coronary artery disease are atherosclerosis, coronary artery spasm or a coronary thrombosis.

Gangrene

Gangrene is the death and decay of tissue due to a lack of blood supply. It can be caused by many conditions including diseases such as diabetes mellitus, injury, frostbite or severe burns.

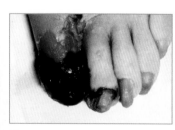

Gangrene

Haemorrhoids (Piles)

Haemorrhoids, or piles, are enlarged, dilated and often twisted veins located in the rectum and anus. They can be caused by repeated straining during defecation (usually due to constipation), frequent heavy lifting or additional pressure resulting from pregnancy. The main symptom of haemorrhoids is small amounts of bleeding after a bowel movement.

Hypertension

Hypertension is abnormally high blood pressure. Normal blood pressure is 120/80mm Hg and hypertension is defined as a systolic pressure higher than 140mm Hg or a diastolic pressure higher than 90 mm Hg or both (these pressures are taken in a resting adult). Hypertension can be caused by a number of factors but risk factors include being older than 75 years of age, obesity, poor diet, lack of exercise, stress, metabolic defects and genetics. Hypersecretion of aldosteronism by the adrenal cortex and kidney disease can also cause hypertension.

Hypotension

Hypotension is blood pressure that is low enough to cause symptoms such as dizziness and fainting. Although low blood pressure is generally healthy, it can cause an insufficient supply of blood to the brain resulting in fainting and dizziness or an insufficient supply of blood to the heart resulting in shortness of breath or chest pain. Hypotension can be caused by a number of factors including heart disease, infections and excess fluid or blood loss.

Myocardial infarction (Heart attack)

Myocardial infarction, or a heart attack, is the death of an area of heart muscle due to an interruption in the supply of blood to the heart. It is most commonly caused by a blood clot in a coronary artery that has already been partially narrowed by atherosclerotic plaques. The main symptom of a heart attack is severe pain in the middle of the chest, back, jaw or left arm. This pain is not alleviated by rest. Occasionally, no pain is felt at all. Other symptoms include sweating, nausea, shortness of breath, faintness and a heavily pounding heart.

Oedema

Oedema is the excessive accumulation of interstitial fluid in body tissues. It results in swelling and puffiness. Oedema may be localised, for example, swelling at a site of injury, or it may be more generalised. Causes of oedema can range from local injuries to heart or kidney disorders.

Palpitations and panic attacks

A palpitation is an awareness of the heart beating. One is not normally aware of the sensation of the heart beating but occasionally it can be felt, for example, during strenuous exercise or an extremely emotional experience. Panic attacks are episodes occurring in anxiety disorders. Symptoms of panic attacks vary and can include palpitations, chest pain, dizziness, chills, nausea and feelings of unreality.

Phlebitis

Phlebitis is inflammation of the walls of a vein and it is characterised by localised pain, tenderness, redness and heat. Phlebitis often occurs in the legs as a complication of varicose veins and it can lead to the development of a thrombosis.

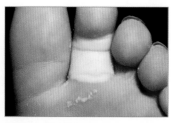

Raynaud's Disease

Raynaud's Disease

Raynaud's disease is a condition in which the arterioles of the fingers and toes constrict abnormally when exposed to cold. It can also be triggered by strong emotions. Symptoms include numbness, tingling or pins-and-needles sensations in the fingers and toes and the ends of the fingers or toes become pale and bluish. Rewarming the affected areas restores normal colour and sensation. The cause of Raynaud's disease is not always known, but it can accompany disorders such as rheumatoid arthritis or atherosclerosis.

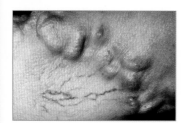

Varicose veins (varices)

Varicose veins (varices)

Varicose veins, or varices, are abnormally enlarged veins. They are most common in the veins of the legs but can occur in other areas of the body and can be inherited, occur during pregnancy or be caused by an obstruction to the flow of blood. Affected veins lengthen and widen and the valve cusps within them separate. This results in the backflow of blood which in turn causes the veins to become even larger and more distended. Varicose veins can result in pain, itchiness, phlebitis and varicose ulcers.

NEW WORDS	
Autorhythmic cells	muscle or nerve cells that generate an impulse without an external stimulus, i.e. they are self-excitable.
Diastole	relaxation of the heart muscle during the cardiac cycle.
Haemostasis	the stopping of bleeding.
Macrophage	large scavenger cell.
Phagocystosis	the engulfment and digestion of foreign particles by phagocytes.
Solvent	a medium (usually a liquid) in which substances (solutes) can be dissolved.
Systole	contraction of the heart muscle during the cardiac cycle.
Vasoconstriction	the constriction of blood vessels.
Vasodilation	the dilation of blood vessels.

Study Outline

Blood

Functions of the blood

Blood functions in transportation, regulation and protection.

Components of blood

1. Blood is composed of blood plasma and formed elements.
2. Blood plasma is composed of water, which acts as a solvent, and solutes that include proteins, electrolytes, nutrients, regulatory substances, gases and wastes.
3. Formed elements include erythrocytes (red blood cells), leucocytes (white blood cells) and thrombocytes (platelets).
4. Erythrocytes transport oxygen.
5. Leucocytes protect the body against foreign microbes and function in the immune response.
6. Thrombocytes function in stopping the bleeding process (haemostasis).

> **Even Red Ladies Will Throw Plates**
>
> **E**rythrocytes, **R**ed blood cells, **L**eucocytes, **W**hite blood cells, **T**hrombocytes, **P**latelets

The heart

The heart pumps blood around the body.

Anatomy of the heart

1. The heart is located in the mediastinum between the lungs in the thoracic cavity.
2. The heart is covered and protected by a triple-layered sac called the pericardium. The pericardium is composed of the outer fibrous pericardium and the inner serous pericardium which forms a double layer (the outer parietal layer and the inner visceral layer).
3. The walls of the heart are composed of three layers: the outer epicardium, the middle myocardium and the inner endocardium.
4. The heart is divided into two halves: the left and right sides.
5. Each side of the heart is composed of two chambers: an atrium for receiving blood into the heart and a ventricle for pumping blood out of the heart.
6. Blood is prevented from flowing backwards by heart valves: atrioventricular valves (AV) separate the atria and ventricles while semilunar valves separate the ventricles and arteries.
7. The right AV valve is the tricuspid valve. The left AV valve is the bicuspid (mitral) valve.
8. The right semilunar valve is the pulmonary semilunar valve. The left semilunar valve is the aortic semilunar valve.

Pulmonary and systemic circulation

1. The heart is divided into two halves: the right and left sides. Each side pumps blood into a different circulation: the pulmonary and systemic circulations.
2. Pulmonary circulation: the right side of the heart receives deoxygenated blood from the body and pumps it to the lungs for oxygen.
3. Systemic circulation: the left side of the heart receives oxygenated blood from the lungs and pumps it to the rest of the body.
4. The principle arteries of the heart are the aorta, coronary artery and pulmonary artery.
5. The principal veins of the heart are the superior vena cava, inferior vena cava, coronary sinus and pulmonary vein.

Coronary circulation

The coronary circulation is the blood supply of the heart tissue.

Physiology of the heart

Regulation of the heart rate

1. Cardiac muscle cells have a myogenic rhythm and contract independently of nervous stimulation.
2. The rhythm of the heart is mainly controlled by the intrinsic conduction system (nodal system).
3. The sinoatrial node (SA node) is the heart's pacemaker. It initiates impulses that are then conducted through the atrioventricular node (AV node), to the atrioventricular bundle of His (AV bundle of His), to the right and left bundle branches and finally to the Purkinje fibres (conduction myofibres).
4. The heart rate is the number of times the heart beats in one minute. In a resting male it is 70 beats per minute (bpm) and in a female it is 75 bpm.
5. The heart rate can be modified by the autonomic nervous system: sympathetic stimulation increases the heart rate and parasympathetic stimulation decreases the heart rate.
6. The heart rate can also be modified by chemicals such as hormones, ions, certain drugs and gases.

Cardiac cycle

1. The cardiac cycle is all the events associated with a hearbeat.
2. Starting from the end of the previous heartbeat, a cardiac cycle involves cardiac diastole (relaxation) in which the chambers begin to fill with blood; atrial systole (contraction) in which the atria contract and empty their contents into the ventricles; ventricular systole in which the ventricles contract and pump blood into the arteries; and then the cycle returns to cardiac diastole.

Blood vessels

1. Arteries and arterioles carry blood away from the heart towards the tissues.

Arteries
A = **A**way

2. Capillaries branch throughout tissues and are the site for the exchange of nutrients, oxygen and waste between the blood and interstitial fluid.
3. Veins and venules carry blood away from tissues towards the heart.
4. In general (except for the vessels of the pulmonary circulation), arteries and arterioles carry oxygenated blood while veins and venules carry deoxygenated blood.

Arteries
1. Arteries are composed of a lumen surrounded by the tunica intima (interna), then the tunica media and finally the outer tunica adventitia (tunica externa).
2. Elastic (conducting) arteries are large arteries that conduct blood from the heart to the medium-sized arteries.
3. Muscular (distributing) arteries are medium-sized arteries that distribute blood to various parts of the body.
4. Arterioles are tiny arteries that deliver blood to capillaries.

Capillaries
1. The walls of capillaries consist of a single layer of endothelium and a basement membrane.
2. Capillaries are extremely thin to allow for the rapid exchange of substances.

Veins
1. The structure of veins is similar to that of arteries except that their walls are thinner and their lumen is larger.
2. Some veins have valves to prevent backflow of blood.
3. Blood is moved through veins by the milking action of skeletal muscles and the movement of the diaphragm.

Blood pressure
1. Blood pressure is the force exerted by blood on the walls of a vessel.
2. Blood pressure is highest in the aorta and lowest in the inferior vena cava.
3. In a normal adult blood pressure is approximately 120/80mm Hg. The first figure is the systolic pressure and the second is the diastolic pressure.
4. Blood pressure is affected by cardiac output and resistance.
5. Cardiac output is affected by the heart rate.
6. Resistance is affected by changes in the tunica media of arterioles, blood viscosity and blood vessel length and radius.

Primary blood vessels of systemic circulation

1. In systemic circulation, oxygenated blood is ejected from the left ventricle into the aorta which branches into arteries and then into arterioles which transport oxygenated blood to the capillaries. At the capillaries the blood loses its oxygen and nutrients and gains carbon dioxide and wastes. Deoxygenated blood then drains into venules that lead into veins that finally drain into the superior vena cava, inferior vena cava and coronary sinus. These three vessels all empty deoxygenated blood into the right atrium of the heart.
2. Please refer to the flowcharts in the main text of this chapter for summaries of the vessels of the body.

Revision

1. Describe the functions of the blood.
2. Identify the main components of blood.
3. Explain the functions of erythrocytes.
4. Explain the functions of leucocytes.
5. Explain the functions of thrombocytes.
6. Describe the pericardium.
7. Describe the walls of the heart.
8. Identify the chambers of the heart and explain their functions.
9. Name the vein in the body that transports oxygenated blood.
10. Identify which vessels transport blood to the heart tissue.
11. Explain the nodal system.
12. Identify where in the body the sinoatrial node is located.
13. Explain what effect parasympathetic stimulation has on the heart rate.
14. Describe the stages of a cardiac cycle.
15. Identify the differences between arteries, capillaries and veins.
16. Explain what blood pressure is.
17. Describe the route blood takes through the body, beginning when it leaves the heart and ending when it returns to the heart.
18. Identify the blood vessel that receives oxygenated blood from the left ventricle.
19. Name the three vessels that deposit deoxygenated blood into the heart.
20. Describe the following disorders:
    - Deep vein thrombosis
    - Angina pectoris
    - Sickle cell anaemia
    - Varicose veins.

Multiple choice questions

1. **Water functions as:**
   a. An electrolyte in the blood.
   b. An ion in the blood.
   c. A solvent in the blood.
   d. A solute in the blood.

2. **Which of the following is the pacemaker of the heart:**
   a. Sinoatrial node.
   b. Atrioventricular bundle of His.
   c. Purkinje fibres.
   d. None of the above.

3. **Which of the following conducts blood from the heart to medium-sized arteries:**
   a. Arterioles
   b. Muscular arteries
   c. Elastic arteries
   d. Capillaries.

4. **The superior vena cava:**
   a. Drains blood from the head, neck, upper limbs and thoracic wall.
   b. Drains blood from the abdominal viscera, most of the abdominal walls and the lower limbs.
   c. Drains blood from the myocardium of the heart.
   d. Drains blood from the lower limbs.

5. **Where in the body is the axillary vein located?**
   a. The head
   b. The neck
   c. The armpit
   d. The abdomen.

6. **Where in the body is the great saphenous vein located?**
   a. The abdomen
   b. The groin
   c. The leg
   d. The foot.

7. **Which of the following statements is correct:**
   a. Anaemia is a reduction in the oxygen carrying capacity of the blood.
   b. Anaemia is characterised by a reduced number of white blood cells.
   c. Anaemia cannot occur in menstruating females.
   d. Anaemia is a blood-clotting disorder.

8. **Which of the following best describes hypotension:**
   a. High clotting ability.
   b. Low clotting ability.
   c. High blood pressure.
   d. Low blood pressure.

9. **Which of the following is not a type of white blood cell:**
   a. Neutrophil
   b. Lymphocyte
   c. Erythrocyte
   d. Basophil.

10. **Which of the following is considered a normal blood pressure in a healthy, resting adult:**
    a. 120/80
    b. 140/100
    c. 160/80
    d. 180/100.

10 The Lymphatic and Immune System

Introduction

To understand the importance of a system it sometimes helps to try and imagine your body without it. Imagine what would happen to you if your body could not return excess fluid to the bloodstream. Or if your body had no means of transporting fats and fat soluble vitamins. Or if you could not defend yourself against invasion from disease. These are sobering thoughts.

Although the lymphatic and immune system often appears to be simple, it is of the utmost importance to the health and maintenance of our bodies. In this chapter you will discover more about this system and how it works closely with the cardiovascular system to maintain your health.

Student objectives

By the end of this chapter you will be able to:

- Describe the functions of the lymphatic and immune system.
- Explain the organisation of the lymphatic and immune system.
- Identify the location of major lymphatic and immune system.
- Describe the organs of the lymphatic and immune system.
- Explain non-specific resistance to disease and immunity.

Functions of the Lymphatic and Immune System

At a glance the lymphatic system does not seem as impressive as some of the other systems of the body. It first appears as a simple system that transports a clear, straw-coloured fluid from the interstitial spaces surrounding cells to the blood. However, on closer examination you see that this system does so much more: it drains interstitial fluid to prevent tissues from becoming waterlogged, transports dietary lipids and protects the body against invasion.

Before we look at the lymphatic system, it is a good idea to ensure you understand the following vocabulary (some of which you may have encountered already):

- Antibody – a specialised protein that is synthesised to destroy a specific antigen.
- Antigen – any substance that the body recognises as foreign.
- Lymphocyte – a type of white blood cell involved in immunity. B cells and T cells are types of lymphocytes.
- Macrophage – a scavenger cell that engulfs and destroys microbes.
- Microbe (micro-organism) – an organism that is too small to be seen by the eye. Microbes include bacteria, viruses, protozoa and some fungi.
- Pathogen – a disease-causing micro-organism.
- Phagocyte – a cell that is able to engulf and digest microbes. Phagocytes include macrophages and some types of white blood cells.

The lymphatic system is a secondary circulatory system that drains excess fluids from tissues and returns the fluid to the cardiovascular system.

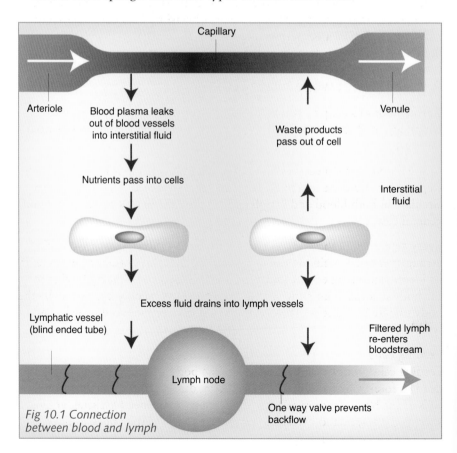

Fig 10.1 Connection between blood and lymph

Drainage of interstitial fluid

Every cell is bathed in a dilute saline solution called interstitial fluid. Everyday approximately 21 litres of blood plasma and its components, such as protein particles, fat molecules and debris, leak out of blood vessels and into the interstitial fluid. Most of this is reabsorbed back into the bloodstream, however, about 3 litres is not reabsorbed. If this fluid is not somehow returned to the bloodstream then the tissues will become waterlogged and blood volume will fall. The lymphatic system plays a vital role in draining this excess fluid and returning it to the bloodstream.

Transportation of dietary lipids

Located inside the lining of the small intestine are finger-like projections called villi. Inside each of these villi are blood capillaries and specialised lymphatic vessels called lacteals. Lacteals absorb lipids and lipid-soluble vitamins (vitamins A, D, E and K) and transport these substances into larger lymphatic vessels that finally deposit their contents into the bloodstream. The fluid inside the lacteals is a creamy white colour due to its fat content and is called chyle.

Protection against invasion: the immune response

In addition to the lymphatic vessels that drain interstitial fluid and transport lipids, the lymphatic system also contains specialised tissues and organs that are involved in protecting the body against invasion. Phagocytic cells and lymphocytes are located in these organs and they function in defending the body through what is known as the immune response.

A closer look at white blood cells...

Throughout this chapter you will hear about macrophages, phagocytes, lymphocytes, B cells, and T cells. So let's get a basic understanding of what these are.

White blood cells (leucocytes) are pale, colourless cells that are derived from stem cells in red bone marrow. Although they are called 'blood' cells, they circulate in both blood and lymph and they are the body's police, army and general cleaners.

Every single cell in the body has its own 'identification papers': a specific arrangement of protein molecules on the surface of its plasma membrane. If its identification papers are faulty, or if it is recognised as being foreign to the body it is called an *antigen* and is destroyed as quickly as possible by the body's white blood cells. These white blood cells include feeding cells and lymphocytes.

Feeding cells
Feeding cells are the general cleaners of the body and they clean up old, dead or foreign material including bacteria, dust particles and even dead cells of the body. Feeding cells are often found in connective tissue, organs such as the spleen and liver, and lymphoid tissue and they flock in their thousands to sites of injury.

In the classroom
Discuss the impact a drop in blood volume can have on your body.

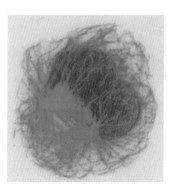

Macrophage

Once a feeding cell comes into contact with an attacker, cytoplasmic projections extend towards the attacker like multiple hands and pull it inwards. Then the feeding cell engulfs the attacker, isolates it in an internal cavity and empties enzymes onto it that break the attacker down. Thus, they literally 'eat' their prey through a process called *phagocytosis*.

Macrophages, granulocytes and monocytes are all types of feeding cells.

Lymphocytes

Lymphocytes are the trained army personnel of the body. They do not eat their attackers, instead they kill them with *antibodies* or poison and in order to do this they first need to be trained or made 'immune competent'. After being born in bone marrow, lymphocytes migrate to either lymphoid tissue or the thymus. These are the body's 'military training academies' and it is here that lymphocytes specialise and mature.

- **Lymphoid tissue** – Lymphocytes that have migrated to the lymphoid tissue and lymph nodes learn to react to antigens and develop receptors on their surfaces that bind to specific antigens. These lymphocytes are now called *B cells* and every B cell is able to bind to only one specific antigen. B cells circulate in blood or lie in wait in lymph nodes and when they come across their specific enemy antigen, they bind to it. In the process of binding to the antigen, a signal is transmitted to the nucleus of the B cell which then divides the cell into many new clones called *plasma cells*. It is these plasma cells that then manufacture antibodies which bind to the antigen and stop it from penetrating body cells and causing disease.
  - *Complement system* – Special molecules in the blood act as dynamite to help antibodies destroy their enemies. These molecules are called *complement factors* and when an antibody binds to an antigen the complement factors flock to the site of battle. Once all the complement factors have arrived, they perforate the membrane of the enemy cell.
- **Thymus** – Those lymphocytes that have migrated to the thymus become trained in killing foreign cells and are called T cells. Unlike B cells, T cells do not produce antibodies that attack their enemy. Instead, they attack the enemy directly. T cells include:
  - *Helper cells* – Helper cells help B cells and feeding cells.
  - *Killer cells* – Killer cells kill the body's own cells that have been invaded by antigens and also kill tumour cells in the body.
  - *Suppressor cells* – Suppressor cells suppress the aggression of some lymphocytes.

Now that you have a basic understanding of the battleground within your body, it is time to get back to the lymphatic system.

Structure of the Lymphatic System

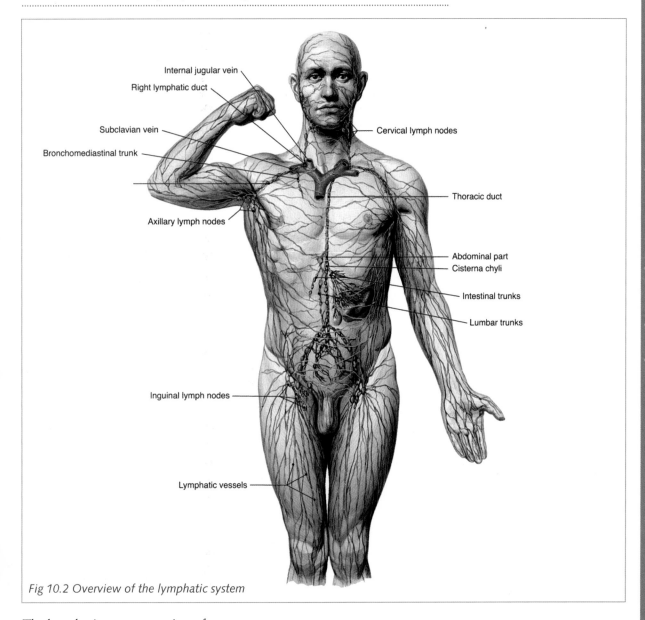

Fig 10.2 Overview of the lymphatic system

The lymphatic system consists of:

- Lymph
- Lymphatic capillaries
- Lymphatic vessels, trunks and ducts
- Lymphatic nodes (glands)
- Lymphatic organs
- Lymphatic nodules (mucosa-associated lymphoid tissue).

Lymph

The word *lymph* means 'clear water' and as mentioned earlier, lymph is a clear, straw-coloured fluid derived from interstitial fluid. It is similar in composition to blood plasma and contains:

- **Protein molecules** that are too large to return to the blood circulation via the capillaries.
- **Lipid molecules** that have been absorbed through lacteals located in the lining of the small intestine.
- **Foreign particles** such as bacteria that can cause disease.
- **Cell debris** from damaged tissues.
- **Lymphocytes** which are a type of white blood cell that functions in the immune response.

Lymph is only found in lymphatic vessels.

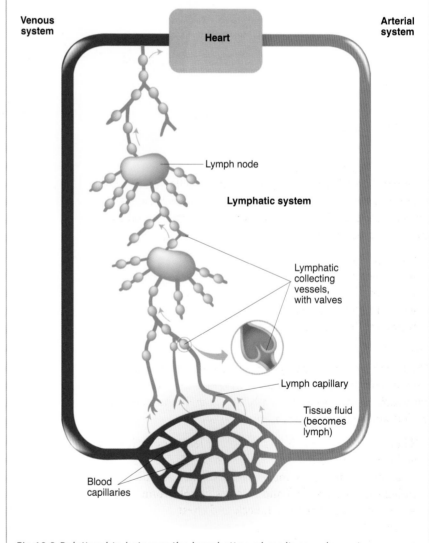

Fig 10.3 Relationship between the lymphatic and cardiovascular systems

Lymphatic capillaries

Lymphatic capillaries are tiny, closed-ended vessels similar to blood capillaries. However, they have a larger diameter than blood capillaries and a unique structure that permits fluid to flow into them but not out of them. Lymphatic capillaries are found throughout the body except in avascular tissue, the central nervous system, splenic pulp (to be discussed shortly) and bone marrow. Interstitial fluid is absorbed into lymphatic capillaries. Once inside the capillaries the fluid is called lymph and it is transported by the tiny lymphatic capillaries into larger vessels called lymphatic vessels.

Lymphatic vessels

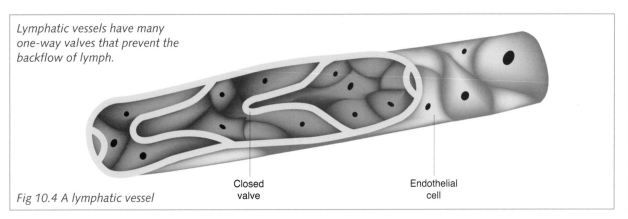

Lymphatic vessels have many one-way valves that prevent the backflow of lymph.

Closed valve

Endothelial cell

Fig 10.4 A lymphatic vessel

Lymphatic vessels and trunks

Lymphatic vessels carry lymph from the capillaries, through a number of lymphatic nodes and into large vessels called lymphatic trunks. These trunks are named after the areas they serve and are the:

- Lumbar trunk
- Intestinal trunk
- Right and left bronchomediastinal trunks
- Right and left subclavian trunks
- Right and left jugular trunks.

Lymphatic ducts

The lymphatic trunks carry lymph into two main channels: the *thoracic duct* (*left lymphatic duct*) and the *right lymphatic duct*. These eventually empty their contents into the left and right subclavian veins respectively.

- **Thoracic duct** – The thoracic duct, or left lymphatic duct, is the main collecting duct of the lymphatic system. It originates near the second lumbar vertebra at a dilation called the *cisterna chyli* and receives lymph from:
  - The left side of the head, neck and chest
  - The left arm
  - The entire body below the ribs.

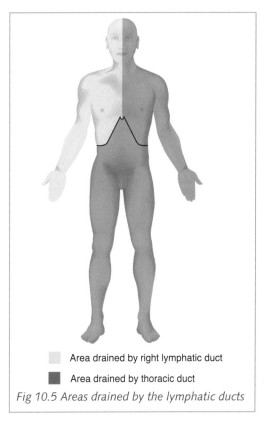

Area drained by right lymphatic duct

Area drained by thoracic duct

Fig 10.5 Areas drained by the lymphatic ducts

- **Right lymphatic duct** – The right lymphatic duct only receives lymph from the upper right side of the body.

Movement of lymph through the lymphatic vessels
Unlike blood, which is pumped around the body by the heart, lymph has no pump to move it through its vessels. Instead, certain mechanisms combine to ensure the flow of lymph:

- **Smooth muscle** found in the walls of lymphatic vessels contracts rhythmically to move lymph.
- **Skeletal muscles** contract to create a milking action.
- **Breathing movements** cause pressure changes in the thoracic cavity.
- **One-way valves** prevent the backflow of lymph in the vessels.

Lymphatic nodes (glands)

As lymph travels through the vessels towards the lymphatic ducts, it passes through a number of nodes, or glands, before it is returned to the blood-stream. The function of these nodes is to filter the lymph and remove or destroy any potentially harmful substances before the lymph is returned to the blood. They also produce lymphocytes that function in the immune response.

Location of lymphatic nodes
Lymphatic nodes are scattered along the length of the lymphatic vessels with higher concentrations of them being strategically placed in sites where there is a greater risk of infection. There are many concentrations of nodes and a few of the primary ones are listed in the chart opposite.

Structure of lymphatic nodes

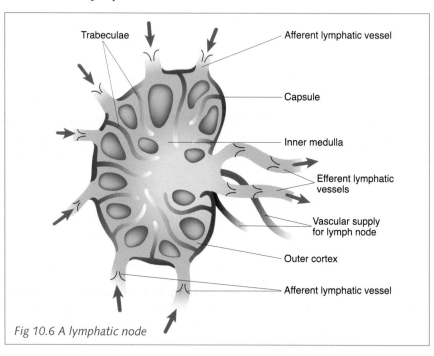

Fig 10.6 A lymphatic node

Trabeculae
Afferent lymphatic vessel
Capsule
Inner medulla
Efferent lymphatic vessels
Vascular supply for lymph node
Outer cortex
Afferent lymphatic vessel

Study tip
Compare the movement of lymph through lymphatic vessels to the movement of venous blood through veins. You will see that they are moved by similar mechanisms.

PRIMARY LYMPHATIC NODES	
NAME	**LOCATION**
Lymphatic nodes of the head and neck	
Superficial parotid nodes (anterior auricular)	In front of ears
Mastoid nodes (posterior auricular)	Behind ears
Submandibular nodes	Beneath mandible
Submental nodes	Beneath chin
Occipital nodes	Base of skull
Deep cervical nodes	Deep within neck
Superficial cervical nodes (includes medial and lateral nodes)	Side of neck
Lymphatic nodes of the body	
Axillary nodes	Armpit
Supratrochlear nodes	Elbow crease
Ileocolic nodes	Abdomen (near the diaphragm)
Iliac nodes	Abdomen
Inguinal nodes	Groin
Popliteal nodes	Knee

Lymphatic nodes are oval-shaped and vary in size from 1mm to 25mm. They are specially structured to allow lymph to flow slowly through them and be filtered of potentially harmful particles and substances.

- Lymph nodes are enclosed in a capsule of dense connective tissue from which strands extend into the node. These strands are called *trabeculae* and they divide the node into compartments. The trabeculae combine together with a network of reticular fibres and fibroblasts to form a framework that supports the continually changing lymphatic tissue within it.
- The lymphatic tissue that makes up lymph nodes consists of:
  - An outer cortex of lymphocytes that are arranged into structures called follicles. The outer rim of these follicles is composed of T cells and macrophages while in the central areas B-cells produce antibody secreting plasma cells.
  - An inner medulla of lymphocytes, macrophages and plasma cells.

Lymph can only flow in one direction: from the capillaries towards the ducts. This means that it also only flows through lymphatic nodes in one direction. The vessels that bring lymph into the nodes are called *afferent vessels* and the vessels that take lymph away from the nodes are called *efferent vessels*.

Lymph flows very slowly through the nodes and as it flows any foreign particles or substances in it are trapped by the reticular fibres of the nodes. Macrophages, antibodies and lymphocytes then all work to protect the body against these foreign substances. How they do this will be discussed shortly under the title Immunity (The Immune Response).

Lymphatic organs

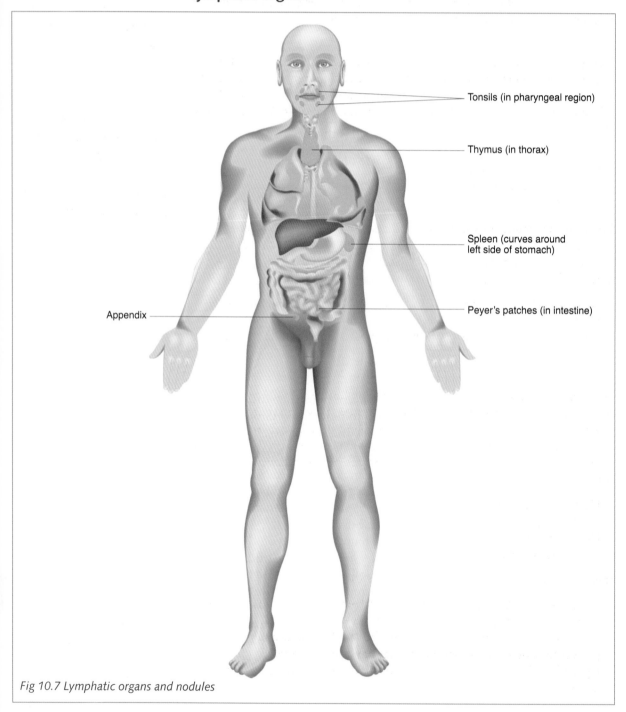

Tonsils (in pharyngeal region)

Thymus (in thorax)

Spleen (curves around left side of stomach)

Peyer's patches (in intestine)

Appendix

Fig 10.7 Lymphatic organs and nodules

Lymphatic tissue, or lymphoid tissue, plays an essential role in protecting the body from invaders and is responsible for the production of lymphocytes and antibodies.

Thymus gland

The thymus gland is located in the mediastinum, behind the sternum and between the lungs. Not all the functions of the thymus gland are yet known, however, it is known that the gland produces hormones such as thymosin that help the development and maturation of T cells.

Spleen

The spleen is the largest single mass of lymphatic tissue in the body. It is located in the abdomen, behind and to the left of the stomach. Although it is an organ of the lymphatic system, it is important to remember that the spleen does not filter lymph. Instead, it filters and stores blood. Functions of the spleen include:

- Filtering and cleaning blood.
- Destroying old worn-out red blood cells.
- Storing platelets and blood.
- Producing lymphocytes.

Similar to lymphatic nodes, the spleen is enclosed in a dense connective tissue from which trabeculae extend and together with reticular fibres and fibroblasts, these trabeculae form the framework of the spleen. The functional part of the spleen consists of two different tissues:

- **White pulp** which functions in immunity and is the site of antibody producing plasma cells.
- **Red pulp** which functions in the phagocytosis of bacteria, red blood cells and platelets.

Lymphatic nodules (mucosa-associated lymphoid tissue)

Lymphatic nodules, or mucosa-associated lymphoid tissue (MALT), are concentrations of lymphatic tissue that are strategically positioned to help protect the body from pathogens that have been inhaled, digested or have entered the body via external openings. Thus, they are scattered throughout the mucous membranes that line systems exposed to the external environment. These include the:

- Gastrointestinal tract
- Respiratory airways
- Urinary tract
- Reproductive tract.

Specific groups of lymphatic nodules include the:

- **Tonsils** – lie at the back of the mouth and help protect the body against pathogens that may have been inhaled or digested.
- **Peyer's patches** – found in the ileum of the small intestine and help protect the body against pathogens that have been digested.
- **Appendix** – located at the end of the caecum. Its function is not yet known.

Did you know?
The thymus gland is at its largest in a 10–12 year old child. Once the child reaches puberty the gland begins to atrophy and thymic tissue is replaced by adipose and areolar tissue.

Did you know?
If damaged, the spleen can be removed and most of its functions can be taken over by other structures such as the liver and red bone marrow.

Fluid flow through the body in a nutshell
Arteries
Blood plasma
⇓
Blood capillaries
Blood plasma
⇓
Interstitial spaces
Interstitial fluid
⇓
Lymphatic capillaries
Lymph
⇓
Lymphatic vessels
Lymph
⇓
Lymphatic ducts
Lymph
⇓
Subclavian veins
Blood plasma

Resistance to Disease and Immunity

Our bodies are constantly at war, fighting off pathogens trying to invade them from all directions. Bacterial attack, fungal attack, viral attack, chemical attack…and so the list continues. The body employs two approaches to defend itself against this multitude of invaders:

- Firstly, the body has a number of defence mechanisms that immediately provide general protection against a variety of invaders. This approach is known as *non-specific resistance to disease.*
- Secondly, the body employs specialised lymphocytes that recognise and combat specific pathogens. This is called the *immune response* or *immunity.*

Non-specific resistance to disease

The body has a number of different defence mechanisms that are in place to firstly ward off invading microbes and secondly help the body deal with any microbes that actually enter into the system. These mechanisms include:

- Mechanical and chemical barriers
- Natural killer cells and phagocytes
- Inflammation
- Fever.

Mechanical and chemical barriers
The body's first line of resistance to invaders is its mechanical and chemical barriers:

Mechanical barriers include:

- **The skin** whose many layers of tough cells provide a strong barrier against invading microbes. The top layer of the skin is also shed regularly and this helps remove any microbes that are on the surface of the skin.
- **Mucous membranes** that trap microbes and also contain hairs or cilia that move microbes and dust away from vital areas.
- **Lacrimal apparatus** which produce tears that wash away microbes and dirt.
- **Saliva** which washes away microbes and dirt from the mouth.
- The processes of **urination, defecation** and **vomiting** which all expel invaders from the body.

Chemical barriers include:

- **Sebum** which acts as a protective film over the skin and helps inhibit the growth of many bacteria and fungi.
- **Perspiration** which helps wash microbes off the surface of the skin and contains an antibacterial substance called lysozyme (discussed shortly).
- **Gastric juice** contains hydrochloric acid which destroys most ingested microbes.
- **Vaginal secretions** whose acidity discourages bacterial growth.

Included in chemical barriers are antimicrobial substances which destroy microbes. Examples of these include:

- **Lysozyme** – an enzyme found in certain body secretions such as tears and saliva. It catalyses the breakdown of the cell walls of certain bacteria.
- **Interferons** – substances produced to protect uninfected cells from viral infection.
- **Complement system** – a group of proteins in blood plasma and on plasma membranes. These proteins are usually inactive, but are activated when antigens and antibodies combine to form an immune complex. When activated, these proteins enhance or 'complement' some immune, allergic and inflammatory processes.

Natural killer cells and phagocytes

If microbes still manage to enter the body, they then face the body's second line of defence: natural killer cells and phagocytes.

- **Natural killer cells** (NK cells) – Natural killer cells are a type of lymphocyte (white blood cell) that can kill a variety of microbes as well as some tumour cells. It is not yet known exactly how they recognise or destroy their targets, yet they are found in the spleen, lymph nodes, red bone marrow and blood.
- **Phagocytes** – The name phagocyte literally means 'eating cell' and phagocytes are a type of lymphocyte that ingest microbes and foreign matter. There are two main types of phagocyte:
    - *Neutrophils* – These white blood cells were discussed in Chapter 9.
    - *Macrophages* – Macrophages are scavenging cells that develop from monocytes (a type of white blood cell). Wandering macrophages are mobile while fixed macrophages are found in the skin, liver, lungs, nervous system, spleen, lymph nodes and red bone marrow.

Inflammation

Despite its formidable defence system, the body can still be injured. If tissue cells are damaged through cuts, burns or microbial invasion, the body responds through a process called inflammation. This is the body's response to injury and is a defensive reaction whose purpose is to help prevent the spread of further damage, to prepare the site for repair and to help clear the area of microbes and toxins.

Inflammation involves four key symptoms:

- Swelling
- Heat
- Redness
- Pain

Occasionally, loss of function of the affected area can be a fifth symptom.

Inflammation is a long process which occurs as follows:

- **Vasodilation and increased permeability of blood vessels** – Immediately after the tissue has been damaged, blood vessels vasodilate and become more permeable. This is partly caused by the release of *histamine*, a chemical present in most tissues. Vasodilation results in more blood flowing to the damaged area and more defensive materials, such as antibodies, phagocytes and clot forming chemicals, leaving the blood and entering the injured site. This flow of blood to the area causes the swelling, heat and redness of inflammation. The pain that usually accompanies inflammation is caused by either injury or irritation to nerve fibres. It can also result from increased pressure caused by the swelling.

- **Phagocyte migration and tissue repair** – About an hour after the inflammatory process has begun, large numbers of phagocytes leave the bloodstream and enter the injured site. Neutrophils are the first type of phagocyte to appear. They engulf microbes and foreign materials but soon exhaust themselves and rapidly die off. They are then followed by wandering macrophages which engulf the damaged tissues, invading microbes and the worn-out neutrophils. As the tissue repairs, pus forms. This is a collection of dead cells and fluid. Pus formation continues until the infection has subsided.

Fever

A fever is an abnormally high temperature and it is the body's way of dealing with infection. Toxins released by microbes can increase the body temperature and this increase in temperature intensifies the effects of the body's own antimicrobial substances, inhibits microbial growth and increases the speed of tissue repair.

Immunity (the immune response)

There are two key differences between immunity and non-specific resistance to disease. Firstly, immunity involves a very specific and focused recognition and response to foreign molecules. Secondly, immunity involves memory.

Recognition and response to foreign molecules

Immunocompetent cells

Any substance that the body recognises as foreign and that provokes a response from the immune system is called an antigen.

In the classroom

Discuss the different types of antigens that exist. These range from microbes such as bacteria, viruses and fungi to pollen, egg white and even transplanted tissues and organs.

The body's lymphocytes that are responsible for responding to antigens are referred to as immunocompetent cells and there are two types:

- **B cells** – develop and mature in red bone marrow throughout life. They develop into plasma cells and are able to synthesise and secrete antibodies.
- **T cells** – develop in red bone marrow and then migrate to the thymus gland where they mature. Two types of T cells exist: CD4+ cells and CD8+ cells. Before T cells are released into the system, they acquire distinctive surface proteins which are capable of recognising specific antigens and are called *antigen receptors*.

B cells and T cells respond differently to pathogens. B cells secrete antibodies while T cells develop antigen receptors. We will now look at how these cells function in the immune response.

Types of immune responses

There are two types of immune responses and pathogens can provoke either one type of response, or both types at the same time.

Cell-mediated (cellular) immune responses (CMI responses) – In cell-mediated immune responses, cells attack antigens directly. CD8+ T cells reproduce into 'killer cells' which leave lymphatic tissues to seek out and destroy antigens. This is the most common response to intracellular pathogens such as fungi, parasites and viruses as well as some cancer cells and tissue or organ transplants.

> Remember:
> In *cell*-mediated responses, *cells* attack cells.

Antibody-mediated (humoral) immune responses (AMI responses) – In antibody-mediated immune responses, antibodies bind to antigens and inactivate them. B cells develop into plasma cells which secrete antibodies. Antibodies then leave the lymphatic tissue to circulate in the blood and lymph and bind to the particular antigen for which they were made. Once bound to this antigen they destroy it. This response is more common against antigens that are dissolved in body fluids and pathogens such as bacteria that have multiplied in body fluids.

> Remember:
> In *antibody*-mediated responses, *antibodies* attack cells.

If CD8+ T cells develop into killer cells and B cells develop into antibody-secreting plasma cells, then you may be wondering what CD4+ T cells do. They become 'helper' T cells that aid both CMI and AMI responses.

Immunological memory (acquired immunity)

Immunity involves immunological memory. This is the body's ability to remember and recognise antigens that have triggered an immune response. Some antibodies and lymphocytes can live for many years, even decades, and are the memory cells of the immune system. The immune system's initial response to antigens, called the *primary response*, is slow and it can take several days before antibodies are detected in blood serum.

Anatomy and physiology in perspective
The AIDS virus uses the CD4+ cells to enter into and destroy other T cells.

These antibodies also gradually decline unless there is another encounter with the same antigen. If the same antigen is encountered again a *secondary response* occurs. This is a quicker, more intense response in which the body recognises and remembers the antigen and so quickly produces more antibodies. Secondary responses can be so quick and successful that you may not even be aware of them.

Immunity can be acquired naturally or artificially as well as actively or passively as detailed in the chart below.

Anatomy and physiology in perspective
Vaccinations (immunisations) are based on the principle of immunological memory. Vaccinations are pre-treated to be immunogenic but not pathogenic. This means that they can activate a primary response in a person while not causing the person to become significantly ill. Thus, when the person next encounters the antigen, they only have a secondary response to it.

TYPES OF IMMUNITY	
TYPE OF IMMUNITY	**HOW IS IT ACQUIRED?**
Naturally acquired immunity	
Naturally acquired active immunity	The body is stimulated to produce its own antibodies through actively having the disease.
Naturally acquired passive immunity	This is the transference of antibodies from mother to foetus across the placenta or from mother to baby through breast-feeding.
Artificially acquired immunity	
Artificially acquired active immunity	Antigens are introduced to the body in the form of vaccinations that induce an active immune response but do not make the recipient ill.
Artificially acquired passive immunity	Ready-made antibodies are injected into the recipient.

Common Pathologies of the Lymphatic and Immune System

Allergy (hypersensitivity)

An allergic, or hypersensitive, reaction is an over-reaction to a substance that is normally harmless to most people. Any substance that invokes an allergic reaction is called an allergen and foods such as milk, peanuts, shellfish and eggs are common allergens as are some antibiotics, vitamins, drugs, venoms, cosmetics, plant chemicals, pollens, dust and moulds.

Allergic reactions can only occur if a person has been previously exposed to the allergen and so developed antibodies to it. Symptoms of allergic reactions can range from a running nose and streaming eyes to anaphylactic shock in which there is swelling, heart and lung failure and possible death.

Lymphoedema

Lymphoedema is the accumulation of lymph in the tissues. It results in swelling and most often affects the legs. Lymphoedema can be caused by a congenital defect in which there is a lack of lymphatic vessels or it can be caused by surgery in which lymphatic nodes and vessels have been damaged or removed. Parasites, obstructing tumours and injuries to the vessels are also known to cause lymphoedema.

> **In the classroom**
> Discuss the differences between oedema (discussed in Chapter 9) and lymphoedema.

Cancer and the lymphatic and immune system

Cancer is the uncontrolled division of cells and it can develop within any tissue in the body. When cells become cancerous, the body can often recognise them as abnormal and so destroy them. However, if the immune system does not recognise the cells as cancerous, or is unable to destroy them, the cancerous cells can replicate to form a mass. This mass is called a tumour and is more difficult for the body to destroy.

Leukaemia

Leukaemia is the abnormal production of leucocytes. In other words, it is cancer of the white blood cells. Instead of forming a lump or tumour, leukaemia cells remain as separate cancerous cells and crowd out healthy blood cells in the bloodstream. They also invade other organs such as the liver, spleen and lymph nodes and this results in an increased susceptibility to infection, bleeding and anaemia.

Lymphomas

Lymphomas are cancers of lymphocytes and they can remain confined to a lymph node, spread to other lymphatic tissues such as the spleen or bone marrow or spread to virtually any other organ in the body. The two main types of lymphoma are Hodgkin's disease and non-Hodgkin's disease:

* **Hodgkin's disease (Hodgkin's lymphoma)** – a malignant lymphoma characterised by the progressive, painless enlargement of the lymph nodes

of the neck, armpits, groin, chest or abdomen. Hodgkin's disease can sometimes be accompanied by fever, night sweats, weight loss, itching and fatigue. The cause of Hodgkin's disease is unknown.

- **Non-Hodgkin's lymphomas** – refers to a diverse group of lymphomas that are classified according to the type of cell involved and the degree of malignancy. They are found to be more common in elderly people and those whose immune systems are not functioning normally and the main symptom is the painless enlargement of lymph nodes.

Metastasis

Although the function of the lymphatic system is to help protect the body from disease, it can sometimes be responsible for spreading diseases. An example of this is the metastasis of cancer. Cancer spreads from its origin to other sites via a process called metastasis which occurs across body cavities, via the bloodstream or via the lymphatic system.

Infectious diseases

Acquired immunodeficiency syndrome (AIDS)

Acquired immunodeficiency syndrome, commonly called AIDS, is an unusual disorder in that the virus that causes it only lowers a person's immunity and makes one more susceptible to other diseases. It is these other diseases that produce the fatal symptoms of AIDS.

AIDS is caused by the human immunodeficiency virus (HIV) and results in a lowered T4 lymphocyte count. The infected person then becomes susceptible to opportunistic infections, such as pneumonia or tuberculosis, which their immune system cannot fight. HIV is transmitted through bodily fluids such as blood, semen, vaginal secretions and breast milk.

Glandular fever (Infectious mononucleosis)

Glandular fever is a contagious disease of the lymphatic system which results in large numbers of white blood cells in the bloodstream. It is caused by the Epstein-Barr virus and occurs mainly in children and young adults. It is often spread through kissing and symptoms include fatigue, headaches, dizziness, sore throat, enlarged and tender lymph nodes and fever.

Tonsillitis

Tonsillitis is inflammation of the tonsils. It can be caused by a virus or bacteria and is characterised by a sore throat, difficulty in swallowing and fever.

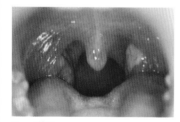

Tonsillitis

Autoimmune diseases and disorders

The role of the immune system is to attack foreign substances that have entered the body. Occasionally, however, the body does attack its own tissues. When this occurs a person is said to have an autoimmune disorder. Some autoimmune diseases and disorders that have already been discussed in other chapters of this book include rheumatoid arthritis, pernicious anaemia, Addison's disease, Graves' disease, insulin-dependent (Type I) diabetes mellitus, myasthenia gravis and multiple sclerosis. Systemic lupus erythematosus is also an autoimmune disease.

Systemic lupus erythematosus (SLE)

Systemic lupus erythematosus, commonly called SLE or lupus, is an autoimmune disease in which the body attacks its own connective tissue. Symptoms vary but can include inflammation, painful joints, fever, fatigue, mouth ulcers, weight loss, hair loss and an enlarged spleen. A 'butterfly' rash across the nose and cheeks is sometimes common. The cause of lupus is unknown, but it affects young women more than any other group of people.

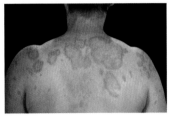

Systemic lupus erythematosus

NEW WORDS	
Antibody	a specialised protein that is synthesised to destroy a specific antigen.
Antigen	any substance that the body recognises as foreign.
Inflammation	the body's response to tissue damage.
Lymphocyte	a type of white blood cell involved in immunity. B cells and T cells are types of lymphocytes.
Macrophage	a scavenger cell that engulfs and destroys microbes.
Metastasis	the spread of disease from its site of origin.
Microbe (micro-organism)	an organism that is too small to be seen by the naked eye. Microbes include bacteria, viruses, protozoa and some fungi.
Pathogen	a disease-causing micro-organism.
Phagocyte	a cell that is able to engulf and digest microbes. Phagocytes include macrophages and some types of white blood cells.

Study Outline

Functions of the lymphatic system

Functions of the lymphatic system include drainage of interstitial fluid, transportation of dietary lipids and protection against invasion.

Structure of the lymphatic system

1. Lymph is derived from interstitial fluid and is only found in lymphatic vessels. Lymph contains protein molecules, lipid molecules, foreign particles, cell debris and lymphocytes.
2. Lymphatic capillaries are closed-ended vessels that permit fluid to flow into them but not out of them. They absorb lymph and transport it to lymphatic vessels.
3. Lymphatic vessels carry lymph from capillaries, through nodes and into lymphatic trunks.
4. Lymphatic trunks empty lymph into the thoracic duct and the right lymphatic duct.
5. The thoracic duct drains lymph from the left side of the head, neck and chest as well as the left arm and the entire body below the ribs.
6. The right lymphatic duct drains lymph from the upper right side of the body only.
7. Lymph is moved through the lymphatic vessels by smooth muscle, skeletal muscle and breathing movements. It is prevented from flowing backwards by one-way valves.
8. Lymph only flows through lymphatic nodes in one direction. The vessels that bring lymph into the nodes are called *afferent vessels* and the vessels that take lymph away from the nodes are called *efferent vessels*.

> Afferent = Arrive
> Efferent = Exit

9. Lymphatic nodes filter lymph of potentially harmful substances and also produce lymphocytes.
10. Lymphatic organs include the thymus gland and the spleen.
11. The thymus gland produces hormones that help the development and maturation of T cells.
12. The spleen filters and cleans blood, destroys worn-out red blood cells, stores platelets and blood and produces lymphocytes.
13. Lymphatic nodules (mucosa-associated lymphoid tissue) are strategically positioned to help protect the body from pathogens that have been inhaled or digested. They are found in mucous membranes lining body cavities.
14. Lymphatic nodules include the tonsils, Peyer's patches and the appendix.

Resistance to disease and immunity

1. Defence mechanisms that give general protection against invaders are classified as non-specific mechanisms. These include mechanical and chemical barriers, natural killer cells and phagocytes, inflammation and fever.
2. Defence mechanisms in which the body produces specific responses to particular organisms fall under the classification of immunity.
3. Immunity involves the body's ability to recognise microbes, respond to them appropriately and remember them so that a secondary response will be quicker and more powerful.

Immunity

1. Immunocompetent cells include B cells and T cells.
2. B cells develop and mature in the red bone marrow and develop into plasma cells which synthesise and secrete antibodies. B cells participate in antibody-mediated immune responses (AMI responses).

> **B** cells = anti**B**odies

3. In AMI responses, antibodies bind to antigens and destroy them.
4. T cells develop in red bone marrow and then migrate to the thymus gland where they mature. T cells participate in cell-mediated immune responses (CMI responses).

> **T** cells = **T**hymus

5. In CMI responses, killer cells attack and destroy antigens.
6. Acquired immunity is the body's ability to remember and recognise antigens that have caused an immune response. The primary response is slow while the secondary response is quick and more successful.

Revision

1. Identify the functions of the immune system.
2. Compare blood plasma to lymph.
3. Compare blood capillaries to lymphatic capillaries.
4. Explain how lymph is moved through the lymphatic vessels.
5. Describe a lymphatic node.
6. Explain the function of Peyer's patches and describe where in the body they are found.
7. Explain the function of the thymus gland.
8. Explain the functions of the spleen.
9. Compare non-specific resistance to disease with immunity.
10. Explain how the body protects itself from general invasion.
11. Explain the inflammatory process.
12. Describe the immune response.

Multiple choice questions

1. **The axillary lymph nodes are located in the:**
   a. Neck
   b. Shoulder
   c. Armpit
   d. Elbow.

2. **Which of the following statements is correct?**
   a. After puberty the thymus gland begins to atrophy.
   b. The thymus gland produces antibodies.
   c. The thymus gland filters and stores blood.
   d. The thymus continues to grow until old age.

3. **Lymphoedema is:**
   a. An accumulation of blood plasma in the tissues.
   b. An accumulation of lymph in the tissues.
   c. An accumulation of interstitial fluid in the lymphatic vessels.
   d. None of the above.

4. **Where is the spleen located?**
   a. Superior to the thymus.
   b. Behind and to the left of the stomach.
   c. Inferior to the large intestine.
   d. In front of and to the right of the liver.

5. **Which of the following is not a lymphatic node of the pelvis and lower limbs?**
   a. Popliteal
   b. Inguinal
   c. Iliac
   d. Supratrochlear.

6. **Where in the body are worn-out erythrocytes destroyed?**
   a. Peyer's patches
   b. Appendix
   c. Axillary node
   d. Spleen.

7. **Which of the following is a function of the lymphatic system?**
   a. Transportation of dietary lipids.
   b. Transportation of haemoglobin molecules.
   c. Transportation of carbon dioxide.
   d. Transportation of bilirubin.

8. **Where in the body is mucosa-associated lymphoid tissue found?**
   a. Liver
   b. Kidneys
   c. Urinary tract
   d. Ovaries.

9. **Where in the body do T cells mature?**
   a. Liver
   b. Lungs
   c. Spleen
   d. Thymus.

10. **Bacteria, viruses and fungi are all:**
    a. Antibodies
    b. Lymphocytes
    c. Microbes
    d. Lacteals.

11 The Digestive System

Introduction

An average person eats approximately 500 kg of food per year[i] and this food gets converted into energy that the body can use to move, breathe, function, build and repair itself. Then the waste is excreted. This amazing transformation of food takes place in the gastrointestinal tract, a 7.6m long tube running from the mouth to the anus, and in this chapter we will follow the journey of our food through this tract.

Student objectives

By the end of this chapter you will be able to:

- Describe the functions of the digestive system.
- Describe the organisation of the gastrointestinal system.
- Identify and describe the main organs and structures of digestion.
- Explain the chemical breakdown of carbohydrates, proteins and lipids.
- Identify the common pathologies of the digestive system.

Functions of the Digestive System

The food we eat contains carbohydrates, fats, proteins, vitamins and minerals – all essential nutrients for the life of every cell. However, none of these nutrients can be used by the body in the form in which they are eaten. They must be broken down into molecules that are small enough to cross the plasma membranes of cells. These foods also carry dangerous microbes and toxins that can harm our bodies and so they need to be prevented from passing into our cells. These are the primary functions of the digestive system: to break down the foods we eat and convert them into a usable form and to destroy or eliminate dangerous substances within them.

The functions of the digestive system can be broken down into the following six basic processes: ingestion, secretion, mixing and propulsion, digestion, absorption and defecation.

Ingestion
Ingestion is the process of taking food into the mouth.

Secretion
Some organs and structures of the digestive system secrete mucous, water, acid, buffers and enzymes. All of these secretions function in helping move and digest food.

Mixing and propulsion
The gastrointestinal tract is a tube that runs from the mouth to the anus and is where digestion and absorption take place. Food is mixed with digestive secretions and is propelled from the mouth to the anus.

Digestion
Digestion is the process by which large molecules of food are broken down into smaller molecules that can enter cells. Food is digested both mechanically and chemically:

- **Mechanical digestion** – In mechanical digestion, food is physically broken down and ground into smaller substances by the teeth, tongue and physical movements such as peristalsis (to be discussed shortly). It is then mixed with fluids until it finally becomes a liquid. Once in a liquid state, it is easier for chemical digestion to take place.
- **Chemical digestion** – In chemical digestion, food molecules are broken down into smaller molecules by enzymes. The chart below outlines the breakdown of carbohydrates, fats and proteins.

Did you know?
Approximately 9 litres of fluid are secreted by the digestive system every day.

CHEMICAL DIGESTION OF CARBOHYDRATES, FATS AND PROTEINS			
NUTRIENT	**ENZYMES**	**FROM**	**TO**
Carbohydrates	Amylases	Starches and polysaccharides into disaccharides	Monosaccharides
Fats	Lipases	Triglycerides	Fatty acids and glycerol
Proteins	Proteases	Peptones and polypeptides	Amino acids

A closer look at enzymes...

The digestion of food relies on the presence of enzymes. Enzymes are catalysts. They speed up reactions but do not actually become involved in the reactions themselves. Under optimal conditions, enzymes can speed up the rates of reactions to up to 10 billion times faster than they would take place without enzymes[ii].

Enzymes are very specific in what they catalyse and a particular enzyme can only affect a particular substrate (molecule on which the enzyme is acting). There are over 1000 known enzymes and each one of these has a specific three-dimensional shape with a characteristic surface area. This means it can only bond to specific substrates that will 'fit' its surface area. In some cases, enzymes are said to fit a substrate like a key fits a lock, in other cases enzymes are said to surround the substrate.

Absorption

Once the larger food molecules have been digested into smaller molecules they enter the lining of the gastrointestinal tract by either active transport or passive diffusion (refer to Chapter 2 to revise these processes). The molecules are absorbed into the bloodstream and lymphatic vessels and are then distributed to the rest of the body.

Defecation

Defecation is the process by which indigestible substances and some bacteria are eliminated from the body. They are eliminated as faeces by the anus.

Overview of the Digestive System

Before studying the digestive system in detail, it will help you to have an overview of the basic structure and function of this system. As mentioned earlier, the digestive system is composed of a long tube that passes from the mouth to the anus. This is the gastrointestinal tract, or alimentary canal. This continuous tube forms the following organs and structures:

- Mouth
- Pharynx
- Oesophagus
- Stomach
- Small intestine – composed of the duodenum, jejunum and ileum
- Large intestine
- Anus

In addition to the gastrointestinal tract, the digestive system includes accessory structures which help with the digestion of food. Most of them produce and/or store secretions that help with the chemical breakdown of food. These accessory structures are the:

- Teeth, tongue and salivary glands – all located in the mouth.
- Liver
- Gall bladder
- Pancreas

Study tip

The names of enzymes usually end in –ase and it is often easy to work out what they catalyse by their names. For example, *proteases* break down proteins while *lipases* break down lipids.

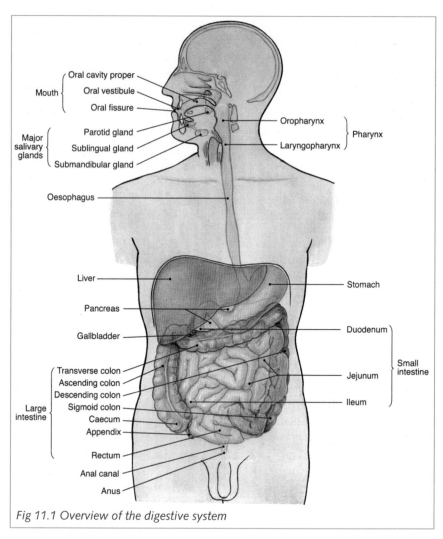

Fig 11.1 Overview of the digestive system

Wall of the gastrointestinal tract

The walls of the gastrointestinal tract (GI tract) are composed of four layers of tissue. This basic arrangement of tissue does differ slightly with some of the organs, but in general it is composed of the:

Mucosa – the deepest layer of the GI tract. Its inner lining is a *mucous membrane* which functions in protection, secretion and absorption. This mucous membrane is attached to the *lamina propria* which is a loose connective tissue layer that supports the blood and lymph vessels into which digested molecules are absorbed. The lamina propria also contains mucosa-associated lymphoid tissue (M.A.L.T) which protects the body against the many microbes and toxins that may have been ingested. The third layer of the mucosa is the *muscularis mucosa* which is a layer of smooth muscle.

Submucosa – this is composed of areolar connective tissue and binds the mucosa to the muscularis layer. It contains many blood and lymphatic vessels and nerves and houses the *submucosal (Meissner's) plexus*. This is a network of nerves that serves the smooth muscle cells of the mucosa and it forms part of the autonomic nervous system.

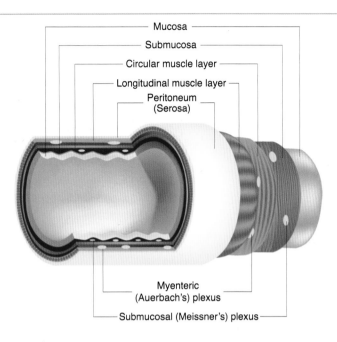

The wall of the gastrointestinal tract is composed of four layers: the inner mucosa lines the lumen of the tract and layered on top of it is the submucosa, the muscularis and the serosa.

Mucosa
Submucosa
Circular muscle layer
Longitudinal muscle layer
Peritoneum (Serosa)
Myenteric (Auerbach's) plexus
Submucosal (Meissner's) plexus

Fig 11.2 Structure of the wall of the gastrointestinal tract

Muscularis – a layer of muscle tissue. It includes some skeletal muscle tissue (especially in the mouth, pharynx and upper portion of the oesophagus as these are the voluntary muscles of swallowing); and smooth muscle tissue. The smooth muscle tissue is composed of an inner sheet of circular fibres and an outer sheet of longitudinal fibres. These two sheets work together to physically break down and propel food along the GI tract. The muscularis also contains the *myenteric plexus (plexus of Auerbach)*. This is a major nerve supply to the GI tract.

Serosa – this is the most superficial layer of the wall of the GI tract and it is a serous membrane. The oesophageal portion of the serosa is called the *adventitia* and it is composed of areolar connective tissue. The rest of the serosa is found below the diaphragm and is called the *visceral peritoneum*.

Peritoneum

The peritoneum is a large serous membrane lining the abdominal cavity. Like the pleural and pericardial serous membranes, the peritoneum has two layers:

* The **parietal peritoneum** lines the walls of the abdominopelvic cavity.
* The **visceral peritoneum** is also called the serosa and it covers the organs of the digestive system. Between the two layers of the peritoneum is a space called the peritoneal cavity. This is filled with serous fluid.

However, unlike the pleural and pericardial serous membranes which smoothly cover the lungs and heart, the peritoneum is composed of large folds. These folds weave between the organs, binding them to each other and to the walls of the abdominal cavity. These folds also contain many blood and lymphatic vessels and nerves and include the:

Did you know?
Together the submucosal and myenteric plexuses have as many neurons as does the spinal cord.

Did you know?
The peritoneum is the largest serous membrane in the body.

375

- **Mesentery** – binds the small intestine to the posterior abdominal wall.
- **Mesocolon** – binds the large intestine to the posterior abdominal wall.
- **Falciform ligament** – binds the liver to the anterior abdominal wall and diaphragm.
- **Lesser omentum** – links the stomach and duodenum with the liver.
- **Greater omentum** – covers the colon and small intestine.

Anatomy and physiology in perspective
The word *omentum* means 'fat' and the greater omentum contains large amounts of adipose tissue. It looks like a 'fatty apron' draping over the colon and small intestine.

Gastrointestinal Tract and its Accessory Organs

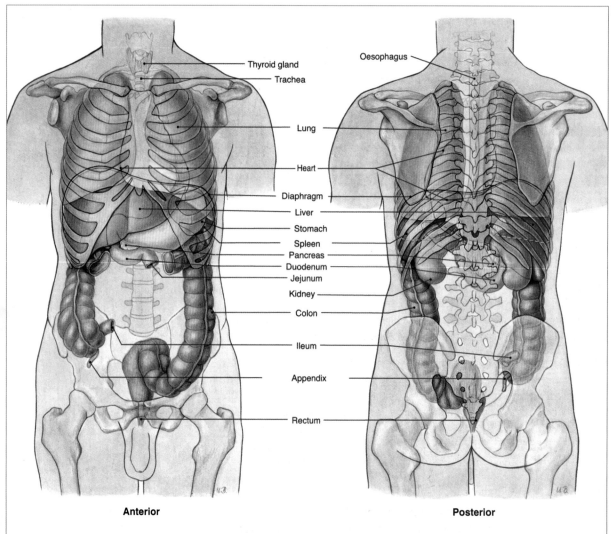

Fig 11.3 Projection of the internal organs onto the surface of the body

The best way to learn the digestive system is to start by having something to eat. So go and get something to eat – make sure it includes carbohydrates, fats and proteins! Now, let's follow the journey of your food.

Mouth (oral or buccal cavity)

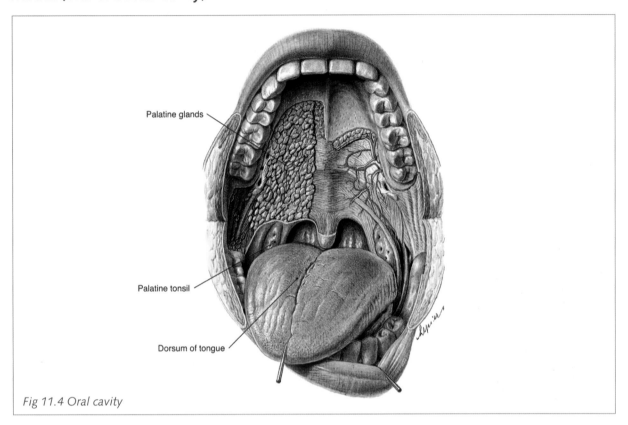

Fig 11.4 Oral cavity

Your food's journey begins in your mouth. You use your teeth to mechanically digest your food by biting and chewing it. Your tongue helps push the food around your mouth and also contains taste buds so that you can appreciate the food. Together with the sensation of the food in your mouth, your senses of taste, smell and sight stimulate the release of saliva which moistens the food in your mouth and begins the chemical breakdown of it.

Structure of the mouth

The mouth is a mucous-membrane lined cavity formed by the cheeks, hard and soft palates and the tongue. The opening to the mouth is protected by the lips (*labia*) and at the back of the mouth is a finger-like projection hanging from the soft palate. This is called the *uvula* and together with the soft palate it closes off the nasopharynx during swallowing. Thus, food is directed into the oesophagus rather than the respiratory tract.

The mouth includes the following structures:

The tongue – moves the food around the mouth.
- It is composed of skeletal muscle covered by a mucous membrane.
- It is attached to the hyoid bone, the styloid process of the temporal bones and the mandible and is secured to the floor of the mouth by the *lingual frenulum*.

- The tops and sides of the tongue are covered by small projections called *papillae* and some of these papillae house the taste buds (for more information on the sense of taste please refer to Chapter 6).
- The surface of the tongue also contains glands that secrete the digestive enzyme *lingual lipase*. This enzyme begins the breakdown of fats.

The teeth (Dentes) – cut and chew food.

- They are composed of a calcified connective tissue called *dentin* which encloses a cavity filled with *pulp*. Pulp is a connective tissue that contains blood and lymph vessels and nerves.
- They are located in sockets of the alveolar processes of the mandible and maxillae and these sockets are covered with *gums (gingivae)* and lined by the *periodontal ligament* which anchors the teeth in the gums and acts as a shock-absorber during chewing.
- Each tooth has three regions: a visible *crown*, one to three *roots* embedded in the socket, and a junction between the crown and roots called the *neck*.
- The dentin of the crown is covered by a layer of calcium phosphate and calcium carbonate. This is called *enamel*. Enamel is an extremely hard substance that protects the teeth from being worn down and also acts as a barrier against acids. The dentin of the root is covered by a substance called *cementum*. Cementum attaches the root to the periodontal ligament.
- In your life you have two sets of teeth. The first set is your *deciduous/primary/milk/baby* teeth. There are twenty in total and they start showing around the age of six months and then begin to fall out between six to twelve years. The second set is your *permanent/secondary* teeth. There are thirty-two in total.
  - There are four types of teeth: *incisors* and *canines* are the sharp pointed teeth used for cutting and biting while *premolars* and *molars* are flatter, broader teeth used for grinding and chewing.

The salivary ducts and glands – produce and secrete saliva.

- Saliva is an alkaline liquid that is continually secreted into the mouth. It helps keep the mouth moist, clean the mouth and teeth, lubricate food and dissolve food molecules. It also contains an enzyme which begins the chemical digestion of carbohydrates and when it is swallowed it helps lubricate the oesophagus. Saliva is:
  - 99.5% water – water acts as a dissolving medium.
  - 0.5% solutes – these solutes include ions which buffer acidic foods; urea and uric acid which help remove bodily wastes; mucous which lubricates food; the chemical lysozyme which helps destroy some bacteria; and the enzyme salivary amylase which begins the breakdown of carbohydrates.

Did you know?
Enamel is the hardest substance in the body.

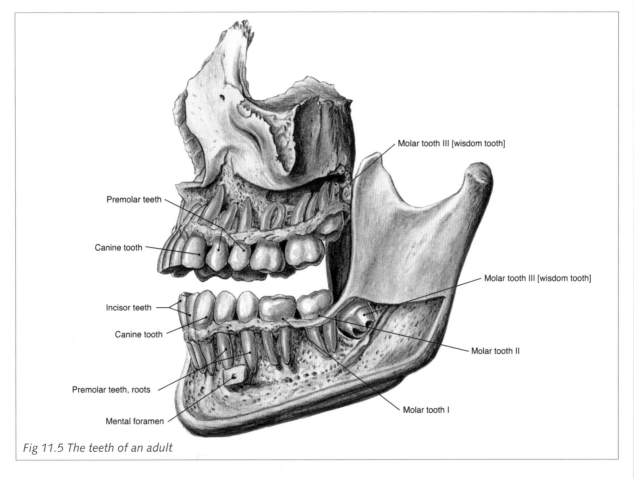

Molar tooth III [wisdom tooth]

Premolar teeth

Canine tooth

Molar tooth III [wisdom tooth]

Incisor teeth

Canine tooth

Molar tooth II

Premolar teeth, roots

Molar tooth I

Mental foramen

Fig 11.5 The teeth of an adult

- Saliva is secreted by the mucous membrane lining the mouth, the buccal glands and the salivary glands. There are three pairs of salivary glands:
  - The parotid glands – These are located below and in front of the ear, near the masseter muscle. They secrete saliva into salivary ducts which open into the mouth at the level of the second upper molar.
  - The submandibular glands – These are located under the angle of the jaw and their ducts open into the mouth on either side of the lingual frenulum.
  - The sublingual glands – These lie in front of the submandibular glands and have numerous ducts that open into the floor of the mouth.

Anatomy and physiology in perspective
Mumps is inflammation of the parotid salivary glands and because these glands are located near the masseter muscle it hurts to open the mouth or chew food when you have mumps.

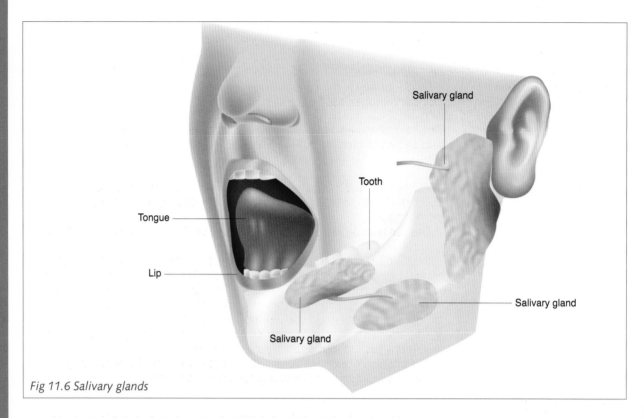

Fig 11.6 Salivary glands

Digestion in the mouth
Please refer to the chart opposite for details on digestion in the mouth.

Oesophagus

Having been swallowed, your food then travels down a long, collapsible tube running behind your trachea. This is your oesophagus and it functions in transporting the bolus from the laryngopharynx to the stomach. No digestion takes place in the oesophagus.

Structure and function of the oesophagus
The oesophagus is a muscular tube approximately 25cm long. It secretes mucous which helps facilitate the movement of the bolus as it pushes it down towards the stomach through a process called peristalsis.

Peristalsis is a wavelike alternating contraction and relaxation of the involuntary circular and longitudinal fibres of the gastrointestinal tract. The circular fibres above the bolus contract and constrict the walls of the tract. This squeezes the bolus downward. Simultaneously, the longitudinal fibres of the tract beneath the bolus contract and shorten, pushing the walls of the tract outwards and thus making room to receive the bolus.

DIGESTION IN THE MOUTH	
DIGESTION/ENZYME	**DESCRIPTION**
Mechanical digestion The mechanical digestion of food begins in the mouth.	
Chewing (mastication)	The tongue moves food, the teeth grind it and saliva mixes with the food and begins to dissolve it. Finally, it is reduced to a soft, flexible mass called a *bolus*.
Swallowing (deglutition)	Swallowing is the mechanical process by which the bolus is moved from the mouth into the stomach. It involves the mouth, pharynx and oesophagus: • The tongue forces the bolus to the back of the oral cavity and into the oropharynx. • Breathing is then temporarily interrupted as the respiratory passages are closed by the upward movement of the soft palate and the uvula. As this movement occurs, the larynx comes forwards and upwards over the tongue and the epiglottis moves backwards and downwards, thus closing off the space between the vocal folds (rima glottidis). • This enables the bolus to pass through the laryngopharynx and into the oesophagus without entering the respiratory tract. • The entire process takes only 1-2 seconds and then the respiratory passage reopens and breathing continues as normal.
Chemical digestion The digestion of both carbohydrates and lipids begins in the mouth. However, because the food is in the mouth for such a short time, the digestion of these continues in the stomach.	
Salivary amylase (in saliva)	Salivary amylase begins the breakdown of large carbohydrate molecules such as starch or polysaccharides into disaccharides and then monosaccharides.
Lingual lipase (secreted by glands on the tongue)	Lingual lipase begins the breakdown of fats (lipids) from trigyclerides into fatty acids and glycerol.

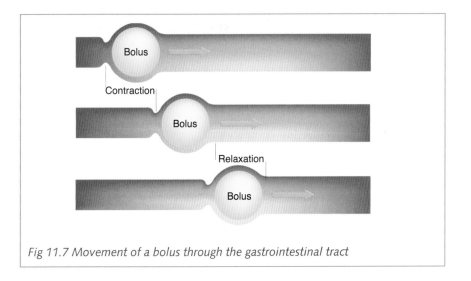

Fig 11.7 Movement of a bolus through the gastrointestinal tract

The stomach

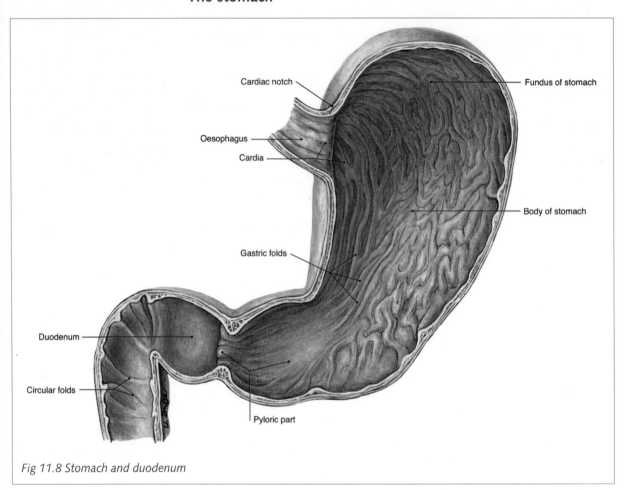

Fig 11.8 Stomach and duodenum

Within seconds of biting into and chewing your food, it arrives in your stomach where it accumulates in layers, the last bit you have eaten being layered on top of your previous mouthful. Once in your stomach, food is gradually churned and mixed with gastric juice which contains a variety of substances that help digest it.

Structure of the stomach

The stomach is a J-shaped organ continuous with the gastrointestinal tract. It lies below the diaphragm and in the left-hand side of the abdominal cavity and is connected to the oesophagus above it by the *cardiac (cardio-oesophageal) sphincter* muscle and the duodenum below it by the *pyloric sphincter* muscle. Because food is eaten more quickly than it can be digested, the stomach is specially structured to act as a mixing area and holding reservoir for food. The stomach can be divided into four regions:

- The *cardia* which surrounds the superior opening to the stomach and receives food from the oesophagus.
- The *fundus* which is an expanded area to the side of the cardia and where food can be held for up to one hour before it comes into contact with gastric juices.

- The *body* which is the large, central mid-portion of the stomach.
- The *pylorus* which is the area closest to the duodenum.

The stomach has the exceptional ability to stretch and when full it can hold up to 4 litres of contents. When fully stretched, its walls are smooth, however, when empty, the stomach shrinks to about the size of a large sausage and its mucous membrane is thrown into large folds called *rugae*. The walls of the stomach have a similar arrangement of tissue to the rest of the gastrointestinal tract. However, there are two differences:

- The inner lining of the mucosa is dotted with deep gastric pits which are narrow channels containing gastric glands that secrete gastric juice. Many different cells are also present in the mucosa that secrete a number of different digestive substances which will be discussed shortly.
- The muscularis includes a third layer of involuntary muscle that is composed of obliquely arranged fibres. This enables the stomach to mix, pummel and churn its contents and so reduce them to a liquid.

Did you know?
It takes approximately four hours for the stomach to be emptied after a meal and if the meal had a very high fat content, then it can take up to six hours before the stomach is properly emptied.

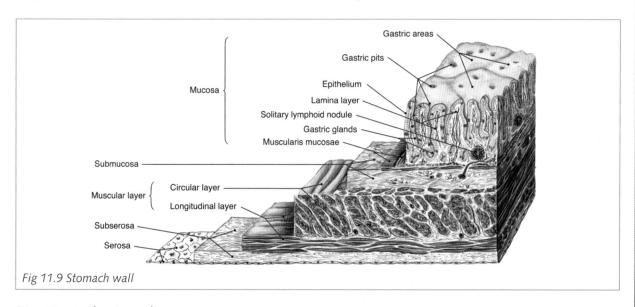

Fig 11.9 Stomach wall

Digestion in the stomach

DIGESTION/ENZYME	DESCRIPTION
Mechanical digestion In the stomach food is mixed, pummelled and churned into a thin liquid called chyme.	
Mixing waves	Every 15–25 seconds a mixing wave passes over the stomach, pushing its contents backwards and forwards and compressing and pummelling it into a liquid state.
Chyme is 'squirted' into the duodenum	The pyloric sphincter muscle between the stomach and duodenum is never completely closed and as the stomach contents are pushed against the muscle, small amounts of chyme (approximately 3 mls) are forced or squirted out of the stomach and into the duodenum. The remaining contents continue to be mixed and churned until the next wave forces a little more chyme into the duodenum.

DIGESTION/ENZYME	DESCRIPTION
Chemical digestion In the presence of food, endocrine cells in the walls of the stomach secrete the hormone gastrin which stimulates the production of gastric juice. Gastric juice is secreted by gastric glands and contains: • **Water** – liquefies the food. • **Hydrochloric acid** (HCL) – kills microbes that have been ingested; partially denatures proteins; stimulates the secretion of hormones that promote the flow of bile and pancreatic juice; stops the action of salivary amylase and lingual lipase; and is needed for the conversion of pepsinogen into pepsin. • **Intrinsic factor** – necessary for the absorption of vitamin B12 from the ileum. • **Pepsinogen** – an enzyme precursor that is converted into pepsin in the acidic environment of gastric juice. Pepsin is an enzyme that begins the breakdown of protein. • **Gastric lipase** – an enzyme that acts on lipids, breaking triglycerides down into fatty acids and monoglycerides. A 1–3 mm thick layer of mucous forms a protective barrier between the acidic gastric juice and the stomach wall and also protects the epithelial cells of the wall from pepsin which digests proteins.	
Salivary amylase	The digestion of carbohydrates by salivary amylase continues in the stomach for about one hour before salivary amylase is denatured by hydrochloric acid.
Lingual lipase	Similarly, the digestion of fats by lingual lipase continues for about one hour before the lingual lipase is denatured by hydrochloric acid.
Gastric Lipase	Continues the breakdown of fats. However, gastric lipase is soon denatured by the acidity of gastric juice.
Pepsin	Pepsin begins the breakdown of proteins.
Rennin	Rennin is an enzyme found only in the stomachs of infants. It begins the digestion of milk by converting the protein caseinogen into casein.

Did you know?

The stomach is not essential to life and a person can survive without it.

From the stomach your food passes into the duodenum, the first part of your small intestine. However, before we look closely at the small intestine, we are going to take a detour in this journey and look at three accessory organs that secrete their substances into the duodenum. These are the pancreas, liver and gall bladder.

Pancreas

The pancreas is a long, thin gland lying behind the stomach and connected to the duodenum by two ducts, the larger *pancreatic duct (duct of Wirsung)* which joins the common bile duct from the liver and the smaller *accessory duct (duct of Santorini)*. The pancreas has a head, body and tail and is 12–15cm long.

Ninety-nine per cent of pancreatic cells are exocrine cells that secrete *pancreatic juice*. This is a clear liquid composed of mostly water, some salts, sodium bicarbonate and some enzymes. It is slightly alkaline and so buffers the acidic chyme coming from the stomach, stops the action of pepsin and creates the correct pH in the small intestine in which the enzymes here can work. There are many different enzymes in pancreatic juice, including:

• Pancreatic amylase – continues the breakdown of carbohydrates.
• Trypsin – continues the breakdown of proteins.
• Pancreatic lipase – continues the breakdown of lipids.
• Enzymes that digest nucleic acids.

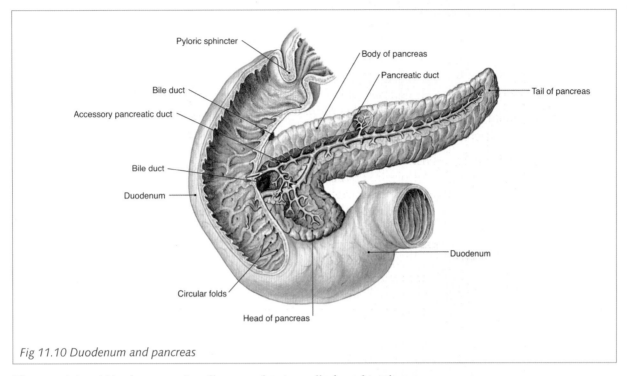

Pyloric sphincter

Bile duct

Accessory pancreatic duct

Bile duct

Duodenum

Circular folds

Head of pancreas

Body of pancreas

Pancreatic duct

Tail of pancreas

Duodenum

Fig 11.10 Duodenum and pancreas

The remaining 1% of pancreatic cells are endocrine cells found in clusters called *pancreatic islets (islets of Langerhans)*. These secrete hormones such as glucagon and insulin and are discussed in more detail in Chapter 7.

Liver

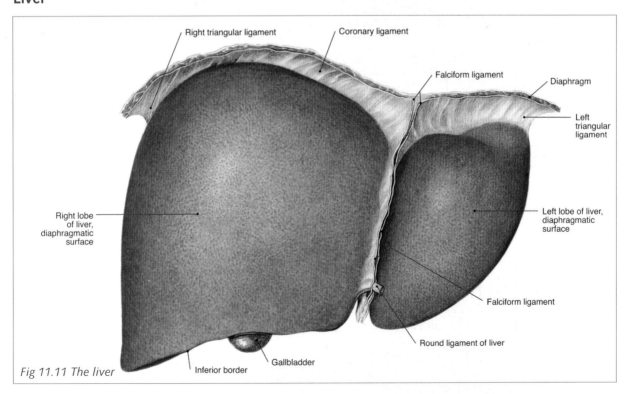

Right triangular ligament

Coronary ligament

Falciform ligament

Diaphragm

Left triangular ligament

Right lobe of liver, diaphragmatic surface

Left lobe of liver, diaphragmatic surface

Falciform ligament

Round ligament of liver

Inferior border

Gallbladder

Fig 11.11 The liver

Structure of the liver

The liver is a large organ located in the top right portion of the abdominal cavity, below the diaphragm. It is composed of two lobes separated by the *falciform ligament*. The right lobe is larger than the left and both are covered by visceral peritoneum under which is a layer of dense irregular connective tissue. Each lobe is composed of smaller functional units called *lobules* and each lobule is made up of specialised epithelial cells called *hepatocytes*. These cells are arranged around a central vein.

The liver does not have capillaries in it. Instead, it has spaces through which blood passes. These spaces are called *sinusoids* and they are partly lined by phagocytes which destroy microbes and potentially harmful foreign matter. The liver also has a unique double blood supply. It receives oxygenated blood via the *hepatic artery*. It also receives deoxygenated blood from the gastrointestinal tract via the *hepatic portal vein*.

This blood contains newly absorbed nutrients as well as drugs, toxins or microbes that may have been absorbed from the gastrointestinal tract. This blood needs to be 'cleaned' or 'made safe' by the liver before it can be circulated to the rest of the body (for more information on hepatic portal circulation please refer to Chapter 9).

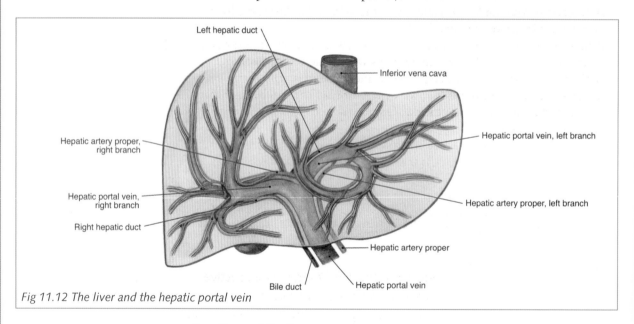

Fig 11.12 The liver and the hepatic portal vein

Functions of the liver

The liver is a vital organ with many important functions in the body. These include:

- **Carbohydrate metabolism** – The liver maintains normal blood glucose levels.
  - If blood glucose levels are too low, the liver converts glycogen into glucose which it then releases into the blood. If necessary, it can also convert some amino acids and lactic acid into glucose.
  - If blood glucose levels are too high, the liver converts the excess glucose into glycogen and triglycerides which it then stores.

- **Lipid metabolism** – The liver stores fats and, when necessary, converts them into a form which can be used by the tissues. It also breaks down fatty acids; synthesises lipoproteins which are necessary for transporting fatty acids, triglycerides and cholesterol; and synthesises cholesterol.
- **Protein metabolism** – The liver synthesises all the major plasma proteins. It also:
  - Converts one amino acid into another through a process called *transamination*.
  - Removes the amino group, NH2, from amino acids so they can be used for ATP production or converted into carbohydrates or fats. This process is called *deamination*.
  - Converts ammonia, which is a toxic by-product of deamination and is also produced by bacteria in the gastrointestinal tract, into urea for excretion.
- **Detoxification** – The liver removes and excretes alcohol and some drugs and also chemically alters and excretes some hormones.
- **Storage of nutrients** – The liver stores glycogen; vitamins A, B12, D, E and K; and the minerals iron and copper.
- **Phagocytosis** – Some liver cells phagocytise worn-out red blood cells, white blood cells and some bacteria.
- **Activation of vitamin D** – Together with the skin and kidneys, the liver participates in activating vitamin D.
- **Production of bile** – The liver produces a yellow, brownish (sometimes greenish) liquid called bile which contains water, bile acids, bile salts, cholesterol, phospholipids, bile pigments and some ions. The liver secretes bile into the gall bladder where it is stored. The gall bladder then intermittently releases it into the duodenum where it plays both excretory and digestive roles.
  - As an excretory product, bile contains bilirubin which is a pigment from worn-out red blood cells that have been broken down.
  - As a digestive product, bile contains bile salts which function in the emulsification and absorption of fats. In this process, large lipid molecules are broken down into a suspension of droplets which makes them more soluble and able to be absorbed.

Did you know?
One of the many amazing facts about the liver is that it can regenerate itself. In 1989 the first living-donor liver transplant took place. In this operation, a 21-month old child received part of her mother's liver. Two months later the mother's liver had returned to its normal size and the child's liver was growing healthily with the child[iii].

Anatomy and physiology in perspective
Bile is broken down into a substance called stercobilin which is excreted in faeces and gives them their brown colour.

Gall bladder

The gall bladder is a pear-shaped, green sac that is located behind the liver and attached to it by connective tissue. The gall bladder receives bile from the liver and concentrates and stores it. It then releases bile into the duodenum via the common bile duct.

Small intestine

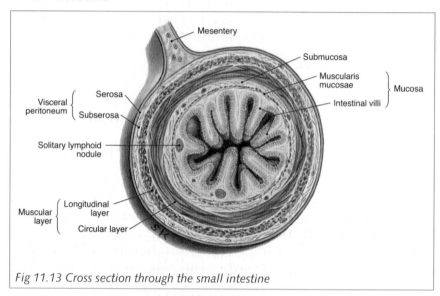

Fig 11.13 Cross section through the small intestine

From the stomach your food passes into a long, coiled and looped tube where it spends the next 3–5 hours. This tube is the small intestine and almost all the digestion and absorption of nutrients occurs here (bacteria in the large intestine complete the digestion of any nutrients that have not been fully broken down in the small intestine).

Structure of the small intestine

The small intestine is made up of three segments:

- **The duodenum** – the first segment of the small intestine and is also the shortest. It is approximately 25cm long and it receives food from the stomach, bile from the gall bladder and pancreatic juice from the pancreas.
- **The jejunum** – approximately 1m long and lies between the duodenum and the ileum.
- **The ileum** – the longest segment of the small intestine and approximately 2m long. It receives food from the jejunum and passes it into the large intestine via the *ileocaecal valve*.

The structure of the wall of the small intestine is similar to that of the rest of the gastrointestinal tract in that it has the same basic four layers of tissues (mucosa, submucosa, muscularis and serosa). However, because most digestion and absorption takes place in the small intestine, it is uniquely structured to ensure a large surface area:

- **Plicae circulares (circular folds)** – The mucosa of the small intestine has ridges that are approximately 10 mm high. These ridges are called the

Did you know?

Your small intestine is approximately 3m long and has a diameter of 2½ cm. However, the small intestine of a dead person has no muscle tone and is actually almost 6½ m long.

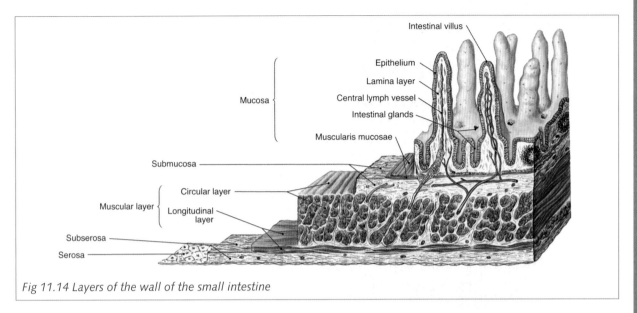

Fig 11.14 Layers of the wall of the small intestine

plicae circulares, or circular folds, and they increase the total surface area of the small intestine and also cause chyme to spiral through the intestine.

- **Villi** (singular = villus)– The mucosa of the small intestine forms finger-like projections called villi that greatly increase the total surface area of the small intestine. These are ½–1 mm long and there are approximately 20–40 of them per square millimetre. Each villus contains an arteriole, venule, capillary network and lacteal and so enable absorbed nutrients to enter the bloodstream and lymphatic system.
- **Microvilli** (singular = microvillus) – Every villus in the mucosa of the small intestine contains a variety of cells including absorptive cells. Each absorptive cell has many tiny membrane-covered projections called microvilli which extend into the small intestine and further increase the surface area for absorption.
- **Brush border** – Together, the microvilli form the brush border across which large amounts of digestive nutrients diffuse into the absorptive cells. The brush border contains enzymes that have been secreted into the plasma membranes of the microvilli instead of into the lumen of the small intestine. Therefore, digestion also occurs at the surface of the cells and not just in the lumen.
- **Brush border enzymes** – Brush border enzymes continue the breakdown of double sugars into simple sugars, complete the digestion of protein and help digest nucleotides. These enzymes are:
  - μdextrinase, maltase, sucrase and lactase which all act on carbohydrates.
  - Aminopeptidase and dipeptidase which together are referred to as peptidases and which act on proteins.
  - Nucleosidases and phosphatises which digest nucleotides.
- **Intestinal glands** (crypts of Lieberkühn) – The mucosa of the small intestine also contains cavities called the intestinal glands. They secrete intestinal juice and contain a number of specialised cells:
  - *Intestinal juice* – Intestinal juice is a clear yellow fluid that has a slightly alkaline pH of 7.6. It contains water and mucous and helps bring nutrient particles into contact with the microvilli.

Did you know?
The many tiny villi lining the small intestine give it a velvety appearance.

Did you know?
It is estimated that there are about 200 million microvilli per square millimetre of small intestine.

- *Paneth cells* – Paneth cells are specialised cells that secrete a bactericidal enzyme called **lysozyme**. Lysozyme also functions in phagocytosis.
- *Enteroendocrine cells* – Enteroendocrine cells are another type of specialised cell that secretes hormones into the intestinal glands. These hormones include *secretin* and *cholecystokinin* which influence the release of pancreatic juice and bile; and *gastric inhibitory peptide*.
- *Goblet cells* – Goblet cells secrete mucous which helps lubricate the small intestine.
  - **Peyer's patches** – The lamina propria of the mucosa of the small intestine contains many mucous-associated lymphoid tissues, including Peyer's patches.
  - **Duodenal glands** (Brunner's glands) – The submucosa of the duodenum contains duodenal glands which secrete an alkaline mucous that helps to neutralise gastric acid in the chyme that has arrived from the stomach.

Digestion in the small intestine

DIGESTION/ENZYME	DESCRIPTION
Mechanical digestion In the small intestine, the thin liquid chyme is mixed with digestive juices and brought into contact with the villi and microvilli where it can be further digested and absorbed. It is moved in and through the small intestine by two movements: segmentation and peristalsis.	
Segmentation	Segmentation is the main movement in the small intestine. It involves localised contractions that move food back and forth and is a similar movement to squeezing opposite ends of a tube of toothpaste so that the paste is moved back and forth.
Peristalsis	Peristalsis in the small intestine is a weak movement compared to that in the oesophagus or stomach. It slowly propels chyme forward towards the large intestine.
Chemical digestion Chyme entering the small intestine contains partially digested nutrients which are then broken down further by a combination of pancreatic juice, bile, intestinal juice and brush border enzymes. Digestion and absorption of most nutrients is usually completed in the small intestine.	
Carbohydrate digestion:	
Pancreatic amylase	Pancreatic amylase present in pancreatic juice completes the breakdown of starches and glycogen. However, it does not digest cellulose which passes into the large intestine as fibre.
μ-dextrinase	The brush border enzyme μ-dextrinase also acts on starches and breaks them down into glucose units.
Maltase	The brush border enzyme maltase breaks down maltose into glucose.
Sucrase	The brush border enzyme sucrase breaks down sucrose into glucose and fructose.
Lactase	The brush border enzyme lactase breaks down lactose into glucose and galactose.

DIGESTION/ENZYME	DESCRIPTION
Protein digestion:	
Trypsin, chymotrypsin, carboxypeptidase, elastase	These enzymes present in pancreatic juice break down proteins into peptides.
Peptidases (aminopeptidase and dipeptidase)	These brush border enzymes complete the break down of proteins into amino acids.
Lipid digestion:	
Bile salts	Bile salts emulsify lipids. In other words, they break down large globules of triglycerides into smaller droplets. This exposes a greater surface area to the enzyme pancreatic lipase.
Pancreatic lipase	Pancreatic lipase present in pancreatic juice breaks down triglycerides into fatty acids and monoglycerides.
Nucleic acid digestion:	
Ribonuclease	Ribonuclease present in pancreatic juice breaks down RNA into nucleotides.
Deoxyribonuclease	Deoxyribonuclease present in pancreatic juice breaks down DNA into nucleotides.
Nucleosidases, phosphatises	These brush border enzymes then break nucleotides down into pentoses, phosphates and nitrogenous bases.

Absorption in the small intestine

The process of digestion breaks large food particles into progressively smaller particles that can finally pass through the epithelial cells lining the small intestine and into the blood capillaries and lymphatic lacteals found in the villi. Absorption is this movement of digested nutrients and ninety percent of all absorption takes place in the small intestine. The remaining ten percent is absorbed in the stomach or large intestine.

Most substances are absorbed into the blood capillaries of the villi and are then carried in the bloodstream to the liver via the hepatic portal vein. Fatty acids, glycerol and the fat-soluble vitamins (A, D, E and K), however, are absorbed into lacteals (lymphatic vessels) in the villi, before being transported in the lymph to the bloodstream. The small intestine absorbs nutrients as well as water, electrolytes and vitamins. Any remaining undigested or unabsorbed matter passes into the large intestine.

Did you know?
Don't forget... all absorbed substances are transported in the blood to the liver where the blood is 'cleaned' before it can be circulated to the rest of the body.

REGION OF SMALL INTESTINE	NUTRIENTS ABSORBED
Carbohydrates can only be absorbed as monosaccharides (glucose, fructose, galactose); proteins as amino acids, dipeptides and tripeptides; and lipids as fatty acids, glycerol and monoglycerides.	
Duodenum	Micro-minerals
Jejunum	Water-soluble vitamins, amino acids, sugars, water and some minerals.
Ileum	Free fatty acids, cholesterol, fat-soluble vitamins and bile.

Large intestine

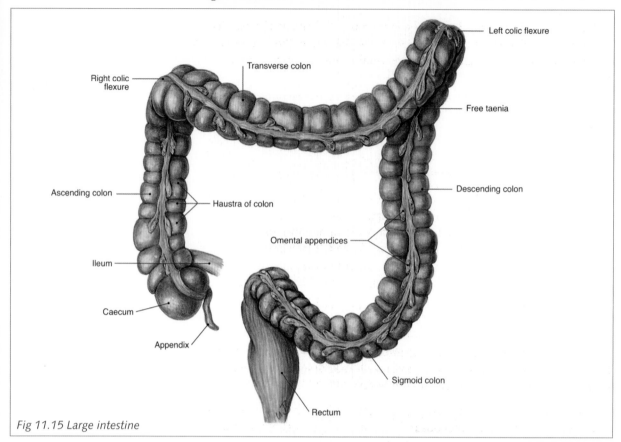

Right colic flexure

Transverse colon

Left colic flexure

Free taenia

Ascending colon

Haustra of colon

Descending colon

Omental appendices

Ileum

Caecum

Appendix

Sigmoid colon

Rectum

Fig 11.15 Large intestine

Did you know?
Your large intestine is approximately 1½ m long and 6½ cm in diameter.

The journey of food from when you first ate it to when you eliminate it is almost over. It has passed through your mouth, oesophagus, stomach and small intestine and almost all of it has been broken down and absorbed into your bloodstream and lymphatic system. Now, any remaining material sits in your large intestine for anywhere from three to ten hours.

Structure of the large intestine

The large intestine is a wide tube running from the ileum to the anus. It is divided into four regions:

- **Caecum** – a 6cm long pouch that receives food from the small intestine via the *ileocaecal valve (ileocaecal sphincter)*. Attached to the caecum is the *vermiform appendix*.
- **Colon** – forms most of the large intestine and is a long tube made up of the:
  - *Ascending colon* – this part of the colon ascends from the caecum up the right side of the abdomen to just beneath the liver. Here it turns to the left, forming the *right colic (hepatic) flexure*.
  - *Transverse colon* – the colon continues across the abdomen to beneath the spleen where it curves downwards at the *left colic (splenic) flexure*.
  - *Descending colon* – the colon then descends to the level of the iliac crest where it turns inwards to form the last part of the colon.
  - *Sigmoid colon* – this is the last part of the colon and it joins the rectum at the level of the third sacral vertebra.

- **Rectum** – approximately 20cm long and lies in front of the sacrum and coccyx.
- **Anal canal** – The last 2–3cm of the rectum is called the *anal canal*. Its external opening is called the *anus* and it is guarded by both internal and external sphincter muscles. These are usually closed except during defecation.

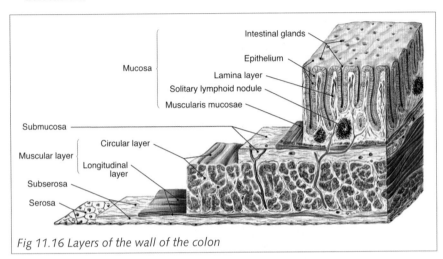

Fig 11.16 Layers of the wall of the colon

The structure of the wall of the large intestine is similar to that of the rest of the gastrointestinal tract in that it contains all four tissue layers (mucosa, submucosa, muscularis and serosa). However:

- The mucosa of the colon contains absorptive cells that absorb water and goblet cells that secrete mucous to lubricate the contents of the large intestine. It also contains a few lymphatic nodules.
- The muscularis has a unique arrangement of muscle fibres. Instead of forming a continuous layer of tissues, the longitudinal muscle fibres collect into three thickened bands at intervals along the colon. These are the *taeniae coli* which are commonly called the 'ribbons' of the colon. These taeniae coli are always slightly contracted (tonic) and so they gather the colon into pouches called *haustra*.

Digestion in the large intestine

DIGESTION/ENZYME	DESCRIPTION
Mechanical digestion: Mechanical digestion in the large intestine is very slow and consists of three movements: haustral churning, peristalsis and mass peristalsis	
Haustral churning	The haustra, or pouches of the colon, are relaxed and distended until they are filled with matter. Then they contract and squeeze their contents into the next haustrum. Thus, the contents of the large intestine are slowly moved from haustra to haustra.
Peristalsis	Peristalsis in the large intestine is a very weak, slow movement.
Mass peristalsis	Three to four times a day, usually during or immediately after a meal, a very strong peristaltic wave begins in the middle of the transverse colon. It quickly travels across the rest of the colon, driving the colonic contents into the rectum.
Chemical digestion: As mentioned earlier, most digestion and absorption of nutrients is completed in the small intestine. However, small amounts of undigested nutrients do pass into the large intestine and these are digested by bacteria living in the lumen.	
Bacteria	Bacteria in the lumen prepare the chyme for elimination by: • Fermenting any remaining carbohydrates. • Converting any remaining proteins into amino acids and breaking down amino acids into simpler substances. • Decomposing bilirubin into simpler substances. • The bacteria also produce some B vitamins as well as vitamin K which are absorbed in the colon.

Absorption in the large intestine

The absorption of water and electrolytes takes place in the ascending and transverse colon.

Faeces

Having spent 3–10 hours in the large intestine and having had most of its water content absorbed by the walls of the colon, the liquid chyme has now become a solid or semisolid mass called *faeces*. Faeces contain water, inorganic salts, sloughed-off epithelial cells from the mucosa of the gastrointestinal tract, bacteria, products of bacterial decomposition and undigested foods. Faeces are eliminated from the body by a process called *defecation*.

Common Pathologies of the Digestive System

Pathologies of the mouth and teeth

Apical abscess (tooth abscess)
An apical abscess is commonly called a tooth abscess and is a collection of pus enclosed by damaged and inflamed tissue. It usually results from an infection that has spread from a tooth into the surrounding tissues.

Gingivitis
Gingivitis is inflammation of the gums. Although gingivitis is not always painful, the gums become red and swollen and bleed easily. It is usually caused by poor oral hygiene, for example not flossing or brushing properly. However, it can also be a side effect of certain drugs, or caused by viral or fungal infections, vitamin deficiencies, an impacted tooth, pregnancy or menopause.

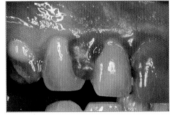

Gingivitis

Halitosis (bad breath)
Halitosis, or bad breath, has a wide range of causes. Most commonly, it is caused by poor oral hygiene, by eating foods that have volatile oils such as garlic or onions, or by nose, throat or lung infections. More extreme causes can include kidney failure, liver failure or uncontrolled diabetes.

Pathologies of the oesophagus and stomach

Dyspepsia (indigestion)
Dyspepsia is commonly referred to as indigestion and is characterised by pain or discomfort in the lower chest or upper abdomen. It usually occurs after eating and can be accompanied by nausea or vomiting. Dyspepsia has many causes ranging from anxiety to gastritis (inflammation of the stomach), stomach or duodenal ulcers.

Heartburn
Heartburn is the sensation of a burning discomfort or pain behind the breastbone. It is often accompanied by a bitter taste in the mouth due to the regurgitation of stomach contents.

Hiatus hernia
In a hiatus hernia the stomach protrudes from the abdominal cavity, through the diaphragm, into the thoracic cavity. It can be accompanied by no symptoms at all, or symptoms of gastro-oesophageal reflux and indigestion. The cause of hiatus hernias is unknown, although it is sometimes present at birth.

Hiccoughs (hiccups)
Hiccoughs are abnormal spasms of the diaphragm accompanied by the quick and noisy closing of the rima glottidis (the space between the vocal folds). Their cause is not always known, but they can be triggered by laughing, talking, eating or drinking.

Gastric ulcer
Please see Peptic Ulcers under the section on Pathologies of the Small and Large Intestines.

Gastro-oesophageal reflux

In gastro-oesophageal reflux the contents of the stomach flow back into the oesophagus where they cause inflammation and pain. Causes include an improperly working oesophageal sphincter muscle whose function can be impaired by smoking and certain foods such as chocolate; extremely high stomach contents, for example, when someone has overeaten; delayed emptying of the stomach contents.

Pathologies of the liver, gall bladder and pancreas

Cholecystitis (inflammation of the gall bladder)

Cholecystitis is inflammation of the gall bladder and it is characterised by severe upper abdominal pain. Choleycystitis usually accompanies gallstones.

Cirrhosis of the liver

Cirrhosis of the liver

Cirrhosis of the liver is the destruction of liver tissue and the replacement of it by scar tissue surrounding areas of healthy tissue. Symptoms can include weakness, nausea, lack of appetite and weight loss. Bile flow can sometimes be obstructed which leads to symptoms such as jaundice and itchiness. Cirrhosis is commonly caused by alcohol abuse or chronic hepatitis.

Gallstones

Gallstones are hard masses of bile pigments, cholesterol and calcium salts that form in the gall bladder. They do not always cause symptoms. However, if they pass into the bile duct and obstruct the flow of bile this can lead to nausea and vomiting and sometimes infection.

Hepatitis

Hepatitis is inflammation of the liver and can be caused by viruses, chemicals, excessive alcohol intake or the use of certain drugs. Hepatitis caused by viruses includes:

- **Hepatitis A** – This infectious form of hepatitis is caused by the hepatitis A virus and is transmitted via contaminated food, drink, faeces or utensils. It is characterised by a loss of appetite, nausea, diarrhoea, fever and chills and most people recover from it within 4-6 weeks.
- **Hepatitis B** – This infectious hepatitis is caused by the hepatitis B virus which is transmitted through sexual contact, contaminated syringes and infusion equipment. Hepatitis B is a chronic disease that can lead to cirrhosis of the liver.
- **Hepatitis C** – This infectious hepatitis is caused by the hepatitis C virus which is usually transmitted through blood transfusions. It is also a chronic form of hepatitis that can lead to liver cirrhosis.

Other types of viral hepatitis exist, including hepatitis D and E.

Jaundice

Jaundice

Jaundice is a yellowing of the skin or whites of the eyes and it is caused by high levels of bilirubin in the bloodstream. It can result from liver disease, blockage of bile ducts or excessive breakdown of red blood cells.

Pathologies of the small and large intestines

Appendicitis
Appendicitis is inflammation of the appendix. It is usually characterised by abdominal pain that starts in the centre and then moves to the lower right. This is then followed by a lack of appetite, nausea and vomiting. Removal of the appendix is usually recommended.

Coeliac disease
In coeliac disease a person is intolerant to gluten which is a protein found in wheat, oats, rye and barley. The consumption of gluten damages the lining of the small intestine and subsequently results in malabsorption. Symptoms of coeliac disease in adults include malnutrition, diarrhoea and weight loss while in children it is characterised by painful abdominal bloating and light-coloured, foul-smelling stools. Coeliac disease is thought to be genetic.

Colitis
Colitis is inflammation of the colon and is characterised by lower abdominal pain and diarrhoea. Blood or mucous can also be present in the stools. There are many different causes and types of colitis, including colitis caused by the use of antibiotics that result in the overgrowth of certain bacteria in the colon; ulcerative colitis which is thought to be genetic or linked to an overactive immune system; and mucous colitis which is also called irritable bowel syndrome (discussed shortly).

Colon cancer (colorectal cancer)
Colon cancer is also called colorectal cancer and is cancer of the large intestine. It usually develops from the lining of the large intestine and then begins to invade its wall as it grows. When the cancer metastasises it often spreads to the surrounding lymph nodes and liver. Colon cancer grows slowly and does not always cause any symptoms until it is in its later stages. Symptoms that can occur include fatigue, weakness and sometimes blood in the stools.

Constipation
Constipation is a condition in which bowel movements are infrequent or uncomfortable. Stools are usually hard and difficult to pass and after passing them the person may feel that they have still not completely emptied the rectum. Constipation may be caused by drugs that slow the movement of matter through the large intestine; a lack of physical exercise; dehydration; a low-fibre diet; ageing; depression; or obstruction of the large intestine.

Crohn's disease
Crohn's disease is the chronic inflammation of the wall of the gastrointestinal tract and, although it can affect any part of the tract, it usually affects the small and large intestines. It is characterised by irregular flare-ups or 'attacks' of abdominal cramping, chronic diarrhoea, fever, loss of appetite and weight loss. Blood can also be present in the stools. The cause of Crohn's disease is unknown although it is thought to be linked to genetics or a dysfunction of the immune system.

Did you know?
People who consume a high fat, low fibre diet are most at risk of developing colon cancer.

Bowel cancer

Diarrhoea

Diarrhoea is the frequent passing of abnormally soft or liquid stools and it can be accompanied by abdominal cramping, large amounts of gas or nausea. Diarrhoea can be caused by infections, inflammation, stress or irritable bowel syndrome (discussed shortly) and prolonged diarrhoea can result in dehydration.

Diverticulosis and diverticulitis

Diverticula are small sac-like pouches that protrude through weak areas of the muscular layer of the gastrointestinal tract (usually the large intestine).

- **Diverticulosis** – The development of diverticula is called diverticulosis and although the causes of diverticulosis are not always known, they are thought to be related to a low fibre diet or inadequate fluid intake. Diverticulosis has few symptoms but can occasionally cause cramping, diarrhoea and bloody stools.
- **Diverticulitis** – Diverticulitis is inflammation or infection of one or more diverticula and is characterised by lower abdominal pain, tenderness and fever. It is usually accompanied by either diarrhoea or constipation.

Enteritis

Enteritis is inflammation of the small intestine and it is usually characterised by diarrhoea. Enteritis can be associated with Crohn's disease or gastroenteritis.

Flatulence

Flatulence is the presence of excess gas in the gastrointestinal tract that is expelled via the mouth (belching) or the anus (flatulence). It is usually accompanied by abdominal pain and bloating. This excess gas comes from air that is swallowed while eating; is produced by bacteria in the large intestine; or can be caused by deficiencies of certain digestive enzymes.

Gastritis

Gastritis is inflammation of the lining of the stomach and is often symptom-free. If there are any symptoms they usually include pain, discomfort, nausea and vomiting. Gastritis can be caused by bacterial, viral or fungal infections or by the ingestion of excess irritants such as alcohol or drugs such as aspirin.

Gastroenteritis

Gastroenteritis is inflammation of the stomach and small intestine and it is characterised by diarrhoea, cramping, vomiting, nausea and a loss of appetite. Gastroenteritis has many different causes including viral or bacterial infections that are transmitted from person to person or through contaminated stools, utensils, food or water. Gastroenteritis is often called 'food-poisoning'.

Haemorrhoids (piles)

Please refer to the pathology section of Chapter 9

Irritable bowel syndrome (IBS, spastic colon)

Irritable bowel syndrome, also called spastic colon, is a condition characterised by recurring flare-ups of abdominal pain, constipation and diarrhoea in an otherwise healthy person. Other symptoms can include fatigue, nausea, headaches, depression, anxiety, difficulty concentrating, abdominal bloating and gas. The flare-ups are usually triggered by eating too quickly or too much, stress, diet, hormones, drugs or minor irritants such as wheat, dairy, tea, coffee or citrus fruits.

Peptic ulcers

Breaks in the mucous lining of the gastrointestinal tract are called peptic ulcers because they are usually caused by the combined action of pepsin and hydrochloric acid. Together these two substances can digest the lining of the tract, causing pain and inflammation.

Symptoms of peptic ulcers include feelings of gnawing, burning, aching, emptiness or hunger and they are usually caused by bacterial infection or the use of drugs that irritate the lining of the stomach (for example, aspirin, nonsteroidal anti-inflammatory drugs and corticosteroids). Peptic ulcers are named according to their location:

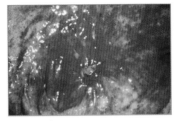

Peptic ulcer

- **Duodenal ulcers** – occur in the duodenum.
- **Gastric ulcers** – occur in the stomach.

Eating disorders

Anorexia nervosa

Anorexia nervosa is a psychological disorder characterised by a distorted body image and a refusal to maintain a minimally healthy body weight. It is also accompanied by a fear of obesity and the absence of menstrual periods. Approximately 95% of people with anorexia nervosa are female.

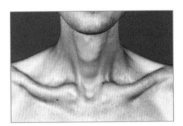

Anorexia nervosa

Bulimia nervosa

Bulimia nervosa is a psychological disorder characterised by bingeing and purging. When a person with bulimia binges they eat large amounts of food rapidly and often in secret. This gives them a sense of having lost control and so they purge themselves by making themselves vomit, taking laxatives, over-exercising or dieting rigorously.

Obesity

Obesity is the accumulation of excessive body fat, mostly in the subcutaneous tissues, and it is diagnosed by assessing one's body mass index (BMI). A person's BMI is their weight in kilograms divided by their height in metres squared and obesity is defined as a BMI of approximately 30 or more. Many factors influence a person's weight and, among other things, obesity can be linked to overeating, a lack of exercise and sometimes genetics.

Obesity

Did you know?
Eating disorders such as anorexia nervosa and bulimia most commonly affect girls or young women in middle and upper socioeconomic classes. They are rare in areas where there is a genuine food shortage.

NEW WORDS

Absorption	the uptake of digested nutrients into the bloodstream and lymphatic system.
Alimentary canal	see gastrointestinal tract.
Catalyst	a substance that alters the rate of a chemical reaction without itself being changed by the reaction.
Defecation	the process by which indigestible substances and some bacteria are eliminated from the body.
Deglutition	the process of swallowing food.
Digestion	the process by which large molecules of food are broken down into smaller molecules that can enter cells.
Enzyme	a protein that speeds up a chemical reaction without itself being used up in the reaction.
Faeces	the waste material of the digestive system that is eliminated through the anus.
Gastrointestinal tract	a tube that runs from the mouth to the anus where digestion and absorption take place.
Ingestion	the process of taking food into the mouth.
Mastication	the process of chewing food.
Peristalsis	an involuntary wave-like movement that pushes the contents of the gastrointestinal tract forward.
Substrate	the substance on which an enzyme acts.

Revision

1. Identify the functions of the digestive system.
2. Explain the following terms:
   - Digestion
   - Absorption
   - Deglutition
   - Peristalsis
   - Enzyme
3. Describe the organisation of the gastrointestinal tract.
4. Explain the function of saliva.
5. Describe the digestion of carbohydrates.
6. Describe the digestion of proteins.
7. Describe the digestion of lipids.
8. Identify the functions of the liver.
9. Explain the following terms associated with the small intestine:
   - Villi
   - Microvilli
   - Brush border
10. Identify the functions of the large intestine.

Study Outline

Digestion of carbohydrates

Polysaccharides such as starch are converted into monosaccharides such as glucose, fructose and galactose. Digestion of carbohydrates begins in the mouth and is completed in the small intestine.

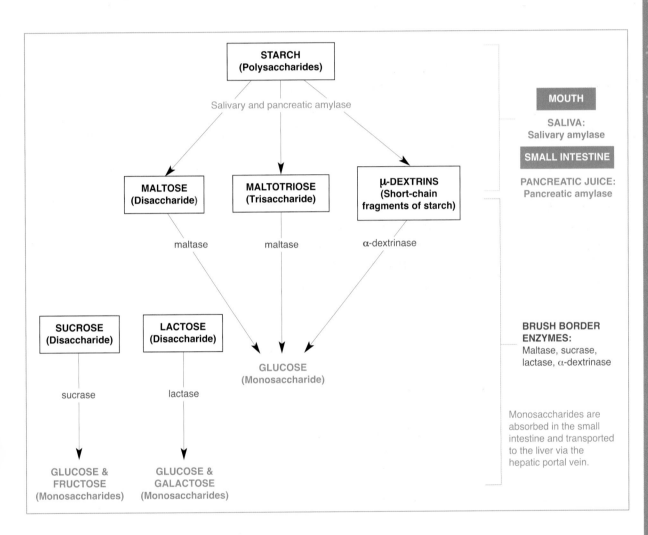

Digestion of proteins

Proteins are converted into amino acids. Digestion of proteins begins in the stomach and is completed in the small intestine.

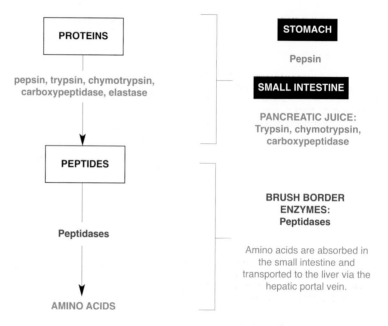

Digestion of lipids

Lipids are converted into fatty acids and monoglycerides. Minimal digestion of lipids begins in the mouth and stomach and most of the digestion of lipids takes place in the small intestine.

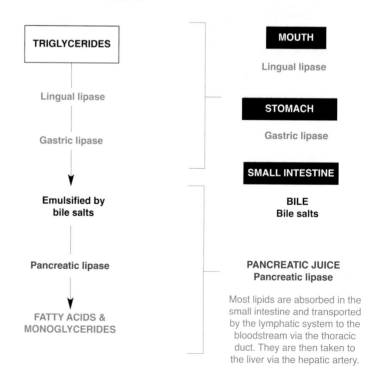

The Journey of Food

Mouth
Food is chewed and mixed with saliva into a bolus. Salivary amylase begins the breakdown of starches and lingual lipase begins the breakdown of fats.

Oesophagus
Bolus is moved to the stomach through peristalsis.

Stomach
Bolus is churned and mixed with gastric juice until it becomes chyme. Salivary amylase and lingual lipase continue to act for about one hour until they are denatured by HCL. Gastric lipase continues the breakdown of fat until it is also denatured by HCL. Pepsin begins the breakdown of proteins and in infants rennin begins the digestion of milk.

Liver
Produces bile which it secretes into the gall bladder. Also has numerous essential functions.

Gall bladder
Stores bile and secretes it into the duodenum. Bile emulsifies fats.

Pancreas
Secretes pancreatic juice into the duodenum of the small intestine. Pancreatic juice contains pancreatic amylase which continues the breakdown of carbohydrates, trypsin which continues the breakdown of proteins, pancreatic lipase which continues the breakdown of lipids and enzymes that digest nucleic acids.

Large intestine
Completes mechanical digestion through the movements of haustral churning, peristalsis and mass peristalsis. Bacteria complete all digestion and produce some B vitamins and vitamin K. Absorbs most of the water content of chyme and turns it into faeces, ready for elimination.

Small intestine
Moves chyme by the actions of segmentation and peristalsis. Completes the digestion of most nutrients by a combination of pancreatic juice, bile, intestinal juice and brush border enzymes. Also absorbs most nutrients: the duodenum absorbs some micro-minerals; the jejunum absorbs water-soluble vitamins, amino acids, sugars, water and some minerals; and the ileum absorbs free fatty acids, cholesterol and fat-soluble vitamins.

Multiple choice questions

1. **Which of the following enzymes is present in the mouth?**
   a. Pepsin
   b. Trypsin
   c. Salivary amylase
   d. Pancreatic amylase.

2. **Where is bile manufactured?**
   a. The pancreas
   b. The liver
   c. The gall bladder
   d. The duodenum.

3. **Which of the following are all parts of the small intestine?**
   a. Duodenum, jejunum, ileum.
   b. Duodenum, ruggae, rectum.
   c. Sigmoid, rectum, anus.
   d. Sigmoid, ruggae, rectum.

4. **Where are brush border enzymes found?**
   a. Mouth
   b. Stomach
   c. Small intestine
   d. Large intestine.

5. **Which of the following enzymes digest lipids?**
   a. Salivary amylase, maltase.
   b. Rennin, pepsin.
   c. Lingual lipase, pancreatic lipase.
   d. Trypsin, chymotrypsin.

6. **Which of the following statements is correct?**
   a. Carbohydrates are broken down into glucose.
   b. Proteins are broken down into lactose.
   c. Fats are broken down into amino acids.
   d. None of the above.

7. **What is the function of bile salts?**
   a. They break down carbohydrates.
   b. They convert proteins into peptides.
   c. They emulsify fats.
   d. They help mix all nutrients together.

8. **Where is the jejunum located?**
   a. At the end of the oesophagus.
   b. Between the duodenum and the ileum.
   c. Between the stomach and the small intestine.
   d. At the end of the small intestine.

9. **Functions of the bacteria of the large intestine include:**
   a. Producing some B vitamins and vitamin K.
   b. Fermenting remaining lipids.
   c. Decomposing minerals such as calcium.
   d. None of the above.

10. **Which form of viral hepatitis is transmitted via contaminated food, drink, faeces or utensils?**
    a. Hepatitis A
    b. Hepatitis B
    c. Hepatitis C
    d. Hepatitis D.

12 The Urinary System

Two small organs, each no bigger than a large bar of soap and buried in fat, filter and clean the blood of toxins, control its volume, composition, pH and pressure. In addition, these organs secrete essential hormones and can, in times of starvation, even manufacture glucose. These intricate organs are the kidneys and each one is composed of around a million tiny filters that would stretch for more than 80 km if unwound and placed end to end[i].

The study of the urinary system is called urology and in this chapter you will discover more about the invaluable role this system plays in your body.

Student objectives

By the end of this chapter you will be able to:

- Describe the functions of the urinary system.
- Explain the organisation of the urinary tract.
- Describe the structure of the kidneys.
- Explain how urine is produced and how the volume, composition, pH and pressure of blood are regulated.
- Describe the common pathologies of the urinary system.

Functions of the urinary system

Anatomy and physiology in perspective

There are many accounts of hunger-strikers who refuse to eat as a form of non-violent protest. Mahatma Gandhi is one such example. Some strikers have been known to survive for many weeks as long as they have received fluids, but anyone who tries to refuse water cannot survive even a few days. Why is water so vital to our bodies?

After oxygen, water is the most important substance to our survival:

- Every cell in our bodies contains water.
- The size and shape of every cell is maintained by water.
- Our blood is mainly water (91.5% of blood plasma is water).
- Water is the solvent in all body fluids.
- Water is the medium for all biochemical processes in the body.
- More than half our body weight is made up of water.
- Everyday we lose about 1.5–1.7 litres of water through our urine, sweat, faeces and the air we exhale.

Water is vital to every cell in our bodies and without it we would not survive. It circulates in blood plasma and as it circulates it gathers toxins, wastes and any other substances that are not used by the cells. Blood, therefore, needs to be continually cleaned and its water content regulated otherwise tissue cells would 'drown' in their own water and blood would become a river of circulating toxins and waste products.

This vital role of cleaning and regulating blood is carried out by the two kidneys. They function in controlling the composition, volume and pressure of blood and in doing this they help to maintain homeostasis of the entire body. The other organs of the urinary system (the ureters, bladder and urethra) transport, store and excrete urine.

Regulation of blood composition and volume

The kidneys filter blood and remove from it any substances that the body no longer needs. For example waste products, toxic substances and excess essential materials such as water. During this filtering process, the kidneys also restore certain amounts of water and solutes to the blood if and when it needs it. Thus, the composition and volume of blood is constantly regulated.

Regulation of blood pH

In addition to regulating the composition and volume of blood, the kidneys also filter out and excrete differing amounts of hydrogen (H^+) ions from the blood. This helps to regulate its pH.

Regulation of blood pressure

The kidneys secrete an enzyme called *renin* which causes an increase in blood pressure and blood volume. Thus, they help to regulate blood pressure.

Other regulatory functions

The kidneys also have other functions in the body:

- **Synthesis of calcitriol** – The kidneys help synthesise the hormone calcitriol which is the active form of vitamin D.
- **Secretion of erythropoietin** – The kidneys secrete the hormone erythropoietin which stimulates the production of red blood cells.
- **Synthesis of glucose** – During periods of starvation the kidneys can synthesise new glucose molecules in a process called gluconeogenesis.

Organisation of the urinary system

The urinary system is composed of:

- **Two kidneys** – these are the functional organs of the urinary system and are the site where blood is filtered and its composition, volume and pressure regulated.
- **Two ureters** – these are long, thin tubes that transport urine from the kidneys to the bladder. One ureter leaves each kidney.
- **The urinary bladder** – this is a collapsible muscular sac where urine is temporarily stored.
- **The urethra** – this is a thin-walled tube that transports urine from the bladder to the outside of the body.

The Kidneys

The kidneys are a pair of reddish, kidney-bean shaped organs located slightly above the waistline, between the levels of the last thoracic and third lumbar vertebrae. They are partially protected by the eleventh and twelfth pairs of ribs and the right kidney is slightly lower than the left because the liver occupies such a large area on the right side of the abdominopelvic cavity.

Anatomy of the kidneys

The kidneys are approximately 10–12cm long, 5–7cm wide and 2.5cm thick and are uniquely structured to filter blood and produce urine. Each kidney is encapsulated in and protected by three layers of tissue:

- **Renal fascia** – This outer, superficial layer of dense irregular connective tissue holds the kidneys in place, binding them to their surrounding structures and the abdominal wall.
- **Adipose capsule (perirenal fat)** – This intermediate layer of fatty tissue protects the kidneys and also holds them firmly in place.
- **Renal capsule** – This inner, deep layer is composed of a smooth, transparent fibrous membrane that is continuous with the outer coat of the ureters and that protects the kidneys and helps to maintain their shape.

Above each kidney sits an *adrenal (suprarenal)* gland.

Each kidney is shaped like a bean, with its concave border facing inward towards the vertebral column. In this concave border is a deep fissure where

Study tip
Don't confuse the digestive enzyme rennin with the renal enzyme renin!

Study tip
Nephros and *renalis* are different names for the kidneys. So if you see any words that are derived from these names, remember that they will be related to the kidneys. For example, 'renal', 'nephron', 'nephritis'.

Did you know?
Although the adrenal glands sit on top of the kidneys, they are part of the endocrine and not urinary system.

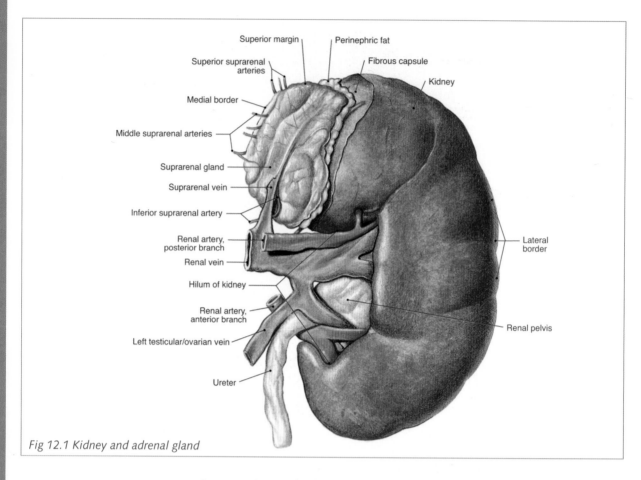

Fig 12.1 Kidney and adrenal gland

Superior margin

Perinephric fat

Superior suprarenal arteries

Fibrous capsule

Kidney

Medial border

Middle suprarenal arteries

Suprarenal gland

Suprarenal vein

Inferior suprarenal artery

Renal artery, posterior branch

Renal vein

Hilum of kidney

Renal artery, anterior branch

Left testicular/ovarian vein

Ureter

Lateral border

Renal pelvis

Study tip
The word *cortex* means 'rind or bark', the word *medulla* means 'inner' and the word pelvis means 'basin'.

the ureter leaves the kidney and where blood vessels, lymphatic vessels and nerves enter and exit the kidneys. This fissure is called the *renal hilus* and is the entrance to the *renal sinus*, a cavity within the kidney.

If you cut a kidney lengthwise in a frontal (coronal) section, you will see that it has three distinct regions:

- **Renal cortex** – The outer, reddish region of the kidney is called the renal cortex. Parts of the cortex extend into their neighbouring region and these extensions are called *renal columns*.
- **Renal medulla** – The middle, reddish-brown region is the renal medulla. It is composed of cone-shaped structures called *renal (medullary) pyramids* whose broad bases face the cortex and thin tips (apexes) point towards the centre of the kidney.
- **Renal pelvis** – The inner, whitish cavity connected to the ureter is called the renal pelvis.

Together the renal cortex and renal pyramids form the functional part of the kidney. They are composed of approximately one million microscopic structures called *nephrons*. These are the functional units of the kidney and are where urine is formed.

Once urine has been formed in the nephron it drains into a large duct, the *papillary duct*, which transports urine into a cuplike structure known as a

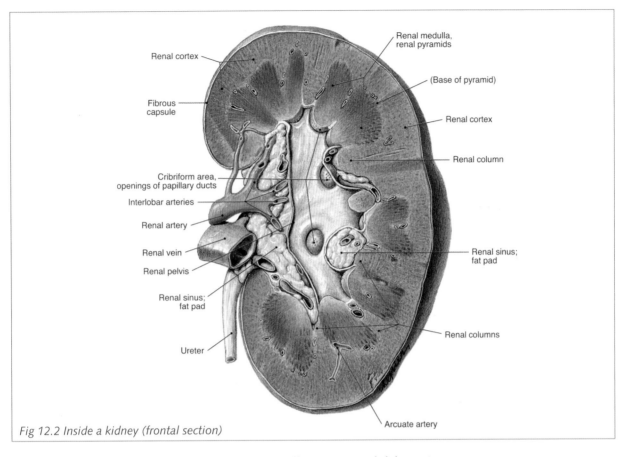

Fig 12.2 Inside a kidney (frontal section)

minor calyx (plural = calyces). The minor calyx collects urine and delivers it to a *major calyx* which then empties the urine into the *renal pelvis*. From here the ureter carries urine to the bladder.

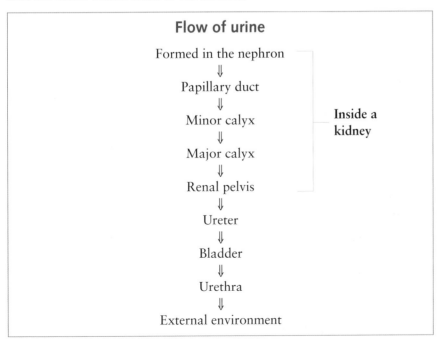

Flow of urine

Formed in the nephron
⇓
Papillary duct
⇓
Minor calyx
⇓
Major calyx
⇓
Renal pelvis

Inside a kidney

⇓
Ureter
⇓
Bladder
⇓
Urethra
⇓
External environment

Nephrons: the functional units of the kidney

Nephrons function in maintaining homeostasis of the blood by:

- Filtering blood through a process called *glomerular filtration.*
- Secreting substances that were not originally absorbed during glomerular filtration and reabsorbing useful materials that the body may need. These processes are called *tubular secretion* and *tubular reabsorption.*

These functions will be discussed in more detail shortly.

Structure of a nephron

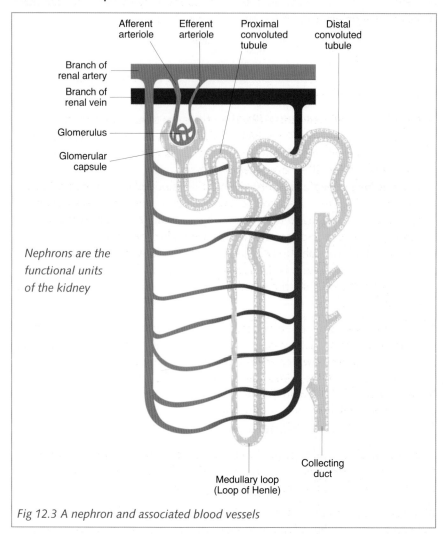

Nephrons are the functional units of the kidney

Fig 12.3 A nephron and associated blood vessels

In order to carry out its functions of filtering, secreting and reabsorbing materials, a nephron is composed of:

- A *renal corpuscle* which filters blood plasma.
- A *renal tubule* in which secretion and reabsorption take place.

Renal corpuscle

Renal corpuscles lie in the renal cortex and are composed of:

- A knotted network of capillaries called the *glomerulus*.
- A cup-like structure surrounding the glomerulus. This cup is called the *glomerular (Bowman's) capsule* and is the closed end of the renal tubule.

Blood flows through the capillaries in the glomerulus and water and most solutes filter from the blood into the glomerular capsule across a membrane called the *filtration membrane*. However, large plasma proteins and formed elements such as red and white blood cells are too big to filter through the walls of the glomerulus and so remain in the blood.

Renal tubule

Filtered fluid passes from the glomerular capsule into the renal tubule which is composed of three sections:

- **Proximal convoluted tubule (PCT)** – this is the area closest to the glomerular capsule. It is a coiled tubule lying in the renal cortex and is the section of the renal tubule where the reabsorption of most substances takes place (this will be discussed shortly).
- **Loop of Henle (nephron loop)** – From the proximal convoluted tubule, the renal tubule extends into the renal medulla where it makes a sharp turn and then returns to the cortex. This section of the tubule is called the loop of Henle and water and salts are reabsorbed here.
- **Distal convoluted tubule (DCT)** – this is the area furthest from the glomerular capsule and is the section of the tube where 'fine-tuning' of the filtrate occurs. Like the proximal convoluted tubule, it also lies in the renal cortex.

Distal convoluted tubules empty their contents into collecting ducts that drain into papillary ducts which then drain into the minor calyces. You will shortly learn about how nephrons function in filtering and regulating the blood. However, before doing this we need to take a quick look at the blood supply to the kidneys and nephrons.

Blood supply to the kidneys and nephrons

The kidneys clean and regulate blood and so it is not surprising that they have an exceptionally rich supply of it: although they make up only 1% of the total body mass, they receive ¼ of total resting cardiac output.

Blood is brought to the kidneys by the right and left *renal arteries* and, once inside the kidneys, they divide into *segmental arteries* which supply segments of the kidney. These arteries then branch into *interlobar arteries* which are found in the renal columns between the renal pyramids. The interlobar arteries then arch at the base of the renal pyramids, where they are referred to as the *arcuate arteries*, before becoming the interlobar arteries once again. They finally branch into *afferent arterioles* in the renal cortex.

Each nephron receives an afferent arteriole which divides into the knot of capillaries known as the *glomerulus*. The glomerular capillaries then reunite

Did you know?
The kidneys receive approximately 1200 ml of blood per minute.

to form an efferent arteriole which takes blood out of the glomerulus. The *efferent arteriole* then divides into a secondary capillary network called the *peritubular capillaries*. These capillaries supply the rest of the nephron with oxygen and nutrients.

Peritubular capillaries drain into *peritubular venules* that drain into *interlobular veins, arcuate veins, interlobular veins* and finally *segmental veins*. Segmental veins drain into the right and left *renal veins* which transport blood away from the kidneys.

Study tip
Note that although blood usually drains out of capillaries into *venules*, blood draining from the glomerulus drains into *arterioles* again.

Flow of blood through the kidneys

Renal artery
⇓
Segmental arteries
⇓
Interlobular arteries
⇓
Arcuate arteries
⇓
Interlobular arteries
⇓
Afferent arterioles
⇓
Glomerulus
⇓
Efferent arterioles
⇓
Peritubular capillaries
⇓
Peritubular venules
⇓
Interlobular veins
⇓
Arcuate veins
⇓
Interlobular veins
⇓
Segmental veins
⇓
Renal vein

Physiology of the kidneys

Did you know?
The kidneys filter approximately 180 litres of fluid every day.

Urine

Urine is a clear to pale yellow fluid that is produced in the kidneys and excreted from the body via the urethra. Although the kidneys filter approximately 180 litres of fluid everyday, only 1–2 litres is excreted as urine and the rest is reabsorbed into the blood. The chart on the next page identifies the characteristics and content of normal urine in a healthy person.

ANALYSIS OF NORMAL URINE	
CHARACTERISTIC	**DESCRIPTION**
Volume	1–2 litres every 24 hours
Colour	Clear to pale yellow. The colour of urine is due to the presence of pigments and can vary depending on the concentration of the urine and on one's diet and health.
Odour	Initially slightly aromatic but it quickly becomes ammonia-like upon standing. As with colour, the odour of urine can vary depending on the concentration of the urine and one's diet and health.
pH	Varies considerably according to diet and ranges between 4.6–8.0. Diets high in protein produce more acidic urine while vegetarian-based diets produce more alkaline urine.
SOLUTE	**DESCRIPTION**
Urea	The main product of protein metabolism.
Creatinine	Product of muscle activity.
Uric acid	Product of nucleic acid metabolism.
Urobilinogen	Bile pigment derived from the breakdown of haemoglobin.
Inorganic ions	These vary with one's diet.

Anatomy and physiology in perspective

The analysis of urine gives many clues to the internal state of the body and substances that should not normally be present in urine include glucose, proteins, pus, red blood cells, haemoglobin and bile pigments. The presence of any of these substances needs to be investigated further.

As mentioned earlier, nephrons have three functions (filtration, secretion and reabsorption) and it is through these functions that the composition and volume of blood is regulated and urine is produced. The remaining areas of the kidneys function as passageways and storage areas. We will now take a closer look at how nephrons function.

Urine production

Glomerular filtration

As blood passes through the capillaries of the glomerulus, fluid and small solutes pass through the filtration membrane into the glomerular (Bowman's) capsule. This fluid is now called the *glomerular filtrate*.

Filtration is assisted by specific structural features of the renal corpuscles:

- The filtration membrane is thin and porous.
- Glomerular capsules have a large surface area.
- Afferent arterioles bringing blood to the glomerulus have a larger diameter

Did you know?
Glomerular filtrate contains all the materials present in blood except the large plasma proteins and formed elements.

413

than efferent arterioles draining blood away from the glomerulus. This means that the pressure in the glomerulus, called the *glomerular hydrostatic pressure*, is higher than the pressure in the glomerular capsule. Thus, fluid is forced into the glomerular capsule. However, there is still some pressure in the glomerular capsule that opposes the capillary hydrostatic pressure. If there were no opposing pressure, fluid would simply stream into the glomerular capsule. This opposing pressure is made up of *blood osmotic pressure* and *capsular hydrostatic pressure* which combine to form the pressure within the capsule.

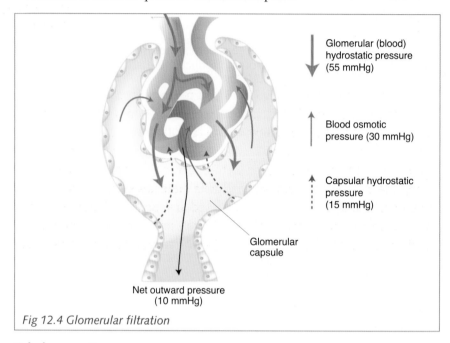

Glomerular (blood) hydrostatic pressure (55 mmHg)

Blood osmotic pressure (30 mmHg)

Capsular hydrostatic pressure (15 mmHg)

Glomerular capsule

Net outward pressure (10 mmHg)

Fig 12.4 Glomerular filtration

Tubular secretion

Blood filters through the glomerulus quite quickly and so there is not enough time for all substances to be filtered into the glomerular capsule. Therefore, the cells of the tubule secrete these remaining substances into the filtrate as it passes through the convoluted tubules. This is to ensure that the blood is cleared of all foreign materials and substances that it no longer needs.

Tubular reabsorption

Once blood plasma has been filtered into the glomerulus, the glomerular filtrate travels through the renal tubule where 99% of it is reabsorbed into the bloodstream. The remaining 1% is excreted from the body as urine.

Reabsorption is carried out by the epithelial cells lining the renal tubule and most reabsorption occurs in the proximal convoluted tubule which is lined by microvilli that greatly increase the surface area for absorption. The distal convoluted tubule functions more in 'fine-tuning' the remaining filtrate.

Substances that are reabsorbed into the bloodstream include: glucose, amino acids, small proteins, and some ions. Other substances which the body no longer needs or cannot use, for example urea and uric acid, remain in the tubules to form urine. The reabsorption of some substances is regulated by hormones as shown in the chart on the next page.

HORMONAL REGULATION OF TUBULAR REABSORPTION		
ENDOCRINE GLAND	HORMONE	REGULATION
Parathyroid glands	Parathyroid hormone (parathormone)	Increases blood calcium and magnesium levels and decreases blood phosphate levels.
Thyroid gland	Calcitonin	Lowers blood calcium levels
Adrenal cortex	Aldosterone	Increases blood levels of sodium and water and decreases blood levels of potassium

Functions of the nephron in a nutshell

It is easy to remember the functions of a nephron if you think about what happens to your household waste: waste disposal vehicles (your blood) transport your household waste (cellular waste products and foreign substances) to a central sorting station (a nephron). At the station it is put onto a conveyor belt (renal tubule) and as it travels along the belt workers remove recyclable materials such as cans (reabsorption) and, at the same time, add any additional rubbish that may be on the floor (secretion). What is left on the conveyor belt is finally put into another vehicle (urine) and transported to a landfill site.

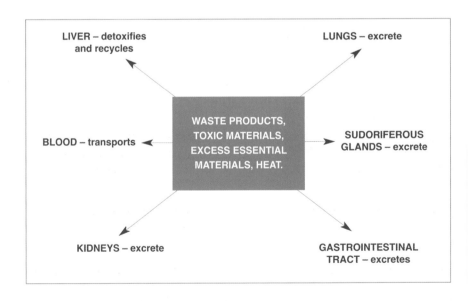

In the classroom
Discuss the different ways in which the body deals with its waste products, toxic materials, excess essential materials and excess heat.

Fluid and electrolyte balance

In the process of producing urine, the kidneys also regulate the fluid and electrolyte balance of the blood and control its pH and pressure.

Water makes up at least 50% of our total body weight and is found in two 'fluid compartments' in the body. Two thirds of the fluid in the body is found in intracellular fluid (ICF). This is the fluid inside cells. The remaining third is found in extracellular fluid (ECF), which is the fluid outside the cells. This includes blood plasma, interstitial fluid, cerebrospinal fluid, serous fluid, aqueous humour and lymph.

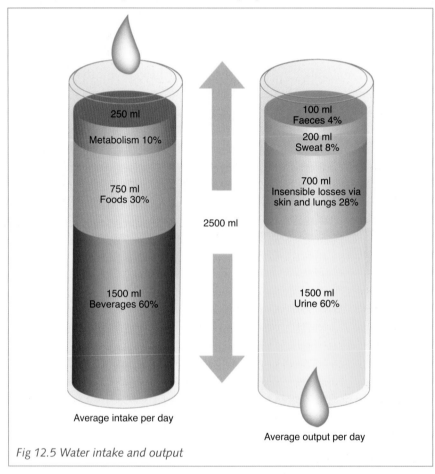

250 ml
Metabolism 10%

750 ml
Foods 30%

1500 ml
Beverages 60%

Average intake per day

100 ml
Faeces 4%

200 ml
Sweat 8%

700 ml
Insensible losses via skin and lungs 28%

1500 ml
Urine 60%

2500 ml

Average output per day

Fig 12.5 Water intake and output

Because water is the universal solvent in the body and electrolytes such as calcium, sodium and potassium dissolve in water, it is vital that the amount of water in each compartment remains constant. This essential role is carried out by the kidneys which regulate the amount of fluids and electrolytes in the body by reabsorbing them or excreting them when necessary. This ensures that the total volume of fluid in the body remains stable despite fluctuating fluid intake.

Two hormones help regulate the reabsorption of water and electrolytes by the kidneys:

- **Antidiuretic hormone (ADH) (vasopressin)** – Specialised cells in the hypothalamus, called osmoreceptors, detect a decrease in water or an increase in solutes in the blood. They send a message to the posterior pituitary gland which then releases antidiuretic hormone. A diuretic is a substance that stimulates an increase in urine production. Thus, 'anti'-diuretic hormone has the opposite effect and it decreases the secretion of urine by causing the collecting ducts of the kidneys to reabsorb more water. When this additional water is returned to the bloodstream, it increases the volume, and therefore pressure, of the blood.
- **Aldosterone** – This is a hormone secreted by the adrenal cortex. It regulates the reabsorption of sodium ions back into the bloodstream and the secretion of potassium ions into the filtrate. Because sodium is responsible for the osmotic flow of water, aldosterone helps regulate the amount of water in the blood. The release of aldosterone is stimulated by:
  - decreased sodium levels or increased potassium levels in the extracellular fluid.
  - the renin-angiotensin mechanism – Low blood pressure in the afferent arterioles of the kidneys or changes in the solute content of the filtrate stimulate the release of renin (an enzyme produced by the kidneys) into the bloodstream. Renin catalyses a series of reactions that produce *angiotensin II* which causes vasoconstriction of blood vessels and stimulates the release of aldosterone.

Ureters

The ureters are two 25–30cm long tubes that carry urine from the kidneys to the bladder. Each ureter drains the renal pelvis of a kidney and inserts into the posterior aspect of the bladder.

The ureters do not have a physical valve that prevents the backflow of urine from the bladder into the ureters. However, when the bladder is full, pressure from within it compresses the ureter openings. This, coupled with the force of gravity, prevents any urinary backflow. The walls of the ureters are composed of three layers of tissue:

- The *outer adventitia* which consists of areolar connective tissue and contains the blood vessels, lymphatic vessels and nerves that supply the ureters. This tissue also holds the ureters in place.
- The *intermediate muscularis* which consists of both longitudinal and circular smooth muscle fibres that move urine via the action of peristalsis.
- The *inner mucosa* which is lined with a mucous membrane of transitional epithelium. These transitional epithelial cells give the ureters their ability to stretch and the mucous secreted by the membrane protects the walls of the ureters from the urine.

Urinary Bladder

The urinary bladder, which is more commonly referred to as the bladder, is a freely movable, hollow muscular organ that is held in its place by folds of the peritoneum. The bladder acts as a reservoir, storing urine until it is excreted out of the body.

The bladder changes its shape depending on how much urine it is holding: when empty it is pear shaped while when full it is more oval in shape. The location of the bladder also varies depending on one's gender: in females the bladder lies in front of the vagina and below the uterus while in males it lies in front of the rectum. Furthermore, the size of the bladder varies depending on one's gender: females have smaller bladders than males because of the uterus above it.

On the floor of the bladder is a triangular shaped area called the *trigone*. The openings to the ureters are located in the posterior two corners of the trigone and the urethral opening, the *internal urethral orifice*, is located in the anterior corner.

Similar to the walls of the ureters, the walls of the bladder consist of three tissues: the outer *adventitia*, the intermediate *muscularis* and the inner *mucosa*. The bladder also has two sphincter muscles that regulate micturition which is the emptying or voiding of the bladder:

- The *internal urethral sphincter* is composed of smooth muscle fibres. When the bladder is approximately half full, stretch receptors in the wall of the bladder send messages to the spinal cord. From here a reflex arc returns to the sphincter causing it to relax and empty the bladder. This is known as the micturition reflex.
- The *external urethral sphincter* is composed of voluntary muscle fibres and although emptying the bladder is a reflex action, the external urethral sphincter is controlled voluntarily. Thus, emptying the bladder becomes a learned, voluntary action.

Urethra

At the end of the urinary system is a passageway that functions in discharging urine from the body. This passageway is the urethra and it is a small tube leading from the internal urethral orifice in the bladder to the external environment.

Just as the size and location of the bladder differs according to one's gender, so does the size and location of the urethra differ. In females the urethra is a 4cm tube lying behind the pubic symphysis and opening to the outside between the clitoris and vaginal opening. In males the urethra is a 15–20cm tube that runs through the prostate gland (prostatic urethra), then through the urogenital diaphragm (membranous urethra) and finally through the penis (spongy urethra), at the head of which it opens to the outside. The urethra also differs between genders in what it transports: in females the urethra only transports urine while in males it transports urine and reproductive secretions.

Common Pathologies of the Urinary System

Ageing and the urinary system

As we age we become more prone to urinary disorders, especially incontinence and urinary tract infections. This is because renal blood flow, glomerular filtration and urea clearance all decrease with age and the filtering mechanism of a healthy 70 year old is half as effective as that of a healthy 40 year old.

Elderly men also often suffer with prostate gland problems which lead to urinary retention and difficulties in urinating. Post-menopausal women, on the other hand, suffer more with incontinence due to a thinning of the urethral lining and a weakening of the urinary sphincters caused by the decrease in oestrogen.

Symptoms of urinary tract disorders differ according to the disorder itself. However, general symptoms can include pain, nausea, vomiting, and fever. Changes in urination patterns or urine composition, colour and odour are also symptomatic of pathologies of the urinary system.

Technical terms for some of the symptoms include:

- **Nocturia** – the need to urinate regularly during the night. It is more common in pregnant women, the elderly or men with enlarged prostate glands.
- **Oliguria** – the production of unusually small amounts of urine and can occur after heavy sweating, diarrhoea or blood loss. It may also result from kidney disease, oedema or poisoning.
- **Anuria** – where the kidneys do not produce any urine at all.
- **Haematuria** – the presence of blood in the urine and may result from injury or disease.
- **Proteinuria (albuminuria)** – the presence of protein in the urine and it is often associated with kidney or heart disease.
- **Renal colic** – a severe cramping pain that occurs in the lower back region and often radiates to the groin. It is a symptom of many kidney or urinary tract disorders.

Calculi (kidney, bladder and ureteral stones)

Calculi are hard masses that can form in the urinary tract (as well as in the gallbladder) and, depending on where they are located, they are commonly referred to as kidney, bladder or ureteral stones. Stones are mainly composed of calcium salts and can form either from an excess of salts in the urine or from a lack of stone inhibitors in the urine. They are more common in the elderly, men and people who eat a high protein diet and have a low water intake. Symptoms of stones include renal colic, back pain, nausea, vomiting, fever and blood in the urine.

Kidney stones

Enuresis (bed wetting)

Enuresis, or bed-wetting, is normal in young children. However, after the age of 5 or 6 years, it may be symptomatic of an infection, narrowing of the urethra, inadequate nerve control or a psychological problem.

Nephritis (Bright's disease)

Nephritis, or Bright's disease, is inflammation of the kidneys and it is characterised by impaired kidney function, fluid and urea retention and blood in the urine. Nephritis can be caused by a variety of factors, including bacterial infection of the kidneys, exposure to toxins and even abnormal immune reactions. Nephritis can also occur anywhere in the kidneys.

- **Glomerulonephritis (glomerular nephritis)** – Glomerulonephritis is inflammation of the glomeruli and when the glomeruli are damaged they are unable to selectively filter blood. Thus, substances such as large proteins and red and white blood cells can pass into the filtrate and be excreted in urine. Urine production also declines and metabolic waste products accumulate in blood. Eventually, scarring of the kidney tissue and impaired kidney function can result.
- **Pyelonephritis** – a bacterial infection of the kidneys and is characterised by fever, shivering and pain.

Nephroblastoma (Wilms' tumour)

Nephroblastoma, also called Wilms' tumour, is a malignant cancer that is usually only found in young children. Symptoms include a rapidly enlarging abdomen, abdominal pain, fever, loss of appetite, nausea and vomiting. High blood pressure and blood in the urine may also occur. The cause is unknown although genetic abnormalities may sometimes be involved.

Pyelitis

Pyelitis is inflammation of the pelvis of the kidney and it is characterised by pain, shivering and fever. Pyelitis is usually caused by a bacterial infection.

Renal failure (kidney failure)

Renal failure is the kidneys' inability to filter blood efficiently and regulate its composition and volume properly. A rapid decline in kidney function is termed acute renal failure and can be caused by kidney disease or any condition that reduces renal blood flow or that obstructs urine flow. Chronic renal failure is a slower, more gradual decline in the functioning of the kidneys and is caused either by acute kidney failure that has been left to develop or by diseases such as diabetes mellitus or hypertension.

Symptoms of renal failure vary depending on the cause, progression and severity of the disorder. They can include: oedema of the feet, ankles, face and hands; dark urine; minimal to no urination; or fatigue, lack of concentration, loss of appetite, nausea and an overall itchiness. In chronic renal failure, blood pressure usually rises and the kidneys' inability to produce sufficient erythropoietin (the hormone that stimulates the production of red blood cells) results in anaemia. If renal failure is caused by an obstruction in the urinary tract, then pain may be an additional symptom.

Anatomy and physiology in perspective

Even if the kidneys are damaged, blood still needs to be cleaned. This can be done through the mechanical process of *kidney dialysis*.

Uraemia

Uraemia is the presence of unusually large amounts of nitrogenous wastes, such as urea, in the blood. Symptoms include drowsiness, lethargy, nausea and vomiting. Uraemia is usually caused by renal failure and it can be fatal if left untreated.

Urinary incontinence

Urinary incontinence, more commonly referred to as incontinence, is the involuntary passing of urine. It is more common in the elderly and also more common in women than in men. Incontinence can be caused by weakened pelvic floor muscles after childbirth or pelvic surgery, menopause, an enlarged prostate, prostate surgery, obesity, constipation or a variety of psychological factors.

Urinary tract infection (UTI)

An infection anywhere along the urinary tract is called a urinary tract infection (UTI) and it is usually caused by microbes (most commonly bacteria) that either enter the urinary tract via the urethra or via the bloodstream. UTIs are more common in women as they have a shorter urethra than men. UTIs are generally classified as upper or lower UTIs. Upper UTIs are infections of the kidneys (pyelonephritis) or ureters (ureteritis). Lower UTIs are infections of the urethra (urethritis) or bladder (cystitis).

- **Pyelonephritis** – inflammation of the kidneys.
- **Ureteritis** – inflammation of the ureters. It often accompanies cystitis.
- **Urethritis** – inflammation of the urethra. It is characterised by painful or difficult urination.
- **Cystitis** – inflammation of the urinary bladder. It is characterised by frequent, burning urination.

NEW WORDS	
Diuretic	a substance that increases urine production.
Electrolyte	a charged particle (an ion) that conducts an electrical current in an aqueous solution.
Micturition	urination.
Urology	the study of the urinary system.

Study Outline

Functions of the urinary system

1. The kidneys are the functional organs of the urinary system and they regulate the volume and composition of blood and produce urine.
2. The kidneys also regulate and balance the pH and pressure of the blood.
3. They secrete calcitriol and erythropoietin and, in times of starvation, produce glucose.
4. They also reabsorb amino acids, small proteins and glucose.

Organisation of the urinary system

The urinary system is made up of two kidneys, two ureters, one urinary bladder and one urethra.

The kidneys

The kidneys are paired organs located between the levels of the last thoracic and third lumbar vertebrae.

Anatomy of the kidneys

1. Each kidney is encapsulated in three layers of tissue: the outer renal fascia, the intermediate adipose capsule and the inner renal capsule.
2. Each kidney is shaped like a bean with its concave border facing medially and it has the following structures:
   - Renal hilus – a deep fissure where the ureter leaves the kidney and the blood vessels, lymphatic vessels and nerves enter and exit.
   - Renal sinus – a cavity within the kidney.
3. Internally, a kidney has three regions:
   - The outer renal cortex whose renal columns extend into the neighbouring region.
   - The middle renal medulla which consists of renal pyramids.
   - The inner renal pelvis which drains into the ureter.

Nephrons: the functional unit of the kidneys

1. The functional units of the kidneys are nephrons. These are where urine is formed.
2. Once the nephrons have produced the urine it drains into papillary ducts, then minor and major calyces, then the renal pelvis and eventually into the ureters which carry it to the bladder.
3. Nephrons consist of a renal corpuscle which filters blood plasma and a renal tubule where secretion and reabsorption take place.
4. Renal corpuscles are composed of a network of capillaries called the glomerulus and a cup-like structure called the glomerular capsule. This is the closed end of the renal tubule.
5. The renal tubule can be divided into three regions: the proximal convoluted tubule, the loop of Henle and the distal convoluted tubule.

Blood supply to the kidneys and nephrons

1. Blood is brought to the kidneys by the right and left renal arteries and, once inside the kidneys, they divide into a series of arteries that eventually branch into afferent arterioles in the renal cortex.
2. Each nephron receives an afferent arteriole which divides into the knot of capillaries known as the glomerulus.
3. The glomerular capillaries then reunite to form an efferent arteriole which takes blood out of the glomerulus.

> Afferent = Arrive
>
> Efferent = Exit

The efferent arteriole then divides into a secondary capillary network called the peritubular capillaries. These capillaries supply the rest of the nephron with oxygen and nutrients.

Peritubular capillaries drain into peritubular venules, that then drain into a series of veins until they finally drain into the right and left renal veins which transport blood away from the kidneys.

Physiology of the kidneys

Urine
Normal constituents of urine include urea, creatinine, uric acid, urobilinogen and varying quantities of inorganic ions.

Urine production
1. Nephrons regulate the volume and composition of blood through the production of urine.
2. Urine is produced through glomerular filtration, tubular secretion and tubular reabsorption.
3. In glomerular filtration, blood passes through the capillaries of the glomerulus and fluid and small solutes pass through the filtration membrane into the capsular space. This fluid is now called the glomerular filtrate.
4. In tubular secretion, the cells of the tubule secrete any remaining substances into the filtrate as it passes through the convoluted tubules.
5. In tubular reabsorption, the tubules reabsorb any substances that the body may need and most of the fluid in the filtrate. What is left passes out of the kidneys as urine.
6. Hormones that regulate tubular secretion include parathyroid hormone, calcitonin and aldosterone.

Fluid and electrolyte balance
1. The kidneys also regulate the fluid and electrolyte balance of the blood and control its pH and pressure.
2. The hormones that help regulate the reabsorption of water and electrolytes by the kidneys are antidiuretic hormone (ADH) and aldosterone.
3. Antidiuretic hormone is secreted by the posterior pituitary gland in response to a decrease in water or an increase in solutes in the blood.

4. Antidiuretic hormone decreases the secretion of urine by causing the collecting ducts of the kidneys to reabsorb more water.
5. Aldosterone is secreted by the adrenal cortex and its release is stimulated by either fluctuations in sodium and potassium levels or the renin-angiotensin mechanism.
6. In the renin-angiotensin mechanism, the enzyme renin (produced by the kidneys) catalyses a series of reactions that produce angiotensin II. Angiotensin II is a molecule that causes vasoconstriction of blood vessels and stimulates the release of aldosterone.

Ureters, bladder and urethra

1. The ureters are long tubes that carry urine from the kidneys to the bladder.
2. The bladder acts as a reservoir, storing urine until it is excreted out of the body.
3. The urethra is a passageway that functions in discharging urine from the body.

Revision

1. Explain the functions of the urinary system.
2. Describe the organisation of the urinary system.
3. Describe the structure of the kidneys.
4. Identify the flow of urine from its formation in the nephrons to the external environment.
5. Describe the nephron and explain its functions.
6. Identify two differences between afferent arterioles and efferent arterioles.
7. Describe how urine is produced.
8. Explain the differences between tubular secretion and tubular reabsorption.
9. Identify the hormones that help regulate the reabsorption of water and electrolytes by the kidneys.
10. Describe the following disorders of the urinary system:
    * Renal failure
    * Nephritis
    * Incontinence

Multiple choice questions

1. **The renal hilus is:**
   a. The outer protective tissue layer of the kidney.
   b. A deep fissure where the ureter leaves the kidney and the blood vessels, lymphatic vessels and nerves enter and exit.
   c. A cavity within the kidney.
   d. None of the above.

2. **Where in the kidneys is urine produced?**
   a. The ureters
   b. The calyces
   c. The nephrons
   d. The urethra.

3. **What is the term used to describe the fluid in the renal tubules?**
   a. Filtrate
   b. Solvent
   c. Solution
   d. Solute.

4. **Another term for kidney failure is:**
   a. Nephrotic failure
   b. Oliguria
   c. Renal failure
   d. Renal colic.

5. **Which of the following statements is correct?**
   a. The pancreas functions in controlling the composition, volume and pressure of blood.
   b. The spleen functions in controlling the composition, volume and pressure of blood.
   c. The liver functions in controlling the composition, volume and pressure of blood.
   d. None of the above.

6. **Functions of the kidneys also include the synthesis of:**
   a. Bilirubin
   b. Calcitriol
   c. Vitamin K
   d. Lingual lipase.

7. **Which of the following statements is correct?**
   a. The renal cortex is the outer region of the kidney, the renal pelvis is the middle region of the kidney and the renal medulla is the inner region.
   b. The renal pelvis is the outer region of the kidney, the renal medulla is the middle region of the kidney and the renal cortex is the inner region.
   c. The renal cortex is the outer region of the kidney, the renal medulla is the middle region of the kidney and the renal pelvis is the inner region.
   d. The renal medulla is the outer region of the kidney, the renal cortex is the middle region of the kidney and the renal pelvis is the inner region.

8. **Which of the following is the correct sequence of the flow of urine:**
   a. Papillary ducts, minor calcyes, nephrons, major calyces, bladder, renal pelvis, ureter, urethra, external environment.
   b. Nephrons, papillary ducts, minor calcyes, major calyces, renal pelvis, urethra, bladder, ureter, external environment.
   c. Papillary ducts, minor calcyes, nephrons, major calyces, renal pelvis, ureter, bladder, urethra, external environment.
   d. Nephrons, papillary ducts, minor calcyes, major calyces, renal pelvis, ureter, bladder, urethra, external environment.

9. **Pyelitis is:**
   a. Inflammation of the pelvis of the kidney.
   b. A malignant cancer that is usually only found in young children.
   c. Bed-wetting, which is normal in young children.
   d. The presence of unusually large amounts of nitrogenous wastes, such as urea, in the blood.
   e. The involuntary passing of urine.

10. **What is the name of the cup-like structure surrounding the glomerulus?**
    a. Distal convoluted tubule
    b. Glomerular capsule
    c. Renal medulla
    d. Papillary duct.

13 The Reproductive System

Introduction

Although it is the last system to be discussed, the reproductive system is one of the most fascinating of all. It is the only system in the body that does not work continually from birth. Instead, it waits until puberty when it bursts into action. It is the only system that is structurally and functionally different in men and women and it is also the only system in which a unique type of cellular division occurs, creating cells with half the number of chromosomes to all other cells in the body.

Student objectives

By the end of this chapter you will be able to:

- Describe the functions of the reproductive system.
- Describe meiosis, spermatogenesis and oogenesis.
- Identify the organs and structures of the male reproductive system.
- Identify the organs and structures of the female reproductive system.
- Explain the female reproductive cycle.
- Explain the effects ageing has on the reproductive systems.
- Identify some of the common pathologies of the reproductive system.

Functions of the Reproductive System

Reproduction occurs in all living organisms and is the process by which a new member of a species is produced. In humans, the reproductive system is unique in that it is the only system in the body that produces cells that have 23 chromosomes instead of 46 chromosomes. These cells are called reproductive cells or haploid cells.

Reproductive cells are called gametes and are produced in the gonads (testes or ovaries) of men and women. Female gametes are produced through the process of oogenesis in the ovaries of a woman and are called ova (eggs). Male gametes are produced through the process of spermatogenesis in the testes of a man and are called spermatozoa sperm.

In the classroom...

Before continuing with this chapter, it may help to revise somatic cell division (Chapter 2). Somatic cell division occurs in most body cells and its function is to replace dead and injured cells or produce new cells for growth. Somatic cell division occurs through a process of nuclear division called mitosis, in which a single diploid (containing 46 chromosomes) parent cell duplicates itself to produce two identical diploid daughter cells.

Reproductive cell division occurs through a process of nuclear division called meiosis. Unlike in mitosis, in which a cell divides only once to produce two identical daughter cells, in meiosis a cell divides twice to produce four daughter cells. These daughter cells are not identical and each one has only one set of chromosomes (23 chromosomes) instead of the usual two sets (46 chromosomes). These new cells, containing only one set of chromosomes, are referred to as haploid cells.

A male gamete formed in the male reproductive system then enters the female reproductive system through sexual intercourse. The male and female gametes unite and fuse in a process called fertilisation. This produces a zygote which is a new cell that now contains two sets of chromosomes (46 chromosomes) – one set from the mother and one from the father. The zygote then begins to divide by mitosis and develops into a new organism.

DIFFERENCES BETWEEN MITOSIS AND MEIOSIS		
CHARACTERISTIC	MITOSIS	MEIOSIS
Description	Somatic cell division	Reproductive cell division
Function	For growth and repair of cells	For reproducing a new organism and the continuation of the species
Number of daughter cells produced	2	4
Are daughter cells identical?	Identical – exact copies	Different – allows for genetic variation
Number of chromosomes in each daughter cell	46 = two sets, diploid number	23 = one set, haploid number

Male Reproductive System

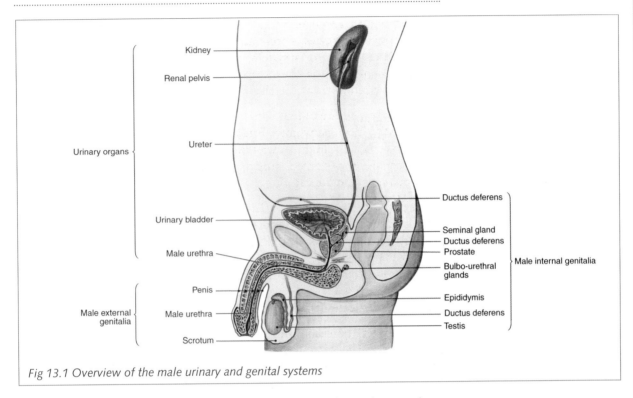

Fig 13.1 Overview of the male urinary and genital systems

The male reproductive system consists of the testes, which are the gonads where sperm are formed, a system of ducts for transporting and storing sperm and accessory organs that produce supporting substances.

Scrotum

Structure of the scrotum
The word *scrotum* means 'bag' and the scrotum is a sac of loose skin and superficial fascia in which lie the testes. The scrotum hangs from the root of the penis and is divided internally into two sacs. Each sac contains one testis.

Function of the scrotum
The function of the scrotum is to house the testes and maintain a temperature approximately 3°Celsius cooler than normal body temperature.

Did you know?
An urologist is a specialist who deals with both the urinary system and the male reproductive system.

Anatomy and physiology in perspective
The internal temperature of the body is too high for the functioning of the testes. Thus, they are kept in a small sac hanging outside the body where it is cooler. This sac is the scrotum.

Testes (testicles)

Structure of the testes

The testes (singular = testis), or testicles, are a pair of oval glands located in the scrotum. They are the gonads of the male reproductive system.

- Each testis is enclosed by a dense white fibrous capsule called the *tunica albuginea*, which in itself is covered by a serous membrane called the *tunica vaginalis*.
- Extensions of the tunica albuginea divide the testis into 200-300 internal compartments called *lobules*.
- Each of these lobules contains tightly coiled tubules called *seminiferous tubules*. It is here, in these seminiferous tubules, that spermatogenesis occurs and sperm are formed.
  - Specialised cells called *sustentacular cells* are also located in the seminiferous tubules. These cells protect, support and nourish the sperm and produce the fluid in which sperm are transported. In addition, these cells help regulate the effects of testosterone and follicle stimulating hormone (FSH).
  - In the spaces between the seminiferous tubules are small clusters of highly specialised cells that secrete *testosterone*. These cells are called *interstitial endocrinocytes (Leydig cells)*.

Did you know?

Women also produce androgens which contribute to their libido. However, they are produced in far smaller quantities than in men and are produced by the adrenal cortex and not by the female gonads.

Study tip

The testes produce male sex hormones called androgens. The principal androgen is testosterone which:

- Stimulates the development of masculine secondary sex characteristics such as the development of pubic, axillary, facial and chest hair; a general thickening of the skin; the skeletal and muscular widening of the shoulders and narrowing of the hips; an increase in sebaceous oil gland secretion; and the enlargement of the larynx and subsequent deepening of the voice.
- Promotes growth and maturation of the male reproductive system and sperm production.
- Promotes male sexual behaviour and stimulates libido (sex drive).
- Stimulates anabolism which is protein synthesis. This results in heavier muscle and bone mass.

Did you know?

Over 300 million sperm cells mature every day.

Function of the testes: spermatogenesis

The testes produce sperm through the process of spermatogenesis. Sperm are highly specialised cells that are able to travel the long journey from the testes, through the male reproductive ducts, into the female reproductive system and finally into an ovum. They are composed of a *head*, which contains the cell's DNA as well as powerful enzymes which help to penetrate into the ovum; a *midpiece* containing many energy-producing mitochondria; and a *tail* which propels the sperm towards the ovum.

Epididymis

Structure of the epididymis

From the seminiferous tubules in each testis, sperm travel down a series of tubules and ducts into a comma-shaped organ lying along the posterior border of each testis. This is the epididymis and is composed of a series of coiled ducts that empty into a single tube called the *ductus epididymis*.

Function of the epididymis

The epididymis is the site of sperm maturation. It stores sperm until they are fully mature and then helps propel them via peristaltic contractions.

Vas deferens (ductus deferens, seminal duct)

Structure of the vas deferens

The end of the ductus epididymis straightens and widens and continues as the vas deferens. This is a very long duct which runs from the epididymis into the pelvic cavity where it loops over the ureter and then over the side and down the posterior surface of the bladder. It finally ends in the urethra.

Function of the vas deferens

The vas deferens transports sperm via peristaltic contractions from the epididymis to the urethra.

Anatomy and physiology in perspective
A vasectomy is the surgical removal of a portion of the vas deferens and it results in sterility.

Spermatic cord

Running alongside the vas deferens is a supporting structure consisting of blood vessels, lymphatic vessels, nerves and muscles. This structure is called the spermatic cord.

Urethra

Structure of the urethra

The urethra is the terminal duct of both the reproductive and urinary systems and is made up of three sections:

- The **prostatic urethra** which runs through the prostate gland (to be discussed shortly).
- The **membranous urethra** which runs through the urogenital diaphragm.
- The **spongy (penile) urethra** which runs through the penis and terminates in the external urethral orifice.

Function of the urethra

The urethra transports both semen and urine to the exterior of the body.

Accessory sex glands and semen

The word *semen* means 'seed' and semen is the fluid in which sperm, their nutrients and other supporting substances are transported. It is a slightly alkaline, milky substance that is sticky to touch.

Closely associated to the urethra are a number of glands whose secretions enter the urethra via small ducts. The fluids of these glands combine with sperm to produce semen:

- **Seminal vesicles** – paired pouch-like structures located at the base of the bladder. They secrete a viscous alkaline fluid that helps neutralise the acidity of the vagina. This fluid contains fructose which the sperm use for energy-production; prostaglandins that help sperm mobility and also stimulate muscular contractions of the female reproductive tract; and clotting proteins that coagulate sperm after ejaculation.
- **Prostate gland** – a doughnut shaped gland that surrounds the prostatic urethra. It secretes a milky, slightly acidic fluid that contributes to sperm mobility and viability.
- **Bulbourethral (Cowper's) glands** – paired pea-sized structures located on either side of the membranous urethra. They secrete a lubricating mucous and an alkaline substance that neutralises the acidity of the urethra.

Penis

Structure of the penis
The penis is a cylindrical shaped organ composed of erectile tissue permeated by blood sinuses. When sexually stimulated, arteries supplying the penis dilate and large quantities of blood enter the sinuses which then expand. In expanding, these sinuses compress any veins that normally drain the penis and so the blood in the penis becomes trapped. This is an erection of the penis.

Three regions make up the penis:

- The *root* is the region attached to the trunk of the body.
- The *shaft* (body) is the main cylindrical region of the penis.
- The *glans penis* is the highly sensitive, acorn-shaped enlargement at the distal end of the penis and it contains a slit-like opening called the *external urethral orifice*. Enclosing the glans penis is a loosely fitting fold of tissue called the *prepuce (foreskin)*.

Infobox

Anatomy and physiology in perspective
Circumcision is the surgical removal of the prepuce (foreskin).

Function of the penis
The function of the penis is to excrete urine and ejaculate semen. During ejaculation, the sphincter muscle at the base of the urinary bladder closes to prevent any urine passing into the urethra.

Female Reproductive System

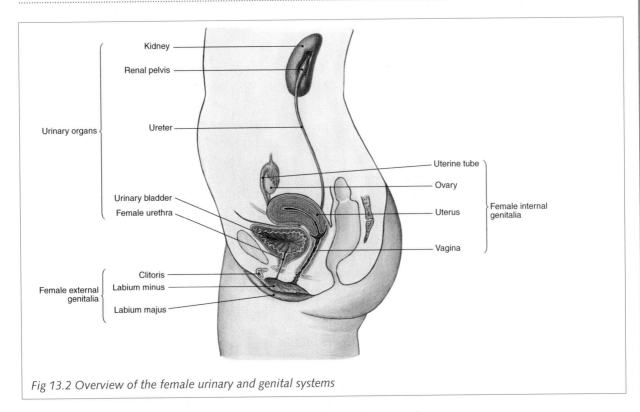

Fig 13.2 *Overview of the female urinary and genital systems*

The female reproductive system is specially structured not only to produce gametes, but also to house, nourish and nurture a growing foetus. It consists of the ovaries, which are the gonads where ova are formed; the Fallopian tubes where an ovum is fertilised and, once fertilised, becomes a zygote; the uterus where a zygote develops into a foetus; and the vagina through which the foetus enters into the world.

Ovaries

Structure of the ovaries

The female gonads, or ovaries, are paired almond-shaped organs located in the superior portion of the pelvic cavity on either side of the uterus. They are held in place by a series of ligaments and a fold of the peritoneum called the *broad ligament*. Inside the ovaries are small sac like structures called *ovarian follicles*.

- Each follicle houses an *oocyte* (an immature ovum) in its differing stages of development. The follicle also contains cells that nourish the oocyte and begin to secrete oestrogens as it enlarges.
- The follicle enlarges as the oocyte matures and eventually develops into a mature *Graafian follicle* (vesicular ovarian follicle). This is a large fluid-filled follicle that soon ruptures and expels the ovum in a process called ovulation (to be discussed later in this chapter).
- After expelling the ovum, the now empty follicle develops into a glandular structure called the *corpus luteum*. For the first 10–14 days

Did you know?

A gynaecologist is a specialist who deals with the female reproductive system.

after ovulation the corpus luteum produces the female hormones *progesterone*, *oestrogen*, *relaxin* and *inhibin*.

- The corpus luteum finally degenerates into a white fibrous tissue called the *corpus albicans*.

Study tip

Oestrogens and *progesterone* are the principal female sex hormones and are produced mainly by the ovaries although small amounts are also produced by the adrenal cortex, testes and placenta. Oestrogens are secreted mainly by the follicle cells that nourish the oocyte and:

- Stimulate the development and maintenance of feminine secondary sex characteristics such as enlarged breasts; the pattern of hair growth on the hair and body; a broadened pelvis; and the distribution of adipose tissue over the abdomen and hips.
- Help regulate fluid and electrolyte balance and lower blood cholesterol levels.
- Increase protein anabolism.

Progesterone is secreted mainly by the corpus luteum and works together with oestrogens to prepare the uterus for pregnancy and the mammary glands for lactation.

Inhibin is a hormone that inhibits the secretion of follicle stimulating hormone (FSH) and relaxin is a hormone that is produced by both the ovaries and placenta during pregnancy. It helps dilate the cervix and increase the flexibility of the pubic symphysis during childbirth.

Functions of the ovaries: oogenesis and ovulation

The ovaries have two functions: oogenesis and ovulation.

Oogenesis is the process through which the ovaries produce ova, or eggs. This occurs in two stages.

- Firstly, during foetal development germ cells differentiate into millions of immature eggs called primary oocytes and at birth a woman has all the oocytes she will ever have.
- Secondly, the release of gonadotropic hormones at puberty stimulates meiosis of a primary oocyte. This occurs in one oocyte in one follicle each month after puberty. The primary oocyte develops into a secondary oocyte, or ovum, and the follicle in which this development occurs is the mature Graafian follicle.

Ovulation is the process by which a mature Graafian follicle releases an ovum. The released ovum then travels down the Fallopian tube towards the uterus. Ovulation will be discussed in more detail later in the chapter.

Did you know?

A baby girl is born with between 200,000 and 2 million oocytes in each ovary. However, many of these degenerate and only about 400 will actually mature and be viable for ovulation.

Fallopian (uterine) tubes

Structure of the Fallopian tubes
The Fallopian tubes are two thin tubes running from the ovaries to the uterus. They are composed of:

- An outer *serous membrane*.
- An intermediate *muscularis* consisting of both circular and longitudinal smooth muscle fibres which contract rhythmically to move the ovum via peristalsis.
- An internal *mucosa* lined with cilia that help move a fertilised ovum towards the uterus. The mucosa also contains secretory cells which provide nutrition for the ovum during its journey.

Functions of the Fallopian tubes
The Fallopian tubes have two functions:

- They are the site of fertilisation and it is here that sperm and ovum unite and fuse to form a zygote.
- They transport the zygote to the uterus. It takes approximately seven days for the zygote to travel down a tube and into the uterus.

Uterus (womb)

Structure of the uterus
The uterus, or womb, is a muscular sac located between the bladder and rectum. It is approximately the size and shape of an inverted pear and is divided into three regions:

- **Fundus** – the dome-shaped region superior to the Fallopian tubes.
- **Body** – the central portion of the uterus.
- **Cervix** – the inferior portion, or neck, of the uterus that forms the narrow opening to the vagina. It is lined with mucous-secreting cells that produce a mixture of water, proteins, lipids, enzymes and inorganic salts. This is called *cervical mucous* and approximately 20–60 ml are produced each day.

Anatomy and physiology in perspective
Cervical cancer is often called the 'quiet killer' because it has few symptoms until it is in its later stages. However, it can be detected early in its development through a *pap smear*. In this procedure, cells are removed from the cervix and from the vaginal area surrounding the cervix and then examined microscopically. Malignant cells have a characteristic appearance that can be easily identified.

The walls of the uterus are composed of three layers of tissue:

- **Outer perimetrium (serosa)** – continuous with the visceral peritoneum and extends laterally to become the broad ligaments which attach the uterus to the pelvic cavity.
- **Intermediate myometrium** – the muscular layer forming the bulk of the uterine wall. It is composed of layers of circular, longitudinal and oblique smooth muscle fibres that, under the influence of the hormone oxytocin, contract powerfully to expel the foetus during childbirth.

- **Inner endometrium** – a highly vascularised layer of tissue that forms the lining of the uterus. It is divided into two layers:
  - *stratum functionalis* – the functional layer of the endometrium and is the layer closest to the uterine cavity. The stratum functionalis is shed during menstruation (to be discussed later in this chapter).
  - *stratum basalis* – the permanent base layer of the endometrium. It produces a new stratum functionalis after each menstruation.

Functions of the uterus

Before fertilisation, the uterus acts as a pathway through which sperm travel into the Fallopian tubes where they attempt to fertilise an ovum.

If fertilisation is successful, then the uterus becomes the site of implantation of a zygote and houses the developing foetus throughout pregnancy. It then contracts forcefully during labour to expel the foetus.

If, however, fertilisation is not successful, the uterus becomes the site of *menstruation*. This is the process by which the lining of the uterus, the *stratum functionalis*, is shed and discarded. Menstruation will be discussed in more detail later in this chapter.

Vagina

Structure of the vagina

The vagina is a muscular tube located between the bladder and rectum and attached to the uterus. The walls of the vagina are composed of three layers:

- **Outer adventitia** – composed of areolar connective tissue and it anchors the vagina to its adjacent organs.
- **Intermediate muscularis** – layers of circular and longitudinal smooth muscle fibres that are able to stretch immensely during childbirth.
- **Inner mucosa** – continuous with that of the uterus and it lies in a series of transverse folds called *rugae*. It secretes an acidic mucous which retards microbial growth but which also harms sperm.

The opening of the vagina to the external environment is called the *vaginal orifice*. It is protected by a fold of mucous called the hymen.

Functions of the vagina

The vagina acts as a passageway for:

- Blood during menstruation.
- Semen during sexual intercourse.
- The foetus during childbirth.

Vulva (external female genitalia)

The word *volvere* means 'to wrap around' and the vulva are the fleshy folds surrounding the opening to the vagina. They are also referred to as the female external genitalia and consist of the:

- **Mons pubis** – an elevation of adipose tissue that cushions the pubic symphysis. It is covered by skin and protected by coarse pubic hair.
- **Labia majora** – two longitudinal folds of skin that are covered by pubic hair and contain adipose tissue, sebaceous glands and sudoriferous

Did you know?
A hysterectomy is the surgical removal of the uterus.

glands. They extend inferiorly and posteriorly from the mons pubis.

- **Labia minora** – two smaller folds of skin running medially to the labia majora. They contain many sebaceous glands but no adipose tissue and very few sudoriferous glands. They are not covered by hair.
- **Clitoris** – a small cylindrical mass of erectile tissue and nerves. It enlarges on tactile stimulation and is found at the anterior junction of the labia minora.
- **Vestibule** – the entire region between the labia minora and it consists of the vaginal orifice and external urethral orifice. It also houses the openings to several ducts, including those from:
  - The mucous-secreting *paraurethral (Skene's) glands* located on either side of the urethral orifice.
  - The mucous-secreting *greater vestibular (Bartholin's) glands* located on either side of the vaginal orifice.
  - Several lesser vestibular glands.

Perineum

The perineum is a diamond-shaped area that contains the external genitals and the anus. It is located between the thighs and buttocks and is present in both males and females.

Mammary glands

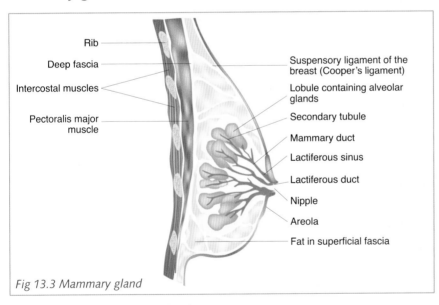

Labels:
- Rib
- Deep fascia
- Intercostal muscles
- Pectoralis major muscle
- Suspensory ligament of the breast (Cooper's ligament)
- Lobule containing alveolar glands
- Secondary tubule
- Mammary duct
- Lactiferous sinus
- Lactiferous duct
- Nipple
- Areola
- Fat in superficial fascia

Fig 13.3 Mammary gland

Structure of the mammary glands

The mammary glands, or breasts, are two modified sudoriferous glands. They are located over the pectoralis major muscles and are attached to them by a layer of dense irregular connective tissue.

Internally, a breast is supported by strands of connective tissue called *suspensory (Cooper's) ligaments* and is composed of compartments separated by adipose tissue. These compartments are called *lobes* and inside each lobe are smaller compartments called *lobules*. Lobules contain grapelike milk-secreting glands called *alveolar glands* that secrete milk into secondary tubules which drain into the *mammary ducts*. It is then stored temporarily in *lactiferous sinuses* and excreted externally through *lactiferous ducts*, via the nipple. Externally, each breast has:

- **A nipple** – This is a pigmented projection of tiny openings leading from lactiferous ducts.
- **An areola** – This is a circular area of pigmented skin surrounding the nipple and containing modified sebaceous oil glands.

Functions of the mammary glands

The mammary glands synthesise and secrete milk through the process of lactation.

The production of milk is stimulated by the hormone *prolactin* (with smaller contributions from progesterone and oestrogens) and the ejection of milk is stimulated by the hormone *oxytocin* whose release is stimulated by the suckling action of the baby on the breast.

Female reproductive cycle

Every month after the onset of puberty a woman's body prepares itself for a possible pregnancy through a series of events called the *reproductive cycle*. The reproductive cycle lasts anywhere from 24–35 days and involves both:

- An *ovarian cycle* in which an oocyte matures until it is ready for ovulation.
- A *uterine (menstrual) cycle* in which the endometrium of the uterus is prepared for the arrival of a fertilised ovum.

If fertilisation does not occur, the stratum functionalis of the endometrium is then shed and *menstruation* occurs. The chart below and on the next page highlights the major phases of these two cycles and is based on an average 28-day reproductive cycle.

FEMALE REPRODUCTIVE CYCLE		
HORMONAL REGULATION	OVARIAN CYCLE	UTERINE (MENSTRUAL) CYCLE
Menstrual phase (menstruation, menses) The word *menses* means 'month' and menstruation marks the beginning of a woman's monthly cycle. Menstruation normally lasts approximately 5 days and the first day of menstruation is termed Day 1 of a woman's cycle. Before menstruation begins, a woman's uterus, especially the stratum functionalis of the endometrium, is prepared to receive a fertilised ovum. If it does not receive a fertilised ovum, levels of oestrogens and progesterone decline and the stratum functionalis dies and is discharged from the body via menstrual flow.		
Declining levels of oestrogens and progesterone cause uterine arteries to constrict and endometrial cells to become deficient in blood. These cells eventually die and the entire stratum functionalis of the endometrium is sloughed off.	Approximately 20 small follicles, some in each ovary, begin to enlarge.	Menstrual flow (consisting of blood, tissue fluid, mucous and epithelial cells derived from the endometrium) is discharged from the vagina.

FEMALE REPRODUCTIVE CYCLE		
HORMONAL REGULATION	**OVARIAN CYCLE**	**UTERINE (MENSTRUAL) CYCLE**

Preovulatory phase

The preovulatory phase is the time between menstruation and ovulation. In a 28-day cycle it can vary between 6-13 days in length. In this phase a mature Graafian follicle forms and the endometrium proliferates.

Follicle stimulating hormone (FSH) from the anterior pituitary gland stimulates follicles to grow. These growing follicles then secrete higher levels of oestrogens and inhibin which in turn decrease FSH secretion	The follicles continue to develop and around day 6, one follicle in one ovary outgrows the other follicles. It becomes the dominant follicle which is now called the mature Graafian follicle and which continues to enlarge until ovulation. The other follicles begin to degenerate. The menstrual and preovulatory phases are called the *follicular phase* of the ovarian cycle.	Oestrogen secreted by the follicles stimulates the repair of the endometrium which now thickens and proliferates. This phase is also called the *proliferative phase* of the uterine cycle.

Ovulation

Ovulation usually occurs around day 14 of a 28-day cycle and involves the rupture of the mature Graafian follicle and the release of an ovum into the pelvic cavity. It is during this phase that a woman can now become pregnant.

High levels of oestrogens stimulate the hypothalamus to release gonadotropin releasing hormone (GnRH) which stimulates the anterior pituitary gland to release follicle stimulating hormone (FSH) and luteinizing hormone (LH). Note that progesterone levels are now low.	LH stimulates the rupture of the mature Graafian follicle and the release of the ovum into the pelvic cavity. The follicle then develops into the corpus luteum which, under the influence of LH, secretes progesterone, oestrogens, relaxin and inhibin.	The prepared endometrium waits for the arrival of a fertilised ovum.

Postovulatory phase

The 14 days after ovulation form the postovulatory phase. This is a 'waiting time' in which the endometrium awaits the arrival of a fertilised ovum. It is now thickened, highly vascularised and secreting tissue fluid and glycogen. Thus, this phase is also called the *secretory phase* of the uterine cycle.

If fertilisation has occurred, the ovum takes approximately a week to arrive at the endometrium where it becomes embedded and develops into a foetus. At this stage a woman is now pregnant.

If fertilisation has not occurred, menstruation and the reproductive cycle begin again.

The corpus luteum survives approximately two weeks and secretes increasing amounts of progesterone and some oestrogens.

If fertilisation has occurred, the corpus luteum and the hormones it secretes remains longer than two weeks and is maintained by human chorionic gonadotropin (hCG), a hormone produced by the embryo approximately 8–12 days after fertilisation. Once the embryo is implanted in the endometrium, the hormones of pregnancy come into effect.

If fertilisation has not occurred, the corpus luteum degenerates and the lack of progesterone and oestrogens causes menstruation.

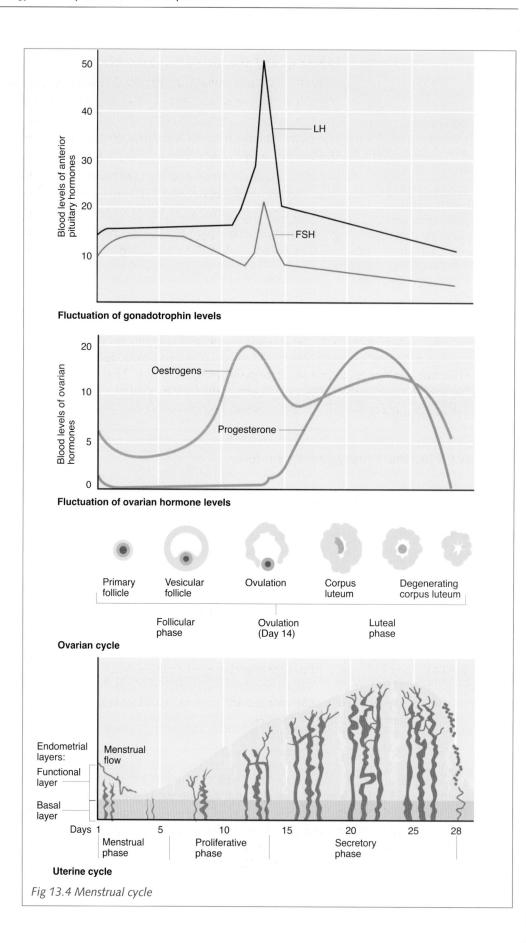

Fluctuation of gonadotrophin levels

Fluctuation of ovarian hormone levels

Ovarian cycle

Uterine cycle

Fig 13.4 Menstrual cycle

Anatomy and physiology in perspective

A woman's body really is a wonder. Ovulation is the only time of the month when a woman can fall pregnant and at this time each month her body makes many small changes to help encourage fertilisation. One such change is to do with cervical mucous. The mucous produced by the cervix generally forms a plug, called the cervical plug, which forms a physical barrier against sperm penetration. However, around the time of ovulation cervical mucous becomes less viscous, more alkaline and easily receptive to sperm. In addition, this mucous now supplements the energy needs of the sperm and protects them from the acidic environment of the vagina and from phagocytes.

Ageing and the Reproductive System

Unlike any other system in the body, the reproductive system appears to be dormant until a child reaches approximately 10 years of age. From this age hormone-directed changes occur and the child goes through puberty. The word *puber* means 'marriageable age' and puberty is the time in which a person develops secondary sexual characteristics and becomes able to reproduce.

Ageing and the male reproductive system

Puberty
Before puberty, which occurs around the age of 14, a boy has low levels of LH, FSH and testosterone and it is only at puberty that the levels of these hormones begin to increase under the influence of GnRH from the hypothalamus. Sustentacular cells in the testes mature and secrete testosterone and spermatogenesis begins.

Increased levels of testosterone bring about the development of secondary sexual characteristics, the enlargement of the reproductive glands and both muscular and bone growth.

Old age
From around the age of 55 years, testosterone levels begin to decline and men lose their muscular strength. Their sperm also become less viable and their libido decreases. However, healthy men are still able to reproduce into their eighties and sometimes even their nineties.

Ageing and the female reproductive system

Puberty
Before puberty, a girl has low levels of LH, FSH and oestrogens. However, under the influence of GnRH at the onset of puberty, LH and FSH stimulate the ovaries to produce oestrogens and girls then develop secondary sexual characteristics and begin menstruating. This event is marked by *menarche* which is a girl's first menses around the age of 12 years.

Pregnancy

Once a woman has begun to menstruate, she is capable of falling pregnant. Pregnancy is the sequence of fertilisation, implantation, embryonic growth and foetal growth.

An average pregnancy lasts 40 weeks from the first day of the woman's last menstrual period (approximately 38 weeks from conception) and is divided into three trimesters, each trimester being made up of approximately three months.

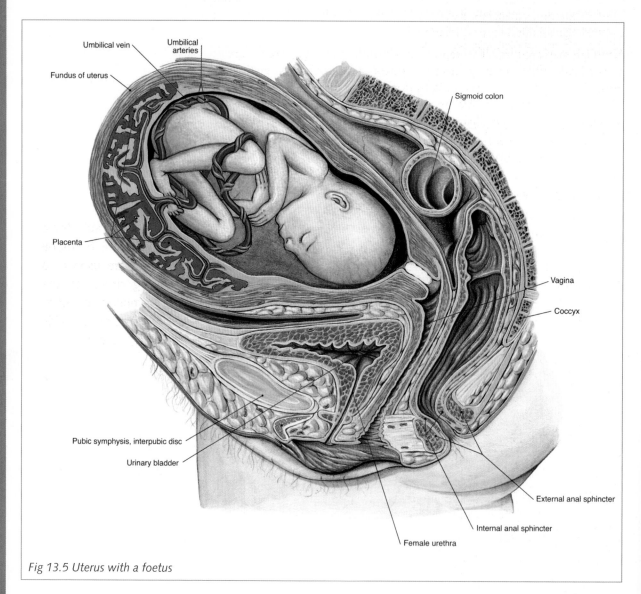

Labels:
Umbilical vein
Umbilical arteries
Fundus of uterus
Sigmoid colon
Placenta
Vagina
Coccyx
Pubic symphysis, interpubic disc
Urinary bladder
External anal sphincter
Internal anal sphincter
Female urethra

Fig 13.5 Uterus with a foetus

Trimester 1, months 0–3

During the first trimester the embryo implants and secures itself in the uterus. It grows from a single cell into a fully formed foetus in only 12 weeks and by the end of the first trimester it has all its organs, muscles, limbs and bones.

Trimester 2, months 4–6
The foetus is now fully formed and just growing and maturing. During the second trimester it develops its individual fingerprints, its toe and finger nails, its eyebrows and lashes and a firm hand grip. It is in this trimester that it even starts grimacing and frowning.

Trimester 3, months 7–birth
During the third trimester the foetus develops its sense of hearing, practises its breathing motions and learns to focus and blink its eyes. It is fully formed and puts on a great deal of weight in the last few weeks.

Menopause
Although men are capable of reproducing well into old age, women are only capable of reproducing into their forties or fifties. At this age they go through menopause which is the cessation of the menses. Menopause is often called the 'change of life' and can be accompanied by a number of symptoms, including hot flushes, headaches, hair loss, sweating, vaginal dryness, insomnia, weight gain and mood swings. After menopause, a woman's reproductive organs begin to atrophy.

Anatomy and physiology in perspective
The symptoms of menopause can sometimes be unbearable for a woman and one form of treatment for these symptoms is hormone replacement therapy (HRT). Oestrogenic hormones are used or a combination of oestrogens and progestin (a drug similar to the hormone progesterone) can be used. HRT has a number of benefits, but it also has risks and it is up to a woman and her doctor to decide on the best course of treatment for her menopausal symptoms.

Common Pathologies of the Reproductive System

Male reproductive system

Benign prostatic hyperplasia (BPH)

Benign prostatic hyperplasia is the benign (non-cancerous) enlargement of the prostate gland. This enlargement leads to a difficulty in urination, incomplete urination and a consequent susceptibility to kidney stones and urinary tract infections. BPH is more common in older men and is thought to be linked to changes in testosterone levels.

Impotence (erectile dysfunction)

Impotence, or erectile dysfunction, is the continual inability to either achieve or maintain an erection. It is sometimes caused by diseases affecting the circulation, for example atherosclerosis or diabetes, or it can be due to nerve damage or illness, fatigue or stress.

Prostate cancer

Prostate cancer is a very slow growing cancer that begins as a small bump in the prostate gland. It usually causes no symptoms until it is in its advanced stages and then symptoms can include difficulty in urinating, the need to urinate frequently or urgently and, in advanced cases, blood in the urine or the inability to urinate. The causes of prostate cancer are not yet known.

Prostatitis

Prostatitis is inflammation of the prostate gland and it is characterised by spasms of the muscles in the bladder and pelvis; pain in the lower back, perineum, penis and testes; and the urge to urinate frequently and burning or painful urination. Prostatitis is usually caused by a bacterial infection.

Testicular cancer

Testicular cancer is cancer of the testes. It is characterised by an irregularly shaped testis with a solid, growing lump either in the testis or, sometimes, elsewhere in the scrotum. The cause of testicular cancer is not known, but it is more common in men whose testes did not descend into the scrotum in early childhood.

Female reproductive system

Breast cancer

Breast cancer is a malignant tumour of the breast and is far more common in women than men. There are different types of breast cancer, but they usually present with a lump in one breast which feels distinctly different from the surrounding tissue. If the cancer is in its more advanced stages, symptoms can also include swollen bumps and sores developing on the breasts and the skin over the lump taking on a dimpled and leathery appearance. Some of the risk factors for developing breast cancer include old age, a family history of breast cancer and never having had a baby.

Candidiasis (candidosis, thrush, yeast infection)
Please refer to Chapter 3.

Cervical cancer
Cervical cancer is cancer of the cervix and it is caused by the human papillomavirus which is transmitted through sexual intercourse. Thus, women at risk of cervical cancer are those who have had a number of different sexual partners. Cervical cancer begins on the surface of the cervix and does not usually have any symptoms until it is in its later stages. Symptoms can then include unusually heavy bleeding during menstruation, bleeding between periods or bleeding after intercourse.

Ectopic (extrauterine) pregnancy
In an ectopic pregnancy, a foetus develops outside of the uterus. The most usual location of an ectopic pregnancy is a Fallopian tube and it usually occurs if the Fallopian tube in which the ovum was fertilised is narrowed or blocked and the ovum becomes stuck. The foetus of an ectopic pregnancy cannot survive. Risk factors for an ectopic pregnancy include having a disorder of the Fallopian tubes, a history of pelvic inflammatory disease or a previous ectopic pregnancy and symptoms include cramping with unexpected vaginal bleeding.

Endometriosis
Endometriosis is a disorder in which endometrial tissue, which is usually only found lining the uterus, develops outside of the uterus on other pelvic organs such as the ovaries or large intestine. Symptoms of endometriosis include lower abdominal and pelvic pain, an irregular menstrual cycle and often severe bleeding and cramping during menstruation. Its cause is still unknown.

Fibroids
Fibroids are non-cancerous (benign) tumours consisting of muscle and fibrous tissue. One or more fibroids can develop in the muscular wall of the uterus and cause symptoms such as pain, a sense of pressure or heaviness in the pelvic area and excessive menstrual bleeding.

Infertility
Infertility is the inability for a couple to conceive a baby after trying for at least one year. Either the man or the woman may be infertile and infertility can be caused by a number of factors, including problems with ovulation, sperm or the Fallopian tubes. If a couple is infertile they can try to conceive with the help of fertility drugs or fertilisation techniques such as in vitro fertilisation (IVF). In IVF, eggs are taken out of a woman's body and fertilised in a laboratory. Once they have been fertilised successfully, the resultant embryos are implanted into the woman's uterus.

Mastitis
Mastitis is inflammation of the breast and is usually caused by a bacterial infection. It is rare except for around the time of childbirth or after an injury or surgery to the breast. Symptoms of mastitis include swelling, redness, warmth and tenderness.

Menstrual disorders

Although every woman has a slightly different menstrual cycle, it is not normal to have extremely heavy or painful periods or to have no periods at all. There are many different menstrual disorders, including:

- **Abnormal Uterine Bleeding** – includes menstruation of excessive duration or amount; diminished menstrual flow; intermenstrual or too frequent menstrual bleeding; and postmenopausal bleeding. Abnormal uterine bleeding can be caused by a number of factors, including hormonal disorders, emotional factors or fibroids.
- **Amenorrhea** – is the absence of a menstrual period and can be due to obesity or extreme weight loss; or abnormal levels of oestrogens due to deficiencies of pituitary or ovarian hormones.
- **Dysmenorrhea** – is severe pain associated with menstruation. It can also be accompanied by headaches, nausea, diarrhoea or constipation and the urge to urinate frequently. Dysmenorrhea is often caused by another disorder such as pelvic inflammatory disease, endometriosis or fibroids.
- **Menorrhagia** – is abnormally heavy bleeding during menstruation. It is usually associated with other conditions such as fibroids, pelvic inflammatory disease or endometriosis.
- **Premenstrual syndrome (PMS), Premenstrual tension (PMT)** – this is a common syndrome that affects many woman. It refers to both a physical and emotional distress that occurs late in the postovulatory phase of the menstrual cycle and sometimes extends into the menstrual phase. Symptoms can include oedema, weight gain, breast swelling and tenderness, abdominal distension, backache, joint pain, constipation, skin eruptions, fatigue and lethargy, depression or anxiety, irritability, mood swings, food cravings, headaches, poor coordination and clumsiness.

Pelvic inflammatory disease (PID)

Pelvic inflammatory disease (PID) is a collective term for any infection of the pelvic organs. It is usually caused by bacteria and is more common in sexually active women. Symptoms of PID tend to be cyclical, usually occurring around the end of menstruation, and can include fever, abdominal pain, irregular vaginal bleeding and a foul-smelling vaginal discharge.

Polycystic ovary syndrome (Stein-Leventhal syndrome)

Polycystic ovary syndrome

Polycystic ovary syndrome (PCOS), or Stein-Leventhal syndrome, is a hormonal disorder in which follicles fail to ovulate and collect as cysts over the ovaries. Women with PCOS have unusually high levels of androgens and often develop male secondary sexual characteristics such as chest and face hair, a decrease in breast size and an increase in muscle size. Other symptoms can include acne, weight gain and irregular vaginal bleeding.

Postnatal depression (postpartum depression)

Postnatal, or postpartum, depression is a depression that can occur in the first few weeks or months after childbirth. A few weeks of 'baby-blues' is normal after giving birth, but a feeling of extreme sadness that lasts for weeks or even months is not normal and is referred to as depression. The exact causes of postnatal depression are unknown, but it is linked to the sudden change in hormone levels, the lack of sleep and the stresses of having to care for a new born baby.

Prolapsed uterus

A prolapsed uterus is a uterus that has dropped downwards from its normal position. Either only the cervix can have dropped downwards, or the entire uterus can have dropped down through the vagina. The most common symptom of a prolapsed uterus is a feeling of heaviness or pressure in the vagina. A prolapsed uterus can be caused by weakness or injury to the ligaments, connective tissue and muscles of the pelvis and can result from pregnancy, vaginal delivery of a foetus, chronic coughing, obesity, straining during bowel movements or even lifting an extremely heavy object.

Sexually transmitted diseases (STDs, venereal disease)

The term sexually transmitted diseases (STDs) encompasses all diseases that are transmitted through sexual intercourse. For example, AIDS, syphilis, gonorrhoea, genital herpes and Chlamydia infection. Symptoms will differ according to the disease contracted.

Toxic shock syndrome

Toxic shock syndrome is a state of shock due to poisoning, usually caused by toxins produced by staphylococci bacteria on foreign objects, such as tampons, that have been put into the body. The symptoms of toxic shock syndrome are sudden and severe and include a high fever, headache, fatigue, sore throat, red eyes, confusion, low blood pressure, vomiting and diarrhoea.

Vaginitis

Vaginitis is inflammation of the lining of the vagina and is characterised by itching, painful urination and increased vaginal discharge. Vaginitis can be caused by a bacterial or fungal infection and can also be the result of poor hygiene, not cleaning the genital area properly or wearing tight and non-absorbent underwear.

NEW WORDS

Fertilisation	the union and fusion of an ovum and a spermatozoa to form a zygote.
Gamete	a mature sex cell (ovum or spermatozoon).
Gonad	a male or female reproductive organ in which gametes are produced.
Haploid cell	a cell with a single set of chromosomes (23 chromosomes).
Lactation	the secretion of milk by the mammary glands.
Meiosis	reproductive cell division in which four daughter cells are produced and each one has only 23 chromosomes.
Menopause	the time in a woman's life when she stops menstruating and ovulating and is no longer able to bear children.
Oogenesis	the production of mature ova in the ovaries.
Ova (singular = ovum)	mature female sex cells (commonly called eggs).
Puberty	the time in a person's life when they become capable of reproducing children and their bodies develop secondary sexual characteristics.
Semen (seminal fluid)	fluid containing sperm and a mixture of fluids secreted by the reproductive glands.
Spermatogenesis	the production of spermatozoa in the testes.
Spermatozoa (singular = spermatozoon)	mature male sex cells (commonly called sperm).

Study Outline

Functions of the reproductive system

1. The human reproductive system functions in reproducing life and continuing the species.
2. Reproductive cells are produced through a type of cell division called meiosis.
3. Meiosis results in four daughter cells that each has only one set of chromosomes (23 chromosomes).
4. Reproductive cells are called gametes and are produced in reproductive organs called gonads.
5. Female gonads are ovaries and they produce ova through the process of oogenesis.
6. Male gonads are testes and they produce spermatozoa through the process of spermatogenesis.
7. The male reproductive system functions in producing spermatozoa and ejaculating it into a woman's reproductive tract.
8. The process in which an ovum and spermatozoon unite and fuse is called fertilisation and it results in a zygote.
9. The female reproductive system functions in producing ova, receiving spermatozoa, being the site of fertilisation and housing the fertilised zygote so that it can grow and develop into a foetus. This is called pregnancy. The female reproductive system also functions in delivering the foetus into the external world through the process of childbirth (labour).

Male reproductive system

The male reproductive system consists of the:

1. **Scrotum** – a paired sac of loose skin hanging externally and housing the testes and epididymis. It functions in maintaining the correct temperature of the testes.
2. **Testes** – a pair of oval glands located in the scrotum. They are the gonads where sperm are formed through spermatogenesis. They also secrete testosterone.
3. **Epididymis** – a series of ducts that act as the site of sperm maturation.
4. **Vas deferens** – a long duct that transports sperm from the epididymis to the urethra. It is supported by a structure called the spermatic cord.
5. **Urethra** – the terminal duct of both the reproductive and urinary systems. It transports semen and urine out of the body.
6. **Accessory glands** – the seminal vesicles, prostate gland and bulbourethral glands all produce fluids that combine with sperm to form semen.
7. **Penis** – a cylindrical organ that excretes urine and ejaculates sperm.

Female reproductive system

The female reproductive system consists of the:

1. **Ovaries** – a pair of almond-shaped organs that are the gonads where ova are produced through oogenesis. They also function in ovulation.
   - The ovaries contain small sac like structures called ovarian follicles. Each follicle houses an immature ovum called an oocyte.
   - One follicle becomes the dominant follicle and is called the mature Graafian follicle. At ovulation this follicle expels the ovum into the uterine cavity.
   - The now empty follicle develops into the corpus luteum, a glandular structure that secretes female hormones.
   - The corpus luteum finally degenerates into the corpus albicans.
2. **Fallopian tubes** – two thin tubes running from the ovaries to the uterus. They are the site of fertilisation and they transport a zygote to the uterus.
3. **Uterus** – a muscular sac composed of three layers of tissue: the outer perimetrium, the middle myometrium and the inner endometrium. The endometrium contains the stratum functionalis which is shed during menstruation. The uterus is a passageway for sperm travelling to an ovum. If fertilisation then occurs, the uterus houses a foetus until birth. If fertilisation does not occur, the uterus is the site of menstruation.
4. **Vagina** – a muscular tube which acts as a passageway for menstrual flow, semen and a foetus.
5. **Vulva** – the external genitals of a woman and include the mons pubis, labia majora, labia minora, clitoris and vestibule.
6. **Mammary glands** – modified sudoriferous glands that function in lactation.

Female reproductive cycle

1. A woman's reproductive cycle lasts anywhere from 24–35 days and involves both:
   - An ovarian cycle in which an oocyte matures until it is ready for ovulation.
   - A uterine (menstrual) cycle in which the endometrium of the uterus is prepared for the arrival of a fertilised ovum.
2. The cycle consists of a menstrual phase in which the stratum functionalis of the endometrium is shed via menstrual flow, a preovulatory phase in which the Graafian follicle forms and the endometrium proliferates, ovulation, and the postovulatory phase which is a 'waiting time'. If fertilisation has occurred the zygote implants in the uterus and a woman becomes pregnant. If it has not occurred, the cycle returns to the menstrual phase.

Ageing and the reproductive system

1. Puberty is the time in which a person develops secondary sexual characteristics and becomes able to reproduce.
2. In females, girls reach puberty when they begin to menstruate. From their first period until they go through menopause they are able to become pregnant. Menopause is the end of menstruation and a woman's capability to bear children.

Revision

1. Explain the functions of the reproductive system.
2. Describe the organisation of the male reproductive system.
3. Identify the functions of the following:
   - Testes
   - Epididymis
   - Urethra
4. Identify the principal male hormone.
5. Describe the organisation of the female reproductive system.
6. Identify the functions of the following:
   - Ovaries
   - Fallopian tubes
   - Vagina
7. Explain the structure of the uterus.
8. Describe the female reproductive cycle.
9. Identify the main female hormones.
10. Describe the following disorders:
    - Pelvic inflammatory disease
    - Breast cancer
    - Prostatitis

Multiple choice questions

1. **The principal male hormone is:**
   a. Oestrogen
   b. Progesterone
   c. Progestin
   d. Testosterone.

2. **Which of the following is not an STD?**
   a. AIDS
   b. Chlamydia
   c. Endometriosis
   d. Syphilis.

3. **The function of the epididymis is to be:**
   a. The site of spermatogenesis.
   b. The site of sperm maturation.
   c. The site of oogenesis.
   d. The site of ova maturation.

4. **Which of the following statements is correct?**
   a. The mature Graafian follicle ruptures to release an ovum into the pelvic cavity.
   b. The mature Graafian follicle secretes oestrogen, progesterone, relaxin and inhibin.
   c. The mature Graafian follicle is the site of fertilisation.
   d. The mature Graafian follicle is found as part of the menstrual flow.

5. **How many chromosomes are found in a haploid cell?**
   a. 13
   b. 23
   c. 43
   d. 46.

6. **How many trimesters are in a pregnancy?**
   a. 1
   b. 3
   c. 6
   d. 9.

7. **Which of the following is the correct definition of a gamete?**
   a. An organ where sex cells are produced.
   b. A fertilised ovum.
   c. A cell derived through somatic cellular division.
   d. A mature sex cell.

8. **Which of the following statements is correct?**
   a. Fibroids are benign tumours consisting of muscle and fibrous tissue.
   b. Fibroids are malignant tumours consisting of muscle and fibrous tissue.
   c. Fibroids are cancerous tumours consisting of muscle and fibrous tissue.
   d. None of the above.

9. **What is menopause?**
   a. The time in which a person becomes capable of reproducing.
   b. The time in which a girl has her first menstrual period and becomes capable of bearing children.
   c. The time in which a man stops producing sperm.
   d. The time in which a woman stops menstruating and becomes incapable of bearing children.

10. **Which layer of the uterus is shed during menstruation?**
   a. Perimetrium
   b. Myometrium
   c. Stratum basale
   d. Stratum functionalis.

Multiple choice answers

Chapter 1
1.b 2.d 3.b 4.c 5. d 6. b 7. c 8. a 9. b 10. c
11. a 12. b 13. a 14. d 15. b

Chapter 2
1.c 2.a 3.c 4.c 5.b 6.a 7.d 8.b 9.c 10.a 11.b
12.d 13.b 14.a 15.d 16.d 17.b 18.c 19.a 20.c.

Chapter 3
1.b 2.c 3.b 4.c 5.b 6.a 7.d 8.d 9.b 10.a 11.b
12.b 13.c 14.d 15.b.

Chapter 4
1.c 2.a 3.b 4.b 5.d 6.a 7.c 8.d 9.a 10.b.

Chapter 5
1.b 2.a 3.c 4.d 5.b 6.c 7.d 8.c 9.a 10.b 11.c
12.d 13.c 14.a 15.c.

Chapter 6
1.c 2.c 3.a 4.a 5.d 6.c 7.d 8.a 9.b 10.a.

Chapter 7
1.c 2.b 3.a 4.d 5. a 6.d 7.b 8.d 9.d 10.c.

Chapter 8
1.b 2.d 3.a 4.b 5.a 6.d 7.a 8.b 9.d 10.c.

Chapter 9
1.c 2.a 3.c 4.a 5.c 6.c 7.a 8.d 9.c 10.a.

Chapter 10
1.c 2.a 3.b 4.b 5.d 6.d 7.a 8.c 9.d 10. c.

Chapter 11
1.c 2.b 3.a 4.c 5.c 6.a 7.c 8.b 9.a 10.a.

Chapter 12
1.b 2.c 3.a 4.c 5.d 6.b 7.c 8.d 9.a 10.b.

Chapter 13
1.d 2.c 3.b 4.a 5.b 6.b 7.d 8.a 9.d 10.d.

Bibliographical references
Chapter 2
i Elaine N. Marieb, *Essentials of Human Anatomy and Physiology* (7th Ed.), p.47
ii Dr Christiaan Barnard, *The Body Machine*, p.10
iii Gerard Tortora & Sandra Grabowski, *Principles of Anatomy and Physiology* (8th Ed.), p.55

Chapter 3
i Tortora and Grabowski, *Principles of Anatomy and Physiology*, p.124
ii Dr David Presbury, *Know Your Skin*, p.11

Chapter 4
i Gerard Tortora & Sandra Grabowski, *Principles of Anatomy and Physiology* (8th Ed.), p.152
ii Dr Christiaan Barnard, *The Body Machine*, p.33

Chapter 5
i Gerard Tortora & Sandra Grabowski, *Principles of Anatomy and Physiology* (8th Ed.), p.249

Chapter 6
i Dr Christiaan Barnard, *The Body Machine*, p.50
ii Gerard Tortora & Sandra Grabowski, *Principles of Anatomy and Physiology* (8th Ed.), p.391

iii Gerard Tortora & Sandra Grabowski, *Principles of Anatomy and Physiology* (8th Ed.), p.396
iv Dr Christiaan Barnard, *The Body Machine*, p.64

Chapter 8
i Elaine Marieb, *Essentials of Human Anatomy and Physiology*, p.420

Chapter 9
i Gerard Tortora & Sandra Grabowski, *Principles of Anatomy and Physiology* (8th Ed.), p.558
ii Gerard Tortora & Sandra Grabowski, *Principles of Anatomy and Physiology* (8th Ed.), p.624

Chapter 11
i Dr Christiaan Barnard, *The Body Machine*, p.123
ii Tortora and Grabowski, *Principles of Anatomy and Physiology*, p.46
iii Tortora and Grabowski, *Principles of Anatomy and Physiology*, p.780

Chapter 12
i Dr Christiaan Barnard, *The Body Machine*, p.132

Appendix

VITAMINS AND MINERALS			
VITAMIN	**MAJOR DIETARY SOURCES**	**MAJOR FUNCTIONS**	**SIGNS OF DEFICIENCY**
A Retinol and Beta-carotene	Animal sources of retinol: liver, eggs, cheese, butter and milk. Plant sources of beta-carotene: orange or yellow vegetables, fruit and dark green leafy vegetables	Known as the 'vision vitamin' or the 'anti-infection' vitamin. Promotes healthy skin, eyes and bones. Protects against infections and boosts the immune system. Essential for night vision. Antioxidant.	Skin problems such as dry flaky skin and dandruff. Mouth ulcers, poor night vision, acne, frequent colds or infections.
B1 Thiamine	In general nuts, whole grains and pork. Also organ meats, yeast extract, wheat germ, sunflower seeds, peanuts, legumes.	Energy production. Brain function and maintenance of the nervous system.	Fatigue, poor memory, lack of concentration and depression.
B2 Riboflavin	In general organ meats and yeast extract. Also almonds, wheat germ, dairy products, whole grain products.	Helps turn fats, sugars and protein into energy. Important for the skin, hair, nails and eyes.	Burning or gritty eyes, sensitivity to bright lights, cracked skin around the mouth, sore tongue and burning lips.
B3 Niacin	In general meat and fish. Also yeast extract, raw peanuts, chicken, eggs, brown rice and seeds.	Energy production. Brain function and a healthy nervous system. Healthy skin. Helps balance blood sugar and lower cholesterol levels.	Lack of energy, insomnia, memory loss, muscle weakness and skin problems.
B5 Pantothenic Acid	In general organ meats, eggs and yeast extracts. Also whole grains, nuts, seeds, pulses, broccoli and cauliflower.	Energy production. Essential for the brain and nerves and helps one cope with stress. Promotes healthy skin and hair and boosts the immune system.	Poor skin, depression, fatigue, loss of appetite and poor coordination.
B6 Pyridoxine	Seeds, whole grains, pulses, yeast extract, bananas, nuts, potatoes, broccoli, cauliflower.	Essential for a healthy immune system and the formation of red blood cells. Helps relieve the symptoms of PMS and menopause and necessary for a healthy pregnancy. Natural anti-depressant and functions in the metabolism of fats and proteins.	Disorders of the nervous system, fluid retention, oily scaling skin, muscular spasms and PMS.

VITAMINS AND MINERALS			
VITAMIN	MAJOR DIETARY SOURCES	MAJOR FUNCTIONS	SIGNS OF DEFICIENCY
B12 Cyanocobalamin	Mainly in animal products such as liver, oysters and sardines. Rare in plant-based foods. Vegans can obtain B12 from fermented soybean curd or fermented milks.	The feel good vitamin. Essential for a healthy nervous system, memory and concentration. Helps the blood carry oxygen. Essential for energy.	Fatigue, exhaustion, anaemia, menstrual disorders, poor hair condition and a lack of energy.
Biotin	In general organ meats, soybeans and whole grains. Also in cauliflower, eggs, mushrooms, nuts and pulses.	Vital for cell growth and replication. Particularly important in childhood. Helps the body use essential fats and maintain healthy skin, hair and nerves.	Dry skin, poor hair condition, premature greying hair, tender or sore muscles, poor appetite, nausea, eczema, dermatitis.
C Ascorbic Acid	In general green vegetables, fruits and fruit juices. Also in tomatoes and potatoes.	Strengthens the immune system and fights infection. Makes collagen to keep bones, skin and joints firm and strong. Antioxidant. Detoxifies pollutants.	Weakness, frequent colds, lack of energy, frequent infections, bleeding or tender gums, easy bruising, nose bleeds, slow wound healing, red pimples on skin.
D Ergocalciferol Cholecalciferol	Formed by the action of sunlight on the skin. Also present in fatty fish, cottage cheese and eggs.	The 'sunshine vitamin'. Essential for strong and healthy bones.	Joint pain or stiffness, backache, tooth decay, muscular weakness and spasm, hair loss.
E D-alpha tocopherol	In general nuts, seeds and their oils. Also soybeans, whole grains and a wide variety of fruits and vegetables.	Primary antioxidant in the body. Essential for immune function. Slows ageing and protects against pollution. Good for the skin.	Lack of sex drive, exhaustion after light exercise, easy bruising, slow wound healing, varicose veins, loss of muscle tone, infertility.
Folic Acid	In general in green leafy vegetables. Also wheat germ, peanuts, sprouts, asparagus, seeds and nuts. Poor in animal foods.	Critical during pregnancy and also essential for brain and nerve function. Needed for utilising protein and red blood cell formation.	Weakness, lethargy and fatigue, irritability, insomnia, recurrent miscarriages and problems with lactation.
K Phylloquinone	Produced by bacteria in the large intestine. Also present in dark green leafy vegetables, oats and whole wheat.	Controls blood clotting and necessary for the formation of strong bones.	Heavy menstrual bleeding, poor blood-clotting, osteoporosis.

VITAMINS AND MINERALS			
MINERAL	MAJOR DIETARY SOURCES	MAJOR FUNCTIONS	SIGNS OF DEFICIENCY
Calcium	Dark green vegetables, nuts seeds, bean curd. Also dairy products and brewer's yeast.	Essential for healthy skin, bones and teeth. Maintains healthy nerves and necessary for muscular contraction.	Muscle cramps or tremors, insomnia, nervousness, joint pain, arthritis, tooth decay, high blood pressure, osteoporosis and osteomalacia.
Chromium	In general whole grains, eggs and meat. Fruits and vegetables are low in chromium.	Helps control blood-sugar levels and therefore normalise hunger and reduce cravings.	Hypoglycemia, fatigue, mood swings and obesity.
Iron	In general animal products such as liver, kidney and meat. Plant products include brewer's yeast, kelp, blackstrap molasses and pumpkin seeds.	As a component of haemoglobin, iron transports oxygen and carbon dioxide to and from cells. Component of enzymes, vital for energy production and the immune system.	Anaemia, pale skin, sore tongue, fatigue, listlessness, loss of appetite, sensitivity to cold.
Magnesium	In general whole grains, nuts, seeds and bean curd. Also green leafy vegetables. Fish, meat and milk are low in magnesium.	Strengthens bones and teeth. Necessary for the normal functioning of muscles and nerves. Activates many enzymes.	Muscle cramps, PMS, sugar cravings, fatigue, nervousness, difficulty relaxing.
Manganese	In general whole grains, nuts and avocados. Also tea, fruits and vegetables.	Important for blood sugar balance, bone formation and detoxification. Activates several enzymes.	Blood sugar imbalances, muscle twitches, childhood growing pains, dizziness or poor sense of balance, fatigue.
Molybdenum	In general pulses and whole grains.	Helps rid the body of the protein breakdown products and detoxifies the body. Essential for the metabolism of iron, amino acids and fats.	Irritability.
Phosphorus	Present in almost all foods.	Forms and maintains bones and teeth. Essential for the structure of cell membranes. Aids metabolism and energy production.	Deficiency is unlikely, but may occur with long-term antacid use or with stresses such as bone fracture. Signs include general muscle weakness, loss of appetite, mental confusion and osteomalacia.

VITAMINS AND MINERALS			
MINERAL	**MAJOR DIETARY SOURCES**	**MAJOR FUNCTIONS**	**SIGNS OF DEFICIENCY**
Potassium	Raisins, potatoes, avocadoes, bananas and a general diet rich in vegetables and fruit and low in sodium.	Promotes healthy nerves and muscles. Maintains fluid balance. Involved in metabolism, the control of acid/alkaline levels in the blood, the functioning of the heart and general health.	Muscle weakness, pins and needles, irritability, nausea, water retention, depression and fatigue.
Selenium	Tuna, oysters, molasses, mushrooms, herrings, cottage cheese, cabbage, liver, courgettes, cod, chicken.	Antioxidant. Slows premature ageing. Reduces inflammation, stimulates the immune system and promotes a healthy heart. Important in male potency and libido.	Premature ageing, cataracts, frequent infections, poor detoxification.
Sodium	Table salt, cured meats, pickles, soy sauce, olives.	Maintains water balance in body and helps nerve functioning. Used in muscle contraction. Utilised in energy production and helps move nutrients into cells.	Weakness, dizziness, heat exhaustion, low blood pressure and headaches.
Zinc	In general shell fish, fish and red meats. Also seeds, nuts, pulses and whole grains.	Essential for the immune system and the structure and function of cell membranes. Essential for growth and healing, hormones, a healthy nervous system, bones and teeth, energy and healthy hair. Helps one cope with stress.	Regular infections, poor sense of taste or smell, white marks on nails, stretch marks, acne, greasy skin, low fertility, pale skin, tendency for depression, loss of appetite.

Bibliography

Alcamo, I. (2003). *Anatomy Colouring Workbook (2nd Ed.)*. NY: Random House.

Allen, R. (1990). *The Concise Oxford Dictionary of Current English*. Oxford: Clarendon Press

Arnould-Taylor, W. (1998). *A Textbook of Anatomy and Physiology (3rd Ed.)*. Cheltenham: Stanley Thornes Publishers Ltd.

Barnard, C. (ed.) (1981). *The Body Machine*. Willemstad: Multimedia Publications Inc.

Beers, M. (ed.) (2003). *The Merck Manual of Medical Information (2nd Home Ed.)*. New Jersey: Merck & Co., Inc.

Blakey, P. (1992). *The Muscle Book*. Bibliotek Books Ltd.

Cheshire, E. (1998). *Gastrointestinal System*. Barcelona: Mosby International Ltd.

Gould, F. (2005). *Anatomy and Physiology*. Cheltenham: Nelson Thornes Ltd.

Gray, H. (2001). *Gray's Anatomy*. Surrey: TAJ Books.

Jarmey, C. (2003). *The Concise Book of Muscles*. Chichester: Lotus Publishing

Marieb, E. (2003). *Essentials of Human Anatomy and Physiology (7th Ed.)*. San Francisco: Benjamin Cummings.

McFerran, T. (ed.). (1998). *Oxford Dictionary for Nurses (4th Ed.)*. Oxford: Oxford University Press

McGuinness, H. (2006). *Anatomy and Physiology Therapy Basics (3rd Ed.)*. London: Hodder Arnold.

Miranda, J. (ed.) (1992). *Milady's Art and Science of Nail Technology (2nd Ed.)*. NY: Milady Publishing Company

Mortimore, D. (2001). *The Complete Illustrated Guide to Vitamins and Minerals*. London: Harper Collins Publishers Ltd.

Neighbors, M. and Tannehill-Jones, R. (2000). *Human Diseases*. NY: Delmar Thomson Learning.

Nilsson, L. (1987). *The Body Victorious*. London: Dell Publishing Co.

Parsons, T. (2002). *An Holistic Guide to Anatomy and Physiology*. London: Thomson

Presbury, D. (2000). *Know Your Skin*. Randburg: Medpress

Pugliese, P. (1991). *Advanced Professional Skin Care*. Bernville: APSC Publishing

Putz, R. and Pabst, R. (ed.) (2008) *Sobotta Atlas of Human Anatomy*. Munich: Elsevier GmbH.

Tortora, G. and Grabowski, S. (1996). *Principles of Anatomy and Physiology (8th Ed.)*. Harper Collins

Underwood, J. (ed.) (2000). *General and Systematic Pathology*. London: Harcourt Publishers Ltd.

Walker, R. (1994). *Atlas of the Human Body.* London: Quarto Children's Books Ltd.

Walker, R. (2001). *Human Body.* London: Dorling Kindersley Ltd.

Waugh, A. and Grant, A. (2001). *Anatomy and Physiology in Health and Illness (9th Ed.).* London: Harcourt Publishers.

Wingate, P. (1976). *Medical Encyclopedia (2nd Ed.).* Harmondsworth: Penguin Books Ltd.

Glossary

A

Abdomen	region of the body between the diaphragm and pelvis.
Abduction	movement away from midline of body.
Abrasion	damaged area of the skin caused by the skin being scraped or worn away.
Absorption	uptake of digested nutrients into the bloodstream and lymphatic system.
Acetylcholine	neurotransmitter found in both the peripheral and central nervous systems.
Acid mantle	film of sebum and sweat on the surface of the skin that protects against bacteria.
Actin	protein that functions in muscle contraction.
Action potential	electrical charge that occurs on the membrane of a muscle cell in response to a nervous impulse.
Active transport	movement of a substance across a cellular membrane that involves the release of energy. It takes place against a concentration gradient.
Adduction	movement towards the midline of the body.
Adenosine tri-phosphate (ATP)	main energy-transferring molecule in the body.
Adipocyte	fat cell.
Adrenaline	hormone secreted by the adrenal medulla that functions in the fight-or-flight response.
Adrenocorticotropic hormone (ACTH)	hormone secreted by the anterior pituitary gland that stimulates and controls the adrenal cortex.
Aerobic	requiring oxygen.
Afferent	carrying towards a centre.
Afferent neurones	see sensory neurones.
Agonist	see prime mover.
Agranulocytes	group of white blood cells that do not contain granules in their cytoplasm (includes lymphocytes and monocytes).
Aldosterone	hormone secreted by the adrenal cortex that regulates the reabsorption of sodium and water in the kidneys.
Alimentary canal	see gastrointestinal tract.
Alveoli	air sacs inside the lungs.
Amphiarthroses	slightly movable joints that permit a minimal amount of flexibility and movement.
Anaerobic	not requiring oxygen.
Anagen	active growing stage of the hair cycle.

Anal canal	last 2–3 cm of the rectum that opens to the exterior.
Anatomical position	position in which the body is standing erect with the feet parallel, the arms hanging down by the side and the face and palms facing forward.
Anatomy	study of the structure of the body.
Antagonist	muscle that opposes the movement caused by the prime mover or agonist.
Anterior	at the front of the body, in front of.
Antibody	specialised protein that is synthesised to destroy a specific antigen.
Antidiuretic hormone (ADH)	hormone released by the posterior pituitary gland that has an antidiuretic effect and raises blood pressure.
Antigen	any substance that the body recognises as foreign.
Antioxidant	substance that combats or neutralises free radicals.
Apocrine gland	type of sweat gland located in the armpits, pubic region and the areolae of the breasts.
Aponeurosis	flat, sheet-like tendon that attaches muscles to bone, skin or another muscle.
Apoptosis	the normal, ordered death and removal of cells as part of their development, maintenance and renewal.
Appendicular skeleton	part of the skeleton consisting of the upper and lower limbs and their girdles.
Appendix (vermiform)	sac attached to the caecum of the large intestine.
Aqueous humour	fluid that nourishes the lens and cornea and helps produce intraocular pressure.
Arachnoid	middle meninge (covering) of the brain and spinal cord.
Areola	circular area of pigmented skin surrounding the nipple.
Arrector pili muscles	smooth muscles attached to hairs that contract to pull the hairs into a vertical position.
Arteries	vessels that usually carry blood away from the heart towards the tissues.
Arterioles	tiny arteries that deliver blood to capillaries.
Arthrology	the study of joints.
Articulation	point of contact between two bones, commonly called a joint.
Atony	lack of muscle tone.
Atrioventricular valves	valves lying between the atria and ventricles that prevent the backflow of blood.
Atrium	receiving chamber of the heart.
Atrophy	wasting away of muscles.
Auditory ossicles	three tiny bones extending across the middle ear: the malleus, incus and stapes.

Auditory tube	see *Eustachian tube*.
Auricle	part of the ear we see.
Autonomic nervous system (ANS)	part of the nervous system that controls all processes that are automatic or involuntary.
Autorhythmic cells	muscle or nerve cells that generate an impulse without an external stimulus, i.e. they are self-excitable.
Avascular	lacking in blood vessels.
Axial skeleton	part of the skeleton comprising the bones found in the centre of the body.
Axilla	armpit.
Axon	transmitting portion of a cell.
Axon terminal	area found at the end of the axon that contains membrane-enclosed sacs called synaptic vesicles.

B

B cells	cells of the immune system that develop into plasma cells and are able to synthesise and secrete antibodies.
Ball and socket joint	synovial joint in which a ball-shaped bone fits into the cup-shaped socket of another bone.
Baroreceptors	sensory nerve endings that monitor blood pressure changes in the arteries and veins.
Basophils	type of white blood cell that contains histamine and is capable of ingesting foreign particles.
Bicuspid valve	left atrioventricular valve.
Bile	liquid produced by the liver that emulsifies fats. It contains water, bile acids, bile salts, cholesterol, phospho-lipids, bile pigments and some ions.
Blood pressure	force exerted by blood on the walls of a blood vessel.
Bowman's capsule	see *glomerular capsule*.
Brachial	pertaining to the arm.
Brain stem	continuation of the spinal cord that connects the spinal cord to the diencephalon. It consists of the medulla oblongata, pons and midbrain.
Bronchus	branch-like passageway inside the lungs.
Bruise	discolouration of the skin caused by the escape of blood from underlying vessels.
Brush border	area in the small intestine composed of microvilli that contain digestive enzymes.
Buccal cavity	mouth.
Bursa	sac-like structure made of connective tissue, lined with a synovial membrane and filled with synovial fluid.

C

Caecum	6 cm long pouch of the large intestine that receives food from the small intestine via the ileocaecal valve.
Calcitonin (CT)	hormone secreted by the thyroid gland that lowers blood calcium levels.
Calcitriol	active form of vitamin D.
Canaliculus	very small channel or canal.
Cancellous bone tissue	see *spongy bone tissue*.
Capillary	very small blood vessel that connects arterioles to venules.
Carbohydrate	organic compound composed of units of glucose that contain carbon, hydrogen and oxygen.
Cardiac cycle	all the events associated with a heartbeat.
Cardiac muscle	muscle forming most of the wall of the heart. It is composed of striated muscle fibres and is involuntary.
Cardiac output	amount of blood pumped out of the heart by the left ventricle.
Carpal	pertaining to the wrist.
Cartilage	resilient, strong connective tissue that is less hard but more flexible than bone.
Cartilaginous joint	joint in which bone ends are held together by cartilage and do not have a synovial cavity between them.
Catagen	the transitionary stage of the hair cycle.
Catalyst	substance that affects the rate of a chemical reaction without itself being changed by the reaction.
Caudal	away from the head, below.
Cellular respiration	metabolic reaction that uses oxygen and glucose and produces energy in the form of ATP. Also called oxidation.
Central nervous system (CNS)	the brain and spinal cord.
Centriole	structure found near the nucleus of a cell that plays a role cell division.
Centrosome	area near the nucleus of a cell that contains centrioles and forms the mitotic spindle in dividing cells.
Cephalad	towards the head, above.
Cephalic	pertaining to the head.
Cerebellum	region of the brain located behind the medulla oblongata and pons. It functions in producing smooth, coordinated movements as well as posture and balance.
Cerebral cortex	outer, most superficial layer of the cerebrum, consisting of grey matter.
Cerebrospinal fluid (CSF)	fluid that circles the CNS, protecting it and helping to maintain homeostasis.

Cerebrum	largest part of the brain and the area that gives us the ability to read, write, speak, remember, create and imagine.
Cerumen	earwax.
Cervical	pertaining to the neck.
Chemoreceptor	receptor sensitive to chemicals.
Choroid	lining of most of the internal surface of the sclera.
Chromatin	mass of 46 chromosomes all tangled together in a non-dividing cell.
Chromosome	thread-like structure found in the nucleus of a cell. It carries the genes.
Chyle	fluid found inside the lacteals of the small intestine.
Chyme	semifluid contents of the stomach consisting of partially digested food and gastric secretions.
Cilia	tiny hairlike projections on the surfaces of cells that move the cell or substances along the surface of the cell.
Ciliary body	part of the eye between the iris and choroid.
Circumduction	circular movement of the distal end of a body part, e.g. circling the shoulder joint.
Cistern	channel or tubule in a cell.
Clitoris	small cylindrical mass of erectile tissue and nerves that forms part of the female genitalia.
Club hair	fully grown hair that has detached from the hair bulb during the catagen stage of hair growth.
Cochlea	bony, spiral canal in the inner ear that resembles a snail's shell and houses the organ of Corti (for hearing).
Colon	long tube that forms most of the large intestine.
Compact bone tissue	very hard, compact tissue with few spaces within it.
Complement system	group of proteins in the blood that help antibodies during an immune response.
Concentric contraction	contractions towards a centre that result in a movement that shortens the angle at a joint.
Conductivity	ability of cells, such as nerve or muscle cells, to move action potentials along their plasma membranes.
Condyloid joint	joint in which an oval protuberance at the end of a bone fits into an elliptical cavity of another bone.
Cones	photoreceptors that respond to colour.
Connective tissue	one of the basic tissue types in the body. Consists of few cells in a large matrix and functions in support, storage and protection.

Contractility	ability of muscles to contract and shorten.
Cooper's ligaments	see *suspensory ligaments*.
Cornea	avascular, transparent coat that covers the iris. It is curved and helps focus light.
Coronal plane	plane that divides vertically into anterior and posterior portions.
Coronary circulation	circulation that supplies the muscles of the heart with blood.
Corpus luteum	gland formed after a Graafian follicle has discharged its ovary. It secretes progesterone, oestrogen, relaxin and inhibin.
Cortex	outer layer of an organ.
Corticoptropin	See *adrenocorticotropic hormone*.
Costal	pertaining to a rib.
Cough	sudden, explosive movement of air rushing upwards through the respiratory passages.
Cranial	towards the head, above, pertaining to the head.
Cranium	hard bones of the skull.
Creatinine	product of muscle activity.
Cross-infection	transfer of infection from one person to another.
Cross-section	division that divides horizontally into inferior and superior portions.
Crust (scab)	accumulation of dried blood, pus or skin fluids on the surface of the skin. Forms where the skin has been damaged.
Cutaneous	pertaining to the skin.
Cutaneous membrane	the skin. Composed of an epidermis and dermis.
Cuticle	outer layer of cells of a hair or the epidermis of the skin or the base of the nail plate.
Cyst	semi-solid or fluid-filled lump above and below the skin.
Cytokinesis	process by which a cell splits into two new cells during cellular division.
Cytology	the study of cells.
Cytoplasm	cellular material inside the plasma membrane, excluding the nucleus.
Cytosol	thick, transparent, gel-like fluid inside a cell.

D

Deep	away from the surface of the body.
Defecation	process by which indigestible substances and some bacteria are eliminated from the body.
Deglutition	process of swallowing food.
Demineralisation	process through which minerals such as calcium and phosphorous are lost from the bones.

Dendrite	receiving or input portion of a neurone.
Dense connective (fibrous) tissue	contains thick, densely packed fibres and fewer cells than loose connective tissue.
Dentes	teeth.
Depolarisation	process by which an action potential is produced.
Depression (of the shoulders or jaw)	dropping the shoulders or jaw downwards.
Dermatology	study of the skin.
Dermis	deep layer of the skin. Composed of dense irregular connective tissue.
Desquamation	process by which the skin is shed.
Diaphysis	main, central shaft of a long bone.
Diarthrose	freely movable joint that permits a number of different movements.
Diastole	relaxation of the heart muscle during the cardiac cycle.
Diencephalon	region of the brain that lies above the brain stem, enclosed by the cerebral hemispheres. Contains the thalamus, hypothalamus and epithalamus and has a number of different functions including housing the pituitary and pineal endocrine glands.
Diffusion	movement of substances from areas of high concentration to areas of low concentration.
Digestion	process by which large molecules of food are broken down into smaller molecules that can enter cells.
Digit	finger or toe.
Diploid	having two complete sets of chromosomes per cell (i.e. 46).
Distal	farther from its origin or point of attachment of a limb.
Distal convoluted tubule	area furthest from the glomerular capsule. Section of the tube where fine-tuning of the filtrate occurs.
Diuretic	substance that increases urine production.
Dopamine	type of neurotransmitter.
Dorsal	at the back of the body, behind.
Dorsiflexion	pulling of the foot upwards towards the shin, in the direction of the dorsum.
Duodenal glands (Brunner's glands)	glands in the duodenum that secrete an alkaline mucous.
Duodenum	first segment of the small intestine. It connects the stomach to the ileum.
Dura mater	outer covering of the brain.
Dyspnoea	laboured or difficult breathing.

E

Eardrum (tympanic membrane)	very thin, semitransparent membrane between the auditory canal and the middle ear. When sound waves hit it, it vibrates, passing the sound waves on to the middle ear.

Eccentric contraction	contraction away from the centre.
Eccrine gland	sweat glands distributed around the body.
Efferent	carrying away from a centre.
Efferent neurone	see motor neurone.
Elasticity	ability of a tissue to return to its original shape after stretching, contracting or extending.
Electrolyte	charged particle (ion) that conducts an electrical current in an aqueous solution.
Elevation (of the shoulders or jaw)	lifting the shoulders or jaw upwards.
Ellipsoid joint	see condyloid joint.
Enamel	extremely hard substance that protects the teeth from being worn down and acts as a barrier against acids.
Endocardium	thin, smooth lining of the inside of the heart.
Endocrine glands	ductless glands that secrete substances into the extracellular space around their cells. These secretions then diffuse into blood capillaries and are transported by the blood to target cells located throughout the body.
Endocrinology	study of the endocrine glands and the hormones they secrete.
Endometrium	mucous membrane lining of the uterus.
Endomysium	connective tissue that surrounds each individual muscle fibre.
Endoplasmic reticulum	network of fluid-filled cisterns within a cell that provides a large surface area for chemical reactions and also transports molecules within the cell.
Endosteum	membrane that lines the medullary cavity of bones.
Enteroendocrine cell	specialised cell that secretes hormones into the intestinal glands.
Enzyme	protein that speeds up a chemical reaction without itself being used up in the reaction.
Eosinophils	type of white blood cell that can destroy certain parasitic worms, phagocytise antigen-antibody complexes and combat the effects of some inflammatory chemicals.
Epicardium	outer layer of the heart wall. Also called the visceral layer of the serous pericardium.
Epidermis	superficial, outer layer of the skin.
Epididymis	the organ lying along the posterior border of each testis. It is composed of a series of coiled ducts and is the site of sperm maturation.
Epimysium	the outermost layer of connective tissue that encircles an entire muscle.

Epinephrine	See *adrenaline*.
Epiphyseal plate	a layer of hyaline cartilage in a growing bone that allows the diaphysis to grow in length.
Epiphysis	the end of a long bone.
Epithelium	basic tissue type. Forms glands, lines internal cavities and vessels and is the superficial layer of the skin.
Equilibrium	balance.
Erythrocyte	red blood cell that contains a protein called haemoglobin which transports oxygen in the blood.
Erythropoietin	hormone secreted by the kidneys that stimulates the production of red blood cells.
Essential fatty acid	fats that are vital for the proper functioning of the body.
Eustachian tube	tube that connects the middle ear with the upper portion of the throat. It equalises the middle ear cavity pressure with the external atmospheric pressure.
Eversion	turning the sole of the foot outwards.
Excitability	ability of muscle or nerve cells to respond to stimuli.
Excoriation	removal of the skin caused by scratching or scraping.
Excretion	elimination of waste products.
Exocrine glands	glands that secrete substances into ducts that carry these substances into body cavities or to the outer surface of the body.
Extensibility	ability of muscles to extend and lengthen or stretch.
Extension	straightening movement in which a body part is restored to its anatomical position after being flexed.
External auditory canal	curved tube that carries sound waves from the auricle to the eardrum.
External nares	openings to the nose. Commonly called the nostrils.
External respiration (pulmonary respiration)	gaseous exchange between lungs and blood. In external respiration, the blood gains oxygen and loses carbon dioxide.

F

Facilitated diffusion	diffusion in which substances are helped across the plasma membrane by channel or transporter proteins within the membrane.
Faeces	waste material of the digestive system that is eliminated through the anus.
Falciform ligament	fold of the peritoneum that binds the liver to the anterior abdominal wall and diaphragm and separates the two principal lobes of the liver.
Fallopian tubes	two thin tubes running from the ovaries to the uterus.

Fascia	connective tissue that surrounds and protects organs, lines walls of the body, holds muscles together and separates muscles.
Fascicle	bundle of 10–100 muscle fibres.
Fatigue (of muscles)	a muscle's inability to respond to stimulus or maintain contractions.
Fertilisation	union and fusion of an ovum and a spermatozoa to form a zygote.
Fibrosis	replacement of connective tissue by scar tissue.
Fibrous joints	joints in which bone ends are held together by fibrous connective tissue with no synovial cavity between them.
Filtration	movement of a liquid through a membrane or filter.
Fissure	crack in the skin that penetrates into the dermis.
Fixator	muscle that helps the prime mover by stabilising and preventing unneccesary movements in surrounding joints.
Flagella	long whiplike extensions of the cell membrane of certain cells such as sperm or bacteria. They move the cell.
Flat bones	thin bones consisting of a layer of spongy bone enclosed by layers of compact bone.
Flexion	bending of a joint in which the angle between articulating bones decreases. The opposite of extension.
Foetus	unborn child in the uterus from the 8th week of development until birth.
Follicle-stimulating hormone (FSH)	hormone secreted by the anterior pituitary gland that stimulates the development of ova and sperm.
Formed elements	cells and cell fragments found in blood.
Free radical	highly unstable, reactive molecule that damages cells.
Frontal plane	see *coronal plane*.

G

Gamete	sex cell (ovum or spermatozoon).
Ganglion	bundle or knot of nerve cell bodies.
Gastric juice	substance secreted by the gastric glands in the stomach, containing water, hydrochloric acid, intrinsic factor, pepsinogen and gastric lipase.
Gastric lipase	enzyme that acts on lipids in the stomach, breaking down triglycerides into fatty acids and monoglycerides.
Gastrin	hormone produced by the stomach that stimulates gastric secretions.
Gastrointestinal tract	tube that runs from the mouth to the anus in which digestion and absorption take place.
Gene	basic unit of genetic material.

Germinal matrix	region of the nail where cell division takes place and growth occurs.
Gingivae	gums
Gliding joint	joint in which two flat surfaces meet.
Glomerular (Bowman's) capsule	cup-like structure surrounding the glomerulus in a kidney's nephron. Forms the closed end of the renal tubule.
Glomerulus	knotted network of capillaries in a kidney's nephron.
Glucagon	hormone produced by the pancreas that raises blood glucose levels.
Glucocorticoids	group of hormones secreted by the adrenal cortex that stimulate metabolism, help the body resist long-term stressors, control the effects of inflammation and depress immune responses.
Glucose	sugar that is the major energy source for all cells.
Gluteal	pertaining to the buttocks.
Glycogen	carbohydrate consisting of sub-units of glucose. It is the main form in which carbohydrates are stored in the body.
Glycolysis	cellular process through which glucose is split into pyruvic acid and ATP.
Goblet cells	cells that secrete mucous.
Golgi apparatus	see Golgi complex.
Golgi complex (apparatus)	cellular structure located near the nucleus that processes, sorts and packages proteins and lipids for delivery to the plasma membrane. It also forms lysosomes and secretory vesicles.
Gonad	male or female reproductive organ in which sex cells are produced.
Graafian follicle	large fluid-filled follicle that ruptures and releases an ovum during ovulation.
Granulocytes	group of white blood cells containing granules in their cytoplasm (includes neutrophils, oesinophils and basophils).
Gustation	sense of taste.

H

Haemoglobin	substance found in red blood cells that transports oxygen and gives the blood cells their red colour.
Haemopoiesis	production of blood cells and platelets.
Haemostasis	the stopping of bleeding.
Hair matrix	ring of cells which divide to create hair.
Hair root	portion of the hair that penetrates into the dermis.
Hair shaft	superficial end of the hair that projects from the surface of the skin. Commonly called the hair strand.
Haploid cell	cell with a single set of chromosomes (23 chromosomes).

Haustra	pouches on the external surface of the colon.
Haversian system	see osteon.
Heart rate	number of times the heart beats in one minute
Hepatocyte	specialised cell found in the liver. It has many metabolic functions.
Hinge joint	joint in which the convex surface of a bone fits into the concave surface of another bone.
Histology	study of tissues.
Hive	see wheal.
Homeostasis	process by which the body maintains a stable internal environment.
Hormone	chemical messenger regulating cellular activity, produced by an endocrine gland and transported in the blood.
Human growth hormone (hGH)	hormone secreted by the anterior pituitary gland that stimulates growth and regulates metabolism.
Hydrophilic	water-loving.
Hydrophobic	water-hating.
Hydroxyapetite	form of calcium phosphate found in bones and teeth.
Hymen	membrane that partially covers the opening of the vagina.
Hyperextension	occurs when a body part extends beyond its anatomical position.
Hypersecretion	over- or excessive secretion.
Hypertonia	increase in muscle tone. Muscles are described as hypertonic.
Hyposecretion	under-secretion.
Hypotonia	loss of muscle tone. Muscles are described as hypotonic.
Hypoxia	inefficient delivery of oxygen to tissues.

I

Ileum	longest segment of the small intestine that receives food from the jejunum and passes it into the large intestine.
Immunity	ability of the body to resist infection.
Inclusions	diverse group of substances that are temporarily produced by some cells.
Inferior	away from the head, below.
Inflammation	body's response to tissue damage.
Ingestion	process of taking food into the mouth.
Inguinal	pertaining to the groin.
Inner ear	system of cavities and ducts that contains the organs of hearing and balance. Also called the labyrinth.
Insertion	point where a muscle attaches to the moving bone of a joint.
Insulin	hormone produced by the pancreas that lowers blood glucose levels.

Integumentary system	system of the skin and its derivatives (hair, nails and cutaneous glands).
Internal nares	openings that connect the nasal cavity to the pharynx.
Internal respiration	gaseous exchange between the blood and tissue cells. In internal respiration the blood loses oxygen and gains carbon dioxide.
Interstitial endocrinocytes	cells in the testes that secrete testosterone. Also called Leydig cells.
Intestinal juice	clear yellow fluid that has a slightly alkaline pH and contains water and mucous. It helps bring nutrient particles into contact with the microvilli.
Intrinsic factor	substance produced by the stomach and necessary for the absorption of vitamin B12 from the ileum.
Inversion	turning the sole of the foot inwards.
Ion	electrically charged molecule.
Iris	coloured portion of the eye suspended between the cornea and lens. It contains the pupil.
Irregular bones	bones that have complex shapes and varying amounts of compact and spongy tissues.
Irritability	ability to respond to a stimulus and convert it into an impulse.
Islets of Langerhans	clusters of endocrine cells located in the pancreas that secrete hormones including insulin and glucagon.
Isometric contraction	contraction in which the muscle contracts but does not shorten and no movement is generated.
Isotonic contraction	contraction in which muscles shorten and create movement while the tension in the muscle remains constant.

J

Jejunum	portion of the small intestine between the duodenum and the ileum.

K

Keratinisation	process in which cells die and become full of the protein keratin.
Keratinocytes	cells that produce keratin. 95% of the cells of the epidermis are keratinocytes.
Kinesiology	study of the motion of the body.
Kinetic energy	energy of motion.

L

Labia	Lips
Labia majora	two longitudinal folds of skin that extend inferiorly and posteriorly from the mons pubis of the female genitalia.
Labia minora	two smaller folds of skin running medially to the labia majora of the female genitalia.
Labyrinth	see *inner ear*.
Lacrimal gland	gland that secretes tears.
Lactase	brush border enzyme that breaks down lactose into glucose and galactose.
Lactation	secretion of milk by mammary glands.
Lacteal	specialised lymphatic vessels found in the villi of the small intestine.
Lactiferous	pertaining to the breasts.
Lactogenic hormone	see *prolactin*.
Lamellae	concentric rings of calcified matrix found in compact bone.
Lamellated corpuscles	nerve endings that are sensitive to pressure. Also called Pacinian corpuscles.
Langerhans cell	a cell that functions in skin immunity.
Lanugo hair	soft hair that begins to cover a foetus from the third month of pregnancy. Usually shed by the eighth month of pregnancy.
Larynx	short passageway between the laryngopharynx and the trachea. Commonly called the voicebox.
Lateral	away from the midline, on the outer side.
Lens	transparent structure of the eye, located behind the iris and responsible for fine tuning of focusing.
Leucocyte	white blood cell. It functions primarily in protecting the body against foreign microbes and in immune responses.
Ligament	tough band of connective tissue that attaches bones to bones.
Limbic system	region of the brain that controls the emotional and involuntary aspects of behaviour and also functions in memory.
Lingual frenulum	fold of mucous membrane that secures the tongue to the floor of the mouth.
Lingual lipase	enzyme in the mouth that begins the breakdown of lipids from trigyclerides into fatty acids and glycerol.
Lipid	a fat. Fats are organic compounds composed of carbon, hydrogen and oxygen and they are usually insoluble in water.
Long bones	bones that have a greater length than width and usually contain a longer shaft with two ends
Longitudinal plane	plane that divides vertically into right and left sides.
Loop of Henle	portion of the renal tubule of a kidney nephron.
Loose connective tissue	tissue that consists of two or more layers of cells. It is durable and functions in protecting underlying tissues in areas of wear and tear.
Lumbar	lower back region between the thorax and pelvis.

Lumen	hollow space within a tubelike structure such as an artery, vein or intestine.
Lunula	crescent-shaped white area at the proximal end of the nail plate.
Luteinizing hormone (LH)	hormone secreted by the anterior pituitary gland that stimulates ovulation, formation of the corpus luteum and secretion of oestrogens and progesterone in females; also stimulates production of testosterone in males.
Lymph	clear, straw-coloured fluid derived from interstitial fluid.
Lymphocyte	type of white blood cell involved in immunity. B cells and T cells are types of lymphocytes.
Lymphoid tissue	tissue where lymphocytes and antibodies are produced. Found in lymph nodes, the tonsils, the thymus, the spleen and as diffuse cells.
Lysosomes	cellular vesicles containing powerful digestive enzymes that can break down and recycle many different molecules.
Lysozyme	enzyme found in certain body secretions such as tears and saliva. It catalyses the breakdown of the cell walls of certain bacteria.

M

Macrophage	scavenger cell that engulfs and destroys microbes.
Macule	small, flat, discoloured spot of any shape. For example, freckles.
Maltase	brush border enzyme that breaks down maltose into glucose.
Mammary glands	two modified sudoriferous glands. Commonly called the breasts.
Marrow cavity	see *medullary*.
Mast cells	large cells in connective tissue that release substances such as histamine during inflammation.
Mastication	process of chewing food.
Meatus	see *external auditory canal*.
Medial	towards the midline, on the inner side.
Median line	imaginary line through the middle of the body.
Mediastinum	space in the thorax containing the aorta, heart, trachea, oesophagus and thymus gland. Found between the two pleural sacs.
Medulla	inner layer of an organ.
Medullary	space within the diaphysis of a bone. It contains yellow bone marrow and is also called the marrow cavity.
Meiosis	reproductive cell division in which four daughter cells are produced and each one has only 23 chromosomes.

Meissner's corpuscles	nerve endings that are sensitive to touch.
Melanocytes	melanin producing cells.
Melanocyte-stimulating hormone (MSH)	hormone secreted by anterior pituitary gland. Exact actions are unknown, but can cause darkening of the skin.
Melatonin	hormone secreted by the pineal gland that causes sleepiness.
Membrane	thin, flexible sheets made up of different tissue layers. They cover surfaces, line body cavities and form protective sheets around organs.
Meninges	three connective tissue membranes that enclose the brain and spinal cord.
Meniscus	pad of fibrocartilage that lies between the articular surfaces of bones.
Menopause	when a woman stops menstruating and ovulating and can no longer bear children.
Menses	see *menstruation*.
Menstrual cycle	cycle in which the endometrium of the uterus is prepared for the arrival of a fertilised ovum.
Menstruation	cyclical, periodic discharge of menstrual flow from the uterus. It contains blood, tissue fluid, mucous and epithelial cells derived from the endometrium.
Merkel cells	only found in the stratum basale of hairless skin and are attached to keratinocytes. They make contact with nerve cells to form Merkel discs that function in the sensation of touch.
Metabolism	changes that take place within the body to enable its growth and function.
Metastasis	spread of disease from its site of origin.
Microbe (micro-organism)	organism too small to be seen by the naked eye. Microbes include bacteria, viruses, protozoa and some fungi.
Microvilli	tiny membrane-covered projections extending into small intestine and increasing surface area for absorption.
Micturition	urination.
Middle ear	small air-filled cavity found between the outer ear and the inner ear. It contains the three auditory ossicles.
Midline	see *median line*.
Mineralcorticoids	group of hormones secreted by the adrenal cortex that regulate the mineral content of the blood.
Mitochondria	powerhouses of the cell where ATP is generated through the process of cellular respiration.
Mitosis	cellular reproduction in which a mother cell divides into two daughter cells, each containing the same genes as the mother cell.

Mixed nerve	nerve containing both sensory and motor fibres.
Mole	small, dark skin growth. A concentrated area of melanin.
Monocyte	type of white blood cell.
Mons pubis	elevation of adipose tissue that cushions the pubic symphysis.
Motor neurone	neurone that conducts impulses from the CNS to muscles and glands.
Mucosa	see *mucous membrane*.
Mucosa associated lymphoid tissue (MALT)	concentrations of lymphatic tissue that are strategically positioned to help protect the body from pathogens that have been inhaled, digested or have entered via external openings.
Mucous membrane	membrane that lines body cavities that open directly to the exterior. They are wet membranes whose cells secrete mucous.
Muscle tissue	tissue composed of elongated cells that are able to shorten (contract) to produce movement.
Muscularis	muscular layer or coat of an organ.
Myelin sheath	sheath that protects and insulates a neurone. It is composed of a white fatty substance called myelin.
Myocardium	middle layer of the heart wall. It is composed of cardiac muscle tissue and contracts to pump blood.
Myofibre	muscle fibre.
Myofibrils	long thread-like organelles, the contractile elements of a skeletal muscle fibre.
Myofilament	filaments found inside myofibrils. There are two types: actin/thin and myosin/thick filaments.
Myogenic rhythm	inherent rhythmicity of certain muscles that does not rely on nervous stimulation, e.g. cardiac muscle.
Myoglobin	protein that binds with oxygen and carries it to muscle cells.
Myology	study of muscles.
Myosin	protein that functions in muscle contraction.

N

Nail bed	area that lies directly beneath the nail plate and secures the nail to the finger or toe.
Nail free edge	the part of the nail that extends past the end of the finger or toe. Also called the distal edge.
Nail groove	grooves on the sides of the nail that guide it up the fingers and toes.
Nail mantle	the skin that lies directly above the germinal matrix of the nail.
Nail plate	the visible body of the nail.
Nail wall	the skin that covers the sides of the nail plate and protects the nail grooves.
Nasal conchae	bony shelves projecting from the lateral walls of the nasal cavity.
Natural killer cells (NK cells)	type of lymphocyte that can kill a variety of microbes as well as some tumour cells.
Nephron	functional unit of the kidney where filtration occurs.
Nerve fibre	term referring to the processes that project from a nerve body, e.g. dendrites and axons.
Nerve tissue	tissue made up of neurones and neuroglia. Found in the brain, spinal cord and nerves and functions in communication.
Neurofibril node	see *node of Ranvier*.
Neuroglia	supporting cells that insulate, support and protect neurones.
Neurolemmocyte	see *Schwann cells*.
Neurology	study of the nervous system.
Neurone	nerve cell responsible for the sensory, integrative and motor functions of the nervous system.
Neurotransmitter	chemical that transmits impulses across synapses from one nerve to another.
Neutrophil	type of white blood cell that engulfs and digests foreign particles and removes waste through phagocytosis.
Node of Ranvier	gap along a myelinated nerve fibre.
Nodule	solid bump that may be raised.
Noradrenaline	hormone secreted by the adrenal medulla that functions in the fight-or-flight response.
Norepinephrine	see *noradrenaline*.
Nucleic acid	organic compound composed of nucleotides, e.g. DNA or RNA.
Nucleolus	spherical body inside the nucleus made up of protein, some DNA and RNA.
Nucleus	structure in a cell that controls all cellular structure and activities and contains most of the genes.

O

Oblique plane	plane that divides at an angle between a transverse plane and a frontal or sagittal plane.
Occipital	pertaining to the back of the head.
Oestrogens	hormones secreted by the ovaries that stimulate the development of feminine secondary sex characteristics and, together with progesterone, regulate the female reproductive cycle.
Olfaction	sense of smell.
Omentum	double layer of the peritoneum. Covers and links the abdominal organs.
Onyx	nail.

Oocyte	immature ovum.
Oogenesis	production of mature ova in the ovaries.
Ophthalmic	pertaining to the eye.
Organelle	little organ of the cell.
Organic compound	compound containing carbon.
Origin	point where a muscle attaches to the stationary bone of a joint.
Osmoreceptor	receptor sensitive to a decrease in water or an increase in solutes in the blood.
Osmosis	diffusion of water through a selectively permeable membrane from an area of higher concentration to an area of water concentration.
Osseous tissue	bone tissue. It is an exceptionally hard connective tissue that protects and supports other organs of the body.
Ossification	process of bone formation.
Osteoblast	cell that secretes collagen and other organic components to form bones.
Osteoclast	cell found on the surface of bones that destroys or resorbs bone tissue.
Osteocyte	mature bone cell that maintains the daily activities of bone tissue. Derived from osteoblasts and are the main cells found in bone tissue.
Osteology	study of the structure and function of bones.
Osteon	basic unit of structure of an adult compact bone. It consists of a system of interconnecting canals called Haversian canals.
Osteoprogenitor cell	stem cell derived from mesenchyme (the connective tissue found in an embryo) that has the ability to become an osteoblast.
Outer ear	external region of the ear that collects and channels sound waves inwards. It is composed of the auricle, external auditory canal and eardrum.
Ova	mature female sex cells (commonly called eggs).
Ovarian cycle	cycle in which an oocyte matures until it is ready for ovulation.
Ovulation	event occuring around day 14 of a 28-day cycle. It involves the rupture of the mature Graafian follicle and the release of an ovum into the pelvic cavity.
Oxidation	see cellular respiration.
Oxytocin (OT)	hormone released by the posterior pituitary gland that stimulates contraction of the uterus during labour and stimulates the milk let-down reflex during lactation.

P

Pancreatic amylase	enzyme present in pancreatic juice that completes the breakdown of starches and glycogen.
Pancreatic islets	see islets of Langerhans.
Pancreatic juice	pancreatic secretion composed of mostly water, some salts, sodium bicarbonate and some enzymes.
Pancreatic lipase	enzyme present in pancreatic juice breaks down triglycerides into fatty acids and monoglycerides.
Paneth cells	specialised cells that secrete a bactericidal enzyme called lysozyme into the small intestine.
Papillae	small projections covering the tongue. Some papillae house the taste buds.
Papillary layer	undulating membrane that makes up approximately ⅕ of the thickness of the dermis. It is composed of areolar connective tissue and fine elastic fibres and has nipple-shaped fingerlike projections called papillae.
Papule	small, solid bump that does not contain fluid, e.g. warts, insect bites and skin tags.
Parasympathetic nervous system	nervous system that opposes the actions of the sympathetic nervous system by inhibiting activity, thus conserving energy.
Parathormone (PTH)	hormone secreted by the parathyroid glands that increases blood calcium and magnesium levels, decreases blood phosphate levels and promotes formation of calcitriol by the kidneys.
Parathyroid hormone	see parathormone.
Parietal	relating to the wall of the body or any of its cavities.
Pathogen	disease-causing micro-organism.
Pathology	study of the diseases of the body.
Pepsin	enzyme in the stomach that begins the breakdown of proteins.
Pepsinogen	enzyme precursor in the stomach that is converted into pepsin in the acidic environment of gastric juice.
Peptidases	brush border enzymes that complete the breakdown of proteins into amino acids.
Pericardium	membranous sac that surrounds and protects the heart.
Perimysium	connective tissue that surrounds bundles of 10–100 muscle fibres.
Perineum	diamond-shaped area that contains the external genitals and the anus. Located between the thighs and buttocks and is present in both males and females.
Periosteum	connective tissue membrane that covers bones.

Peripheral	at the surface or outer part of the body.
Peripheral nervous system (PNS)	part of the nervous system connecting the rest of the body to the central nervous system (brain and spinal cord).
Peristalsis	involuntary wave-like movement that pushes the contents of the gastro-intestinal tract forward.
Peritoneum	large serous membrane lining the abdominal cavity.
Peroxisomes	cellular vesicles containing enzymes that detoxify any potentially harmful substances in the cell.
Peyer's patches	patches of mucous-associated lymphoid tissue located in the lining of the small intestine.
Phagocystosis	engulfment and digestion of foreign particles by phagocytes.
Phagocyte	cell that can engulf and digest microbes. Phagocytes include macrophages and some types of white blood cells.
Phalanges	bones of the fingers and toes.
Pharynx	funnel-shaped tube whose walls are made up of skeletal muscles lined by mucous membrane and cilia. Commonly called the throat.
Photoreceptors	specialised cells that convert light into nerve impulses.
Physiology	study of the functions of the body.
Pia mater	thin inner covering of the brain. Dips into all the folds and spaces of the brain tissue.
Pivot joint	joint in which a rounded/pointed surface of a bone fits into a ring-shaped bone.
Plane	imaginary flat surface that divides the body or organs into parts.
Plane joint	see gliding joint.
Plantar	pertaining to the sole of the foot.
Plantar flexion	pointing of the foot downwards, in the direction of the plantar surface.
Plaque	large, flat, raised bump or group of bumps.
Plasma	liquid portion of blood.
Plasma cell	cell that develops from a B cell (type of lymphoctye) and produces antibodies.
Plasma membrane	barrier that surrounds a cell and regulates the movement of all substances into and out of it.
Platelet	see thrombocyte.
Plexus	network of nerves or blood vessels.
Popliteal	hollow space behind the knee.
Posterior	at the back of the body, behind.
Postovulatory phase	14 days after ovulation in which the endometrium awaits the arrival of a fertilised ovum.
Pregnancy	sequence of fertilisation, implantation, embryonic growth and foetal growth.
Preovulatory phase	time between menstruation and ovulation. In a 28-day cycle it can vary from 6 to 13 days in length and is the time when a mature Graafian follicle forms and the endometrium proliferates.
Prime mover	muscle responsible for causing a movement.
Process	bony projection or prominence.
Progesterone	hormone secreted by the ovaries which, together with oestrogens, regulates the female reproductive cycle and helps maintain pregnancy.
Prolactin (PRL)	hormone secreted by the anterior pituitary gland which stimulates the secretion of milk from the breasts.
Proliferative phase	see preovulatory phase.
Pronation	movement involving turning the palm posteriorly or inferiorly.
Proprioceptor	specialised nerve receptor located in muscles, joints and tendons that provides sensory information regarding body position and movements.
Prostate gland	gland surrounding the prostatic urethra. It secretes a milky, slightly acidic fluid that contributes to sperm mobility and viability.
Protein	organic compound made up of amino acids and containing carbon, hydrogen, oxygen and nitrogen. The main building material of cells.
Protraction	drawing the shoulders or jaw forwards.
Proximal	closer to its origin or point of attachment of a limb.
Proximal convoluted tubule	portion of the renal tubule closest to the glomerular capsule of a kidney nephron. Where the reabsorption of most substances takes place.
Puberty	time at which a person develops secondary sexual characteristics and becomes able to reproduce.
Pulmonary circulation	circulatory system in which the right side of the heart receives deoxygenated blood from the body and pumps it to the lungs where it is oxygenated.
Pulmonary respiration	see external respiration.
Pulmonary ventilation	process in which air is inspired or breathed into the lungs and expired or breathed out of the lungs.
Pupil	hole in the centre of the iris through which light enters the eye.
Pustule	lump containing pus.

Q

Quadrant	region of the abdominopelvic cavity.

R

Rectum	last portion of the gastrointestinal tract.
Reflex	automatic response to a stimulus.
Remodelling	process through which new bone tissue replaces old, worn-out or injured bone tissue.
Renal	pertaining to the kidneys.
Renin	enzyme produced by the kidneys which functions in raising blood pressure.
Rennin	enzyme found only in the stomachs of infants. Begins the digestion of milk by converting the protein caseinogen into casein.
Respiration	exchange of gases between the atmosphere, blood and cells.
Retina	innermost layer of the wall of the eyeball. It consists of a non-visual pigmented portion and a neural portion.
Retraction	drawing the shoulders or jaw backwards.
Ribosomes	tiny granules that are sites of protein synthesis in the cell.
Rima glottidis	vocal folds.
Rods	photoreceptors which respond to different shades of grey only
Rotation	movement of a bone in a single plane around its longitudinal axis.
Rugae	folds in the mucous lining of a hollow organ. Found in the stomach and vagina.

S

Saddle joint	joint in which a surface shaped like the legs of a rider fits into the saddle-shaped surface of another bone.
Sagittal plane	plane that divides vertically into right and left sides.
Salivary amylase	enzyme present in saliva that begins the breakdown of large carbohydrate molecules.
Sarcolemma	plasma membrane of a muscle fibre.
Sarcomere	basic functional unit of a skeletal muscle.
Sarcoplasm	cytoplasm of a muscle fibre.
Scales	areas of dried, flaky cells, e.g. in psoriasis or dandruff.
Scar	area where normal skin has been replaced by fibrous tissue. Forms after an injury.
Schwann cells	cells that wrap around the axon of a neurone to form a myelin sheath.
Sclera	white of the eye. Made up of dense connective tissue, protects the eyeball and gives it its shape and rigidity.

Scrotum	sac of loose skin and superficial fascia in which lie the testes.
Sebaceous oil gland	exocrine gland that secretes sebum. Usually associated with a hair follicle.
Sebum	oily substance secreted by sebaceous glands.
Secretion	substance released from a gland cell. Secretions are usually useful substances as opposed to waste products.
Secretory phase	see *postovulatory phase*.
Segmentation	main movement in the small intestine. It involves localised contractions that move food back and forth.
Semen (seminal fluid)	fluid containing sperm and a mixture of fluids secreted by the reproductive glands.
Semicircular canals	three semicircular canals that project from the vestibule in the ear and contain receptors for equilibrium.
Semilunar valves	valves lying between the ventricles and the arteries that prevent the backflow of blood.
Seminal fluid	see *semen*.
Seminal vesicles	paired pouch-like structures located at the base of the bladder that secrete a viscous alkaline fluid.
Seminiferous tubules	tightly coiled tubules located in the testes where sperm are formed.
Sensory neurones (afferent neurones)	neurones that conduct impulses from sensory receptors to the CNS.
Serous membrane	membrane lining body cavities that do not open directly to the exterior and covering the organs that lie within those cavities.
Sesamoid bones	oval bones that develop in tendons where there is considerable pressure.
Short bones	cube-shaped bones that are nearly equal in length and width.
Simple epithelium	single layer of cells. Usually very thin and functions in absorption, secretion and filtration.
Sinoatrial node (SA node)	the heart's pacemaker.
Skeletal muscle tissue	muscle tissue attached to bones that is composed of long, cylindrical shaped fibres that are striated and under voluntary control.
Smooth muscle tissue	muscle tissue that contains non-striated (smooth) fibres and that is regulated by the autonomic nervous system.
Solvent	medium (usually a liquid) in which substances (solutes) can be dissolved.
Somatic cell	any cell except the reproductive cells.
Somatic cell division	process of nuclear division called mitosis. A single diploid parent cell duplicates to produce two identical daughter cells.

Somatic nervous system (voluntary nervous system) part of the nervous system that allows us to control our skeletal muscles. Also called the voluntary nervous system.

Somatostatin hormone secreted by the pancreas that inhibits insulin and glucagon release.

Somatotropin see *human growth hormone*

Spermatogenesis production of sperm in the testes.

Spermatozoa mature male sex cells (commonly called sperm).

Spinal nerves nerves emerging from the spinal cord that carry impulses to and from the rest of the body.

Spongy bone tissue light bone tissue with many spaces within it and a sponge-like appearance. Does not contain osteons.

Sputum material coughed up from the respiratory tract.

Stratified epithelium consists of two or more layers of cells. It is durable and protects underlying tissues in areas of wear and tear.

Stratum basale deepest layer of the epidermis and the base from where new cells germinate or sprout.

Stratum corneum outermost layer of the skin consisting of dead, tough cells.

Stratum functionalis deep layer of the endometrium of the uterus. It is shed during menstruation.

Stratum germinativum See *stratum basale*.

Stratum granulosum epidermal layer of degenerating cells that are becoming increasingly filled with little grains or granules of keratin.

Stratum lucidum epidermal waterproof layer of dead, clear cells.

Stratum spinosum epidermal layer of prickly cells that are beginning to go through the process of keratinisation.

Stretchmarks small tears in the dermis caused by the skin stretching beyond its ability.

Striated having the appearance of light and dark bands or striations.

Subcutaneous beneath the skin.

Substrate substance on which an enzyme acts.

Sucrase brush border enzyme that breaks down sucrose into glucose and fructose.

Sudoriferous gland gland that excretes sweat onto the surface of the skin.

Superficial towards the surface of the body.

Superior towards the head, above.

Supination movement involving turning the palm anteriorly or superiorly.

Suspensory (Cooper's) ligaments strands of connective tissue that support a breast.

Sympathetic nervous system part of the nervous system that reacts to changes in the environment by stimulating activity, thereby using energy.

Synapse gap at the end of a nerve fibre which an impulse crosses to pass from one neurone to the next.

Synaptic vesicle sac that stores neurotransmitters and is located in a synaptic end bulb at the distal end of an axon terminal.

Synarthrose immovable joint.

Synergist muscle that helps the prime mover.

Synovial joint freely movable joint in which a cavity is present between the articulating bones.

Synovial membrane membrane composed of areolar connective tissue that lines freely movable joints and secretes synovial fluid.

Systemic circulation type of circulation in which the left side of the heart receives oxygenated blood from the lungs and pumps it to the rest of the body.

Systole contraction of the heart muscle during the cardiac cycle.

T

T cells lymphocyte that matures in the thymus gland and functions in cell-mediated immunity.

Taeniae coli three thickened bands of longitudinal smooth muscle running the length of the colon.

Taste bud a receptor for taste.

Telangiectasis a localised collection of blood vessels in the skin. Characterised by a red spot that can be spidery in appearance and that blanches under pressure.

Telogen resting phase of the hair cycle in which the follicle is inactive until stimulated to develop another hair.

Tendon strong cord of dense connective tissue that attaches muscles to bones, to the skin or to other muscles.

Terminal hair hair found on the head, eyebrows, eyelashes, under the arms and in the pubic area.

Testes gonads of the male reproductive system.

Testosterone hormone secreted by the testes that stimulates the development of masculine secondary sex characteristics and libido.

Thermogenesis generation of heat in the body.

Thorax the chest.

Thrombocyte type of white blood cell that functions in haemostasis and blood clotting.

Thymosin hormone secreted by the thymus gland that promotes the growth of T-Cells.

Thyroid hormone	hormone secreted by the thyroid gland that functions in metabolism, growth and development.
Thyroid-stimulating hormone (TSH)	hormone secreted by the anterior pituitary gland that controls the thyroid gland.
Thyrotropin	see *thyroid-stimulating hormone*.
Thyroxine	see *thyroid hormone*.
Tissue respiration	see *internal respiration*.
Tone (tonus)	partial contraction of a resting muscle.
Trachea	long tubular passageway which transports air from the larynx into the bronchi. Commonly called the windpipe.
Tract	bundles of nerve fibres that are not surrounded by connective tissue.
Transverse plane	plane that divides horizontally into inferior and superior portions.
Tricuspid valve	right atrioventricular valve.
Trimester	period consisting of three months. Pregnancy is divided into three trimesters.
Trypsin	enzyme secreted by the pancreas that continues the breakdown of proteins.
Tubercle	solid lump that is larger than a papule.
Tumour	an abnormal growth of tissue.
Tympanic membrane	see *eardrum*.

U

Ulcer	a deep, open lesion on the skin. Ulcers penetrate the dermis.
Ungus	nail
Urea	main product of protein metabolism.
Uric acid	product of nucleic acid metabolism.
Urobilinogen	bile pigment derived from the breakdown of haemoglobin.
Urology	study of the urinary system.
Uterine cycle	see *menstrual cycle*.
Uterus	the womb.
Uvula	finger-like projection hanging from the soft palate at the back of the mouth.

V

Vacuole	space within the cytoplasm of a cell that contains material taken in by the cell.
Vas deferens	long duct running from the epididymis to the urethra. Transports sperm via peristaltic contractions.
Vascular tissue	blood – a type of connective tissue whose matrix is made of a fluid called blood plasma.
Vasoconstriction	constriction of blood vessels.
Vasodilation	dilation of blood vessels.
Vasopressin	see *antidiuretic hormone*.

Veins	vessels that usually carry blood away from the tissues towards the heart.
Vellus hair	soft and downy hair that is found all over the body except the palms of the hands, soles of the feet, eyelids, lips and nipples.
Ventral	at the front of the body, in front of.
Ventricle (brain)	fluid filled cavity in the brain. There are four of them.
Ventricle (heart)	delivery chambers of the heart that pump blood into the blood vessels.
Venule	small vein. It drains blood away from capillaries.
Vesicle	small, fluid-filled sac.
Vesicular ovarian follicle	see *Graafian follicle*.
Vesicular transport	type of transport across the plasma membrane of a cell in which vesicles (small sacs) carry particles.
Vestibule (ear)	central portion of the bony labyrinth of the ear. Contains receptors for equilibrium.
Vestibule (female genitalia)	entire region between the labia minora, consisting of the vaginal orifice and external urethral orifice.
Villi	finger-like projections of intestinal mucosa cells that increase the total surface area of the small intestine.
Viscera	organs of the abdominal body cavity.
Visceral	relating to the internal organs of the body.
Vitreous body	jelly-like substance that helps produce intraocular pressure in the eye.
Voluntary nervous system	see *somatic nervous system*.
Wheal	common allergic reaction in which there is swelling with an elevated, soft area.
Wheezing	whistling sound produced when the airways are partially obstructed.
Zygote	a fertilised ovum.

Index

NB: *f* = figures, *c* = charts, *b* = boxed information.

Picture Credits

We would like to thank the following for granting permission to reproduce copyright material in this book:

Pages 76–86, Wellcome Trust Photo Library

Page 102, 104, 106, 108, 109, 112, 113, 114, 116, 117, 118, 120, 123, 124, 129, Sobotta: Atlas der Anatomie des Menschen, 22nd edition © Elsevier GmbH, Urban & Fischer Verlag München

Page 130, 132, 133, Wellcome Trust Photo Library

Page 157, 158, 159, 160, 163, 164, 165, 166, 170, 171, 174, 175, 176, 180, 181, 182, Sobotta: Atlas der Anatomie des Menschen, 22nd edition © Elsevier GmbH, Urban & Fischer Verlag München

Page 186, 187, 189 Wellcome Trust Photo Library

Page 194, 195, 196 Sobotta: Atlas der Anatomie des Menschen, 22nd edition © Elsevier GmbH, Urban & Fischer Verlag München

Page 201, 213, 214, 216, 220, 223, 224, 226, 228, 230, 233, 234, 236, Sobotta: Atlas der Anatomie des Menschen, 22nd edition © Elsevier GmbH, Urban & Fischer Verlag München

Pages 241 and 243 Wellcome Trust Photo Library

Page 255 Sobotta: Atlas der Anatomie des Menschen, 22nd edition © Elsevier GmbH, Urban & Fischer Verlag München

Page 264, 265, 266 Wellcome Trust Photo Library

Page 273, 274, 276, 278, 279, 280, 282 Sobotta: Atlas der Anatomie des Menschen, 22nd edition © Elsevier GmbH, Urban & Fischer Verlag München

Page 295 Wellcome Trust Photo Library

Page 302 Wellcome Trust Photo Library

Page 306, 307, 308, 310, 311 Sobotta: Atlas der Anatomie des Menschen, 22nd edition © Elsevier GmbH, Urban & Fischer Verlag München

Page 316 Wellcome Trust Photo Library

Page 321, 324, 325, 334 Sobotta: Atlas der Anatomie des Menschen, 22nd edition © Elsevier GmbH, Urban & Fischer Verlag München

Pages 340 and 342 Wellcome Trust Photo Library

Page 351 Wellcome Trust Photo Library

Page 353 Sobotta: Atlas der Anatomie des Menschen, 22nd edition © Elsevier GmbH, Urban & Fischer Verlag München

Pages 366 and 367 Wellcome Trust Photo Library

Pages 374, 376, 377, 379, 382, 383, 385, 386, 388, 389, 392, 393 Sobotta: Atlas der Anatomie des Menschen, 22nd edition © Elsevier GmbH, Urban & Fischer Verlag München

Page 395, 396, 397, 399 Wellcome Trust Photo Library

Page 403 Sobotta: Atlas der Anatomie des Menschen, 22nd edition © Elsevier GmbH, Urban & Fischer Verlag München

Pages 408 and 409 Sobotta: Atlas der Anatomie des Menschen, 22nd edition © Elsevier GmbH, Urban & Fischer Verlag München

Page 419 Wellcome Trust Photo Library

Page 429, 433, 442 Sobotta: Atlas der Anatomie des Menschen, 22nd edition © Elsevier GmbH, Urban & Fischer Verlag München

Page 446 Wellcome Trust Photo Library